Natural Products in Drug Discovery and Development

Natural Products in Drug Discovery and Development

Guest Editor

Tian-Le Gao

MDPI

Basel • Beijing • Wuhan • Barcelona • Belgrade • Novi Sad • Cluj • Manchester

Guest Editor
Tian-Le Gao
Institute of Materia Medica
Peking Union Medical College
Beijing
China

Editorial Office
MDPI AG
Grosspeteranlage 5
4052 Basel, Switzerland

This is a reprint of the Special Issue, published open access by the journal *International Journal of Molecular Sciences* (ISSN 1422-0067), freely accessible at: https://www.mdpi.com/journal/ijms/special_issues/8U3234Y9B8.

For citation purposes, cite each article independently as indicated on the article page online and as indicated below:

Lastname, A.A.; Lastname, B.B. Article Title. *Journal Name* **Year**, *Volume Number*, Page Range.

ISBN 978-3-7258-7899-4 (Hbk)
ISBN 978-3-7258-7900-7 (PDF)
https://doi.org/10.3390/books978-3-7258-7900-7

Contents

About the Editor

Tian-Le Gao

Tian-Le Gao is an Associate Professor at the Institute of Materia Medica, Chinese Academy of Medical Sciences, where his research focuses on pharmacology, natural product-based drug discovery, and the mechanisms of chronic pain. His interests include applying neg entropy theory to uncover novel therapeutic targets and developing innovative analgesics from traditional Chinese medicine. He has led projects funded by the National Natural Science Foundation of China and collaborated with pharmaceutical companies on first in class drug candidates, including a novel analgesic and a T cell activating immunotherapy platform. Dr. Gao has published over 35 papers in journals such as *Molecular Cancer and Engineering*, holds several authorized patents, and serves as a youth editorial board member for *Acta Pharmaceutica Sinica B* and *Chinese Journal of Pharmaceutical Biotechnology*.

International Journal of
Molecular Sciences

MDPI

Editorial

Special Issue "Natural Products in Drug Discovery and Development"

Tian-Le Gao

Institute of Materia Medica, Chinese Academy of Medical Sciences and Peking Union Medical College, Beijing 100050, China; tianlegao@imm.ac.cn

The Special Issue "Natural Products in Drug Discovery and Development" arrives at a pivotal moment when traditional medicine's enduring wisdom is increasingly converging with modern pharmacological research. As global health challenges intensify, researchers are turning to nature's vast repository of compounds to uncover novel therapeutic solutions. This Special Issue is designed to gather cutting-edge research articles, comprehensive reviews, and insightful case studies that explore the multifaceted potential of natural products in drug discovery and development, leveraging multidisciplinary approaches and innovative research methodologies.

A key focus lies in the isolation and characterization of natural product components. Illustrative work in this area includes the study of bile acids, which examines their bioactivities, mechanisms of action, biosynthesis, and potential applications in preventing and treating infectious diseases [1]. Several promising compounds, such as ursodeoxycholic acid, have been identified for further investigation. A notable advance is the development of artificial bear bile—produced via synthetic chemistry and enzyme engineering—which provides a sustainable and ethical alternative to natural bear bile [2]. This innovation addresses critical medical demands while aligning with animal welfare principles.

Pharmacological actions of natural compounds constitute a major research emphasis. This includes their role in ameliorating microcirculatory dysfunction induced by ischemia/reperfusion injury or lipopolysaccharide exposure, providing mechanistic insights essential for developing novel therapies targeting circulatory disorders [3]. This Special Issue further explores the role of gut microbiota in mediating drug efficacy and the beneficial effects of probiotics. For example, isochlorogenic acid C is shown to mitigate asthma through microbiota-dependent pathways [4], a finding complemented by previous research on rosmarinic acid, which modulates gut microbial composition to reduce bronchoconstriction and inflammation, offering a safer alternative to conventional asthma treatments [5]. This issue also includes a study investigating the multi-faceted benefits of berberine in improving diastolic dysfunction and mitigating metabolic comorbidities in heart failure with preserved ejection fraction (HFpEF) [6].

Innovative formulations are also highlighted, such as the development of a salt formulation combining berberine and quercetin to enhance their anti-fibrotic potential in liver fibrosis [7]. This new formulation improves the dissolution and bioavailability of quercetin, showing stronger anti-fibrotic effects than the physical mixture in both cellular and animal models. Such advancements offer new strategies for optimizing multi-drug combination therapy.

Exploring deeper biological principles, the issue aims to delve into neg-entropy mechanisms in disease treatment and prevention. Entropy, a measure of disorder, increases naturally, yet living organisms maintain their complexity and health, indicating a capacity for negative entropy [7,8]. Effective drugs may accelerate recovery by stimulating these mechanisms, which include metabolism and self-organization, defense, self-healing, wear resistance, and adaptability [8]. These interconnected capacities allow life to resist disorder, adapt to changes, and sustain complexity, providing a theoretical framework for understanding how natural products can promote health and combat disease.

The pharmacological properties of aromatic compounds and the health effects of essential oils are attracting increasing research interest. One study investigated a volatile extract from a Uyghur medicinal plant for the treatment of fluconazole-resistant vulvovaginal candidiasis [9]. The extract exhibited dual antioxidant and antifungal activities, demonstrating promising potential as a broad-spectrum natural agent against resistant fungal infections [9].

Combinational studies of natural medicine drugs are encouraged, as demonstrated by research on the synergistic analgesic effects of ligustrazine and sinomenine in treating neuropathic pain [10,11]. Network pharmacology and metabolomics analysis reveal associated targets and key pathways, supporting the clinical potential of this combination therapy [11]. Another study explores the synergistic effects of combining pregabalin with dexborneol, showing enhanced analgesic outcomes and reduced side effects, thus offering a more effective and tolerable therapeutic strategy for neuropathic pain [12].

In conclusion, this Special Issue invites readers to engage with the research and consider the transformative potential of natural medicines in advancing global healthcare. By harnessing the power of nature, we can pave the way for innovative therapies that improve human health and well-being.

Funding: This work was supported by the National Natural Science Foundation of China (82374061), CAMS Innovation Fund for Medical Sciences (2024-I2M-ZH-012, 2022-I2M-1-016), and Key Project Fund of the Chinese Academy of Engineering (Organized Scientific Research Strategy Study on Anti-Infective Drugs for Emerging Infectious Diseases, 2025-XZ-125).

Acknowledgments: The author expresses special gratitude to Yun-Dan Guo from Institute of Materia Medica, Chinese Academy of Medical Sciences and Peking Union Medical College (Beijing, China), and Zhi-Li Zhang from Miqi Institute of Medicinal Plants & Skin Health Research (Beijing, China), for the insightful discussions.

Conflicts of Interest: The author declares no conflicts of interest.

References

1. Liu, S.; Yang, S.; Blazekovic, B.; Li, L.; Zhang, J.; Wang, Y. Bioactivities, Mechanisms, Production, and Potential Application of Bile Acids in Preventing and Treating Infectious Diseases. *Engineering* **2024**, *38*, 13–26. [CrossRef]
2. Li, Y.; Huang, Y.; Feng, N.; Zhang, H.; Qu, J.; Ma, S.; Liu, Y.; Li, J.; Xu, S.; Wang, L.; et al. Artificial Bear Bile: A Novel Approach to Balancing Medical Requirements and Animal Welfare. *Engineering* **2024**, *38*, 100–112. [CrossRef]
3. Han, J.; Li, Q.; Sun, K.; Pan, C.; Liu, J.; Huang, P.; Feng, J.; Liu, Y.; Meininger, G.A. Natural Products Improve Organ Microcirculation Dysfunction Following Ischemia/Reperfusion- and Lipopolysaccharide-Induced Disturbances: Mechanistic and Therapeutic Views. *Engineering* **2024**, *38*, 77–99. [CrossRef]
4. Xu, J.-Y.; Rong, X.-J.; Shen, Z.; Guo, Y.-D.; Zhang, Y.-X.; Ding, C.-C.; Wang, Y.; Han, Y.-X.; Gao, T.-L.; Tie, C. Isochlorogenic Acid C Alleviates Allergic Asthma via Interactions Between Its Bioactive Form and the Gut Microbiome. *Int. J. Mol. Sci.* **2025**, *26*, 4864. [CrossRef] [PubMed]
5. Guo, H.; Han, Y.; Rong, X.; Shen, Z.; Shen, H.; Kong, L.; Guo, Y.; Li, J.; Xu, B.; Gao, T.; et al. Alleviation of allergic asthma by rosmarinic acid via gut-lung axis. *Phytomedicine* **2024**, *126*, 155470. [CrossRef] [PubMed]
6. Mu, Y.; Geng, J.; Liu, C.; Jiang, S.; Han, Y.; Jiang, J.; Wang, Y. Exploring the Multi-Faceted Effects of Berberine in Ameliorating Diastolic Dysfunction in Rats with Heart Failure with Preserved Ejection Fraction. *Int. J. Mol. Sci.* **2025**, *26*, 4847. [CrossRef] [PubMed]
7. Jiang, J.-D. Editorial for the Special Issue on Natural Medicine. *Engineering* **2024**, *38*, 1. [CrossRef]
8. Gao, T.-L.; Guo, H.-H.; Jiang, J.-D. Neg-Entropy Mechanism as a Target for Natural Medicines. *Engineering* **2024**, *38*, 11–12. [CrossRef]

9. Guo, Y.-D.; Zhang, M.-X.; Yu, Q.-Y.; Wang, L.-L.; Han, Y.-X.; Gao, T.-L.; Lin, Y.; Tie, C.; Jiang, J.-D. *Hyssopus cuspidatus Boriss* Volatile Extract (SXC): A Dual-Action Antioxidant and Antifungal Agent Targeting *Candida albicans* Pathogenicity and Vulvovaginal Candidiasis via Host Oxidative Stress Modulation and Fungal Metabolic Reprogramming. *Antioxidants* **2025**, *14*, 1046. [CrossRef]
10. Gao, T.; Hao, J.; Wiesenfeld-Hallin, Z.; Wang, D.Q.; Xu, X.J. Analgesic effect of sinomenine in rodents after inflammation and nerve injury. *Eur. J. Pharmacol.* **2013**, *721*, 5–11. [CrossRef] [PubMed]
11. Yuan, Z.; Zhao, X.; Zhang, Y.; Jiao, Y.; Liu, Y.; Gao, C.; Zhang, J.; Ma, Y.; Wang, Z.; Li, T. Using Integrated Network Pharmacology and Metabolomics to Reveal the Mechanisms of the Combined Intervention of Ligustrazine and Sinomenine in CCI-Induced Neuropathic Pain Rats. *Int. J. Mol. Sci.* **2025**, *26*, 2604. [CrossRef] [PubMed]
12. Shen, Z.; Guo, Y.-D.; Tang, M.-Z.; Zhou, P.; Su, Y.-X.; Shen, H.-R.; Li, T.; Jiang, W.; Han, Y.-X.; Tie, C.; et al. Dexborneol Amplifies Pregabalin's Analgesic Effect in Mouse Models of Peripheral Nerve Injury and Incisional Pain. *Antioxidants* **2024**, *13*, 803. [CrossRef] [PubMed]

International Journal of
Molecular Sciences

MDPI

Article

Isochlorogenic Acid C Alleviates Allergic Asthma via Interactions Between Its Bioactive Form and the Gut Microbiome

Jing-Yi Xu [1,†], Xiao-Juan Rong [2,†], Zhen Shen [3], Yun-Dan Guo [3], Yi-Xuan Zhang [1], Chen-Chen Ding [1], Yi Wang [1], Yan-Xing Han [3], Tian-Le Gao [3,*] and Cai Tie [1,*]

1 State Key Laboratory for Fine Exploration and Intelligent Development of Coal Resources & School of Chemical and Environmental Engineering, China University of Mining and Technology-Beijing, Ding 11 Xueyuan Road, Beijing 100083, China; 18873361988@163.com (J.-Y.X.); 17788266130@163.com (Y.-X.Z.); 13574319616@163.com (C.-C.D.); sycamorey11@163.com (Y.W.)
2 Xinjiang Institute of Material Medica, Urumqi 830004, China; rxj1125@stu.xjmu.edu.cn
3 State Key Laboratory of Bioactive Substances and Function of Natural Medicine, Institute of Materia Medica, Chinese Academy of Medical Sciences & Peking Union Medical College, Beijing 100050, China; shenzhen@imm.ac.cn (Z.S.); guoyundan@imm.ac.cn (Y.-D.G.); hanyanxing@imm.ac.cn (Y.-X.H.)
* Correspondence: tianlegao@imm.ac.cn (T.-L.G.); tiecai@cumtb.edu.cn (C.T.)
† J.-Y.X. and X.-J.R. contributed equally to this work.

Abstract: The global prevalence of asthma is approximately 4.3%, and current asthma treatments focus on reducing symptoms, maintaining normal activity levels, and preventing the deterioration of lung function, rather than achieving a cure or complete prevention. We identified isochlorogenic acid C (ICGAC) as a potential natural medicine for the treatment of asthma. However, the bioavailability of ICGAC was low, ranging from 14.4% to 16.9%, suggesting the involvement of the gut microbiota. The full spectrum of ICGAC's anti-asthmatic mechanism remains to be elucidated. This study investigated the mechanism by which ICGAC alleviates allergic asthma through the gut–lung axis. We discovered anti-asthma pathways and targets based on the selective regulation of lipid peroxidation and employed pharmacological tools to preliminarily validate their mechanisms and efficacy. To study the role of ICGAC in regulating the gut microbiota, we performed 16S rRNA gene sequencing and metabolite analysis. Furthermore, by combining molecular biology and lipid metabolomics, we elucidated the underlying anti-asthma mechanisms of ICGAC. The effective form of ICGAC varies between single and long-term administration. The oral administration of ICGAC enhances the gut-microbiota-derived production of short-chain fatty acids (SCFAs) as the active substances, modulates immune cell activity, influences the differentiation of T- and B-cells, and reduces airway inflammation. ICGAC also regulates the metabolic network of lipid mediators (LMs) and polyunsaturated fatty acids (PUFAs), thus exerting anti-inflammatory effects by modulating arachidonate lipoxygenase (ALOX) activity and LM levels. In addition, ICGAC enhanced the antioxidant response by upregulating the expression of glutathione peroxidase 4 (GPX4), solute carrier family 7 member 11 (SLC7A11), and nuclear factor erythroid 2-related factor 2 (Nrf2), while inhibiting the release of interleukin-4 (IL-4), thereby suppressing asthma inflammation and IgE production. The anti-asthmatic mechanism of oral ICGAC involves the inhibition of lipid peroxidation by chlorogenic acid (CGA) and SCFAs produced by the gut microbiota. ICGAC suppresses asthma-associated inflammatory and oxidative stress responses through the upregulation of GPX4, SLC7A11, and Nrf2 in lung tissue. This study not only provides a solid foundation for the potential clinical use of ICGAC in asthma treatment but also offers novel insights for future research and therapeutic strategies targeting asthma.

Keywords: isochlorogenic acid C; asthma; airway inflammation; docosahexaenoic acid; lipid metabolomics

1. Introduction

According to the latest data from the World Health Organization in 2019, asthma has a global incidence rate of 3%, affecting over 600 million individuals with symptoms, resulting in 461,000 deaths annually [1]. In China, the incidence rate stands at 4.2% [2]; it is characterized by a large patient base, high incidence, and high deterioration rates. However, challenges such as insufficient diagnosis and treatment persist, leading to reduced quality of life and posing a significant financial burden, particularly for those with poorly controlled asthma and in low-income settings [3].

Marked by airway inflammation, hyperresponsiveness, and airway remodeling, asthma is a condition closely associated with exposure to allergens or diverse physical and chemical stimuli. Its clinical presentation includes wheezing, coughing, shortness of breath, chest tightness, and breathing difficulties, significantly disrupting patients' daily activities and overall quality of life [4]. In more severe cases, asthma can damage airways and impair oxygen flow to the alveoli, potentially leading to life-threatening complications [3]. Influenced by both genetic and environmental factors, including infections, allergens, and irritants, asthma exhibits heterogeneity in its inflammatory and remodeling processes [5]. Given its complexity, asthma treatment aims to alleviate symptoms, preserve normal activity levels, and prevent the deterioration of lung function, rather than achieving a complete cure or absolute prevention [6].

Asthma therapy is guided by a stepwise approach tailored to disease severity and control, as recommended by global guidelines [7]. All patients require immediate access to a rapid-onset bronchodilator (e.g., short-acting β_2-agonists [SABAs]) for symptom relief. For mild intermittent asthma, as-needed SABA use may suffice, while patients with persistent symptoms or exacerbations require controller medications. First-line controllers include low-dose inhaled glucocorticoids (IGCs), which reduce airway inflammation and are the cornerstone of long-term management. For moderate disease, IGCs are often combined with long-acting β_2-agonists (LABAs) in a single inhaler (fixed-dose combination) to improve adherence and efficacy [8]. Alternative options for mild-to-moderate asthma include leukotriene receptor antagonists (LTRAs) or theophylline.

Severe asthma warrants high-dose IGCs/LABA combinations, and patients should be evaluated for add-on therapies such as long-acting muscarinic antagonists (LAMAs) or biologic agents. Systemic glucocorticoids should be avoided whenever possible; instead, patients with eosinophilic or type 2 inflammation (e.g., elevated blood eosinophils or FeNO) may benefit from targeted biologics such as anti-IgE (omalizumab), anti-IL-5/IL-5R (mepolizumab, reslizumab, benralizumab), or anti-IL-4/IL-13 (dupilumab) [9]. These therapies address specific inflammatory pathways and reduce the risk of exacerbation, aligning with the principles of personalized medicine [10].

However, these treatments have limitations, such as the requirement for lifelong use, ineffectiveness in some patients, and potential side effects including suppression of the hypothalamic–pituitary–adrenal axis, growth retardation, osteoporosis, diabetes, and respiratory infections [11]. Therefore, there is a significant unmet clinical demand for asthma therapies that are both highly effective and safe. This has prompted a shift towards exploring natural medicines, which offer a plethora of potential drug candidates with minimal adverse effects. These natural compounds often exhibit multiple therapeutic advantages, despite their complex mechanisms of action and potentially unidentified primary targets [12,13].

In pursuit of more effective and safer asthma therapies, our research has focused on *Hyssopus cuspidatus* Boriss, a perennial herb of the Labiatae family with traditional uses in treating colds, fever, phlegm, and cough. Its anti-allergic inflammation properties have been reported in mouse asthma and anaphylaxis models, with Isochlorogenic acid C (ICGAC) identified as a primary bioactive component [14,15]. This dicaffeoylquinic acid, belonging to the caffeoylquinic class of acids, exhibits a wide range of pharmacological effects such as antioxidant, anti-inflammatory, and antimicrobial activities [14]. Preclinical studies have shown that ICGAC can inhibit eosinophil activation and migration, suppress histamine release from mast cells, and restore the Th1/Th2 immune balance, thereby effectively alleviating asthma symptoms [16]. Although the therapeutic potential of ICGAC in asthma is evident, its precise mechanisms of action remain to be fully elucidated, particularly given its low bioavailability, which ranges from 14.4% to 16.9%. This suggests a potential role for the gut microbiota in mediating some of these effects, given its critical involvement in immune and inflammatory regulation, as well as its emerging implications in respiratory diseases [14].

Lipidomics, which involves the comprehensive analysis of thousands of lipid species and their functions, has revealed that lipids, as crucial signaling molecules, modulate various cellular processes in asthma patients' airways, influencing disease onset and progression [17]. Notably, asthma has been associated with alterations in glycerophospholipid and sphingomyelin metabolism, leading to the discovery of potential lipid biomarkers such as PC(18:1/18:2), PC(16:0/18:1), and PC(18:0/22:5), serving as valuable references in the research and development of asthma treatments [18].

The aim of this study is to elucidate the anti-asthma mechanism of ICGAC by adopting an integrative gut–lung axis approach. Employing an Ovalbumin (OVA)-induced asthma mouse model, we place special emphasis on deciphering how ICGAC interacts with the gastrointestinal system when exerting its anti-asthmatic effects, particularly focusing on the gut microbiota and intestinal epithelial cells. Through the application of lipid metabolomics and network pharmacology, we aim to uncover the multifaceted mechanisms of ICGAC in asthma, including its regulation of the lipid metabolism, modulation of inflammatory mediators, immune cell function, and airway responsiveness. Ultimately, this research aims to lay a scientific foundation for the potential clinical application of ICGAC in asthma management.

2. Result

2.1. Histopathological Analysis of Lungs

We assessed lung tissue morphology using H&E staining and light microscopy to compare the effects of ICGAC and dexamethasone (DEX) on asthma-induced structural alterations in mice. As shown in Figure 1A, the control group displayed normal lung tissue architecture with no notable inflammatory infiltration. In contrast, the model group exhibited marked inflammatory cell infiltration around the bronchi and blood vessels, along with bronchial epithelial degeneration, bronchial obstruction, alveolar wall thickening, and interstitial inflammatory cell infiltration. The peribronchial and perivascular inflammatory infiltrates in OVA-induced asthma mice consisted of eosinophils and mast cells, along with lymphocytic involvement [19]. Treatment with both DEX and ICGAC significantly attenuated these pathological changes, including inflammatory cell infiltration and alveolar wall thickening, highlighting their protective roles in mitigating asthma-induced lung damage. The analysis of the pathological results of lung injury in mice is shown in Table S1. In the L-ICGAC and H-ICGAC groups, the mice still had inflammatory cell infiltration in the bronchi of the lungs. However, in general, lung injury was reversed to a certain extent after ICGAC administration. The treatment with DEX could also reverse lung tissue injury in asthmatic mice. In the asthma model group, one mouse (12.5%) had moderate pathological changes in most of the lung tissue, six mice (75%) had moderate pathological changes in part of the lung tissue, and one mouse (12.5%) had mild pathological changes in a small part of the lung tissue. After ICGAC and DEX treatment, no mice with severe pathological changes in the lung tissue were found in the administration groups. After ICGAC treatment, the lung injury in asthmatic mice was reversed and showed a dose-dependent trend. In the low-dose group, seven mice (87.5%) had moderate pathological changes in part of the lung tissue, and one mouse (12.5%) had mild pathological changes in a small part of the lung tissue. In the high-dose group, only two mice (25%) had moderate pathological changes in part of the lung tissue, and six mice (75%) had mild pathological changes in a small part of the lung tissue.

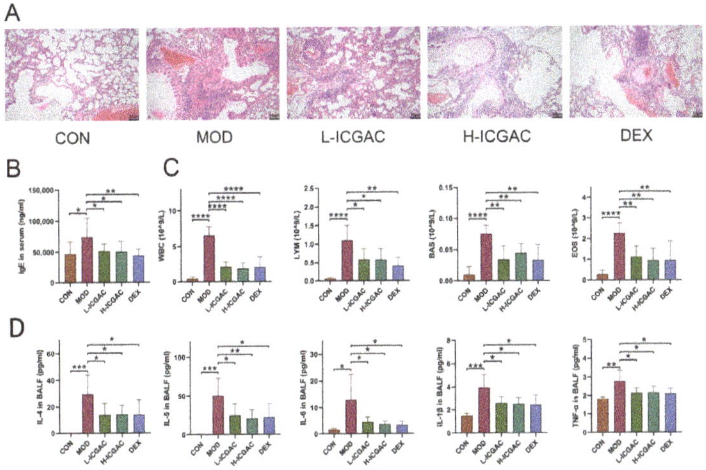

Figure 1. (**A**) Lung histology analysis of different groups. (**B**) IgE levels in the serum of mice. (**C**) The number of white blood cells (WBC), lymphocytes (LYM), basophils (BAS), and eosinophil (EOS) in the BALF. (**D**) IL-4, IL-5, IL-6, IL-1beta, and TNF-alpha levels in the supernatant of BALF. Data are shown as the mean ± SEM (n = 5–8), * $p < 0.05$, ** $p < 0.01$, *** $p < 0.001$, **** $p < 0.0001$.

2.2. Modulatory Effects of ICGAC on Immune Cells and Biochemical Cytokines

As shown in Figure 1B, serum IgE levels in the model group were significantly elevated compared to the control group, confirming the successful induction of the mouse asthma model through OVA sensitization. Treatment with DEX and ICGAC effectively reversed the abnormal increase in IgE levels. Notably, the IgE concentrations in the L-ICGAC (low-dose ICGAC) and H-ICGAC (high-dose ICGAC) groups were comparable, indicating no significant dose-dependent effect of ICGAC on IgE reduction.

The numbers of various immune cells (WBC, LYM, EOS, MON) are shown in Figure 1C. In the model group, the number of immune cells was increased relative to the control group. After treatment with both DEX and ICGAC acid, there was a significant reduction in the number of immune cells. Moreover, we did not observe significant differences between the L-ICGAC and H-ICGAC groups in the assessed indicators. From each group of mice, bronchoalveolar lavage fluid (BALF) was collected. Asthma-associated type 2 cytokines (IL-4, IL-5, IL-6, IL-1β and TNF-α) in BALF are shown in Figure 1D. All cytokines were higher in the model group than in the control group. Inflammatory cytokines (including IL-4, IL-5, IL-6, IL-1β and TNF-α) were substantially reversed after treatment with the ICGAC and DEX.

However, there was no significant difference in cytokine levels between the L-ICGAC and H-ICGAC groups, suggesting a lack of dose-dependent effects in cytokine modulation.

2.3. Differences in Effective Substance Profiles in Serum Between Long-Term and Single-Dose ICGAC Treatments for Asthma in Mice

We quantified the concentrations of ICGAC, chlorogenic acid, and caffeic acid in serum samples from long-term-treated mice using UPLC-MS. ICGAC was not detected in the control group. However, chlorogenic acid was identified in the serum of both the L-ICGAC and H-ICGAC groups. As illustrated in Figure 2A, the levels of chlorogenic acid were comparable between the L-ICGAC and H-ICGAC groups, with concentrations of 22.64 ng/mL and 22.85 ng/mL, respectively. These findings indicate that chlorogenic acid is the effective substance in serum following the long-term administration of ICGAC.

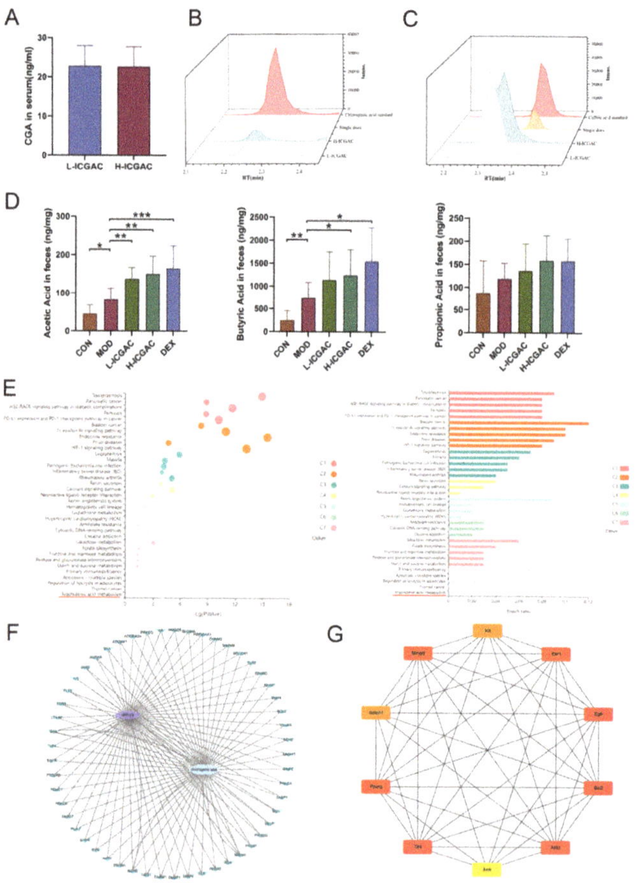

Figure 2. (**A**) CGA quantity in serum. (**B**) Chromatogram of chlorogenic acid (L−ICGAC, H−ICGAC, single does, and chlorogenic acid standard). (**C**) Chromatogram of caffeic acid (L−ICGAC, H−ICGAC, single does, and caffeic acid standard). (**D**) Quantitative results of acetate acid, propionate acid, and butyrate acid concentrations in feces. (**E**) KEGG analysis for the signaling pathways and metabolic pathways of potential target genes. (**F**) Network of CGA and asthma compound (ingredient) target network (**G**) Network of CGA and asthma TOP10 compound (ingredient) target network. Data are shown as the mean ± SEM (n = 6–8), * $p < 0.05$, ** $p < 0.01$, *** $p < 0.001$.

We measured the concentrations of ICGAC, chlorogenic acid, and caffeic acid in serum samples from single-dose-treated mice using UPLC-MS. Neither ICGAC nor chlorogenic acid was detected in the control or single-dose groups. However, caffeic acid was identified in the serum of the single-dose groups (Figure 2B,C). These results suggest that caffeic acid is the primary effective substance in serum following the single-dose administration of ICGAC.

2.4. ICGAC and DEX Influence the Concentration of SCFAs Produced by Gut Microbiota in Asthmatic Mice

In our study, we determined SCFA levels in the feces using GC-MS (Figure 2D). The current study showed that the levels of acetic acid, propionic acid, and butyric acid in feces were higher in the model group of mice compared to the control group. Meanwhile, the DEX, H-ICGAC and L-ICGAC groups had even higher levels of acetic acid, propionic acid, and butyric acid than the model group. Specifically, acetic acid and butyric acid were significantly up-regulated following the ICGAC treatment. The highest level of acetic acid was observed in the H-ICGAC group, while propionic acid levels were comparable between the H-ICGAC and DEX groups. The highest concentration of butyric acid was detected in the DEX group. These findings highlight the role of ICGAC and DEX in enhancing SCFA production, potentially contributing to their therapeutic effects in asthma.

2.5. Gut Microbiota Metabolizes ICGAC into Chlorogenic Acid, Its Active Form, Modulating Asthma Pathways

Chlorogenic acid was identified as the active metabolite form of ICGAC responsible for controlling allergic asthma, as demonstrated through a network pharmacology analysis. Kyoto Encyclopedia of Genes and Genomes (KEGG) pathway enrichment analysis showed that multiple signaling pathways and metabolic pathways, including asthma-related arachidonic acid metabolism pathways, were significantly affected (Figure 2E). Targets related to asthma were obtained from DisGeNET (https://disgenet.com/; access date: 17 January 2025), GeneCards, Home (https://www.genecards.org/; access date: 17 January 2025), OMIM (https://secure.jhu.edu/form/OMIM; access date: 17 January 2025) and e-TSN (http://www.lilab-ecust.cn/etsn/; access date: 17 January 2025). Targets related to chlorogenic acid were obtained from Swiss Target Prediction (http://www.swisstargetprediction.ch/; access date: 17 January 2025) and BATMAN (http://bionet.ncpsb.org.cn/batman-tcm/#/home; access date: 17 January 2025). The correlation analysis between the targets of chlorogenic acid and targets of asthma identified 51 potential targets to construct a PPI network using Cytoscape v3.8.2 (Figure 2F). The top 10 genes were selected as hub genes, including RAC-alpha serine/threonine-protein kinase 1 (Akt1), Peroxisome Proliferator-Activated Receptor Gamma (Pparg), Epidermal Growth Factor Receptor (Egfr), Matrix Metalloproteinase 9 (Mmp9), Toll-Like Receptor 4 (Tlr4), Estrogen Receptor 1 (Esr1), KIT Proto-Oncogene Receptor Tyrosine Kinase (Kit), Neurogenic Locus Notch Homolog Protein 1 (Notch1), Angiotensin-Converting Enzyme (Ace), and Bruton's Tyrosine Kinase (Btk) (Figure 2G).

2.6. Effect of ICGAC on the Composition of the Gut Microbiota

In the study, 16S rRNA gene sequencing was performed on fecal samples from each group to investigate how ICGAC modulates the gut microbiota. Principal coordinate analysis (PCoA) and non-metric multidimensional scaling (NMDS) showed that the ICGAC treatment significantly altered the structure of the gut microbiota. However, these changes did not restore the microbiota to a state similar to that of the control group, indicating a distinct directional shift (Figure 3A).

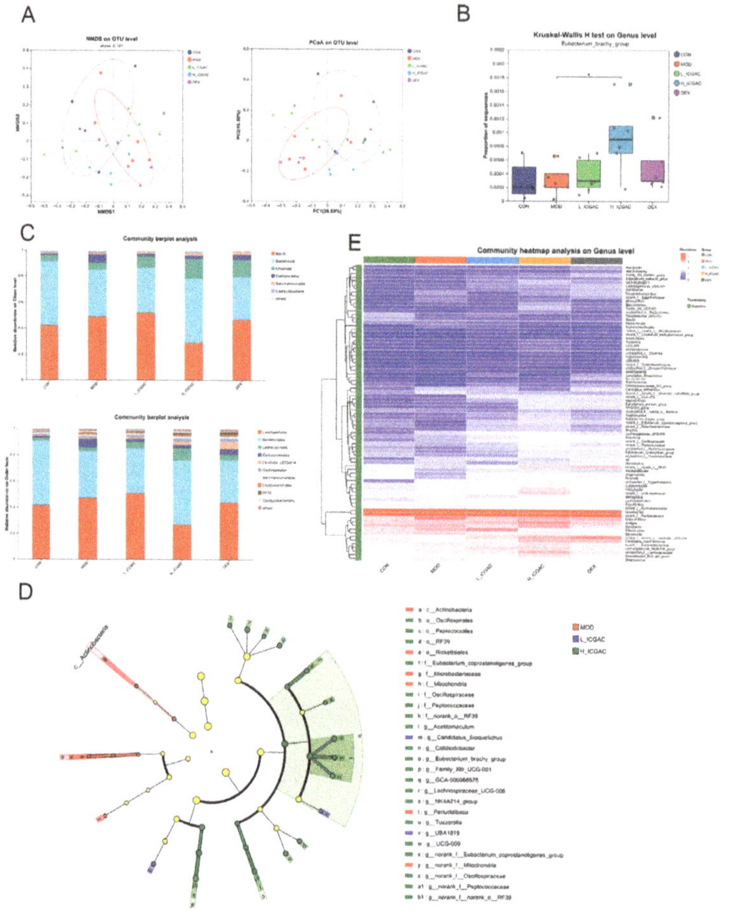

Figure 3. (**A**) Microbiota community analysis based on NMDS and PCoA score plots. (**B**) Quantitative analysis of Eubacterium_brachy_group. (**C**) Community abundance profiling of the gut microbiota on the class (left) and order (right) levels. (**D**) Linear discriminant analysis effect size (LEfSe) analysis for identifying the key enriched bacteria between model group, L−ICGAC group and H−ICGAC group. (**E**) Heatmap analysis of the top 80 differentiated taxa on the genus levels. * $p < 0.05$, when high dose ICGAC group was compared with model group.

To further confirm which bacterium altered by the ICGAC treatment could in turn affect the disease progression of OVA-induced asthma, we performed high-dimensional categorical comparisons using the linear discriminant analysis (LDA) of effect size (LEfSe), which detected marked differences in the predominance of bacterial communities between the model and ICGAC groups (Figure 3D). The results showed that *Eubacterium_brachy_group* displayed an enrichment in the ICGAC group, which might be associated with the transformation of ICGAC into chlorogenic acid and the subsequent amelioration of asthma symptoms (Figure 3B). Studies have shown that chlorogenic acid has remarkable anti-asthmatic efficacy [20]. Specifically, the abundance of *Clostridia* and *Lachnospirales* increased after the ICGAC treatment (Figure 3C). Overall, treatment with ICGAC induced significant changes in the structure and composition of the gut microbiota in asthmatic mice. Those trends were further supported by a heatmap analysis of the top 80 differentiated taxa on the genus levels (Figure 3E).

2.7. ICGAC Regulates Pulmonary Lipid Mediators to Mitigate Asthma Pathogenesis

To confirm the regulation of the phosphatidylcholine (PC) metabolic network by ICGAC, we quantified 206 PCs in the lung using UPLC tandem mass spectrometry. In total, 159 PCs were detected in mouse lung tissue. For the lung tissue, scatter plots based on principal component analysis (PCA) scores showed significant differences in PCs levels between the control and model groups (Figure 4D). PCA showed that there was a significant difference in PCs levels between the control group compared with the model group, with the ICGAC treatment resulting in PCs levels similar to those in the model group. Focusing on PCs containing docosahexaenoic acid (DHA), we observed no significant changes in PC concentrations in the ICGAC group compared to the model group. However, PC concentrations were higher in the DEX group than in the model group (Figure 4G). These trends were further supported by heatmap analysis, as illustrated in Figure 4A.

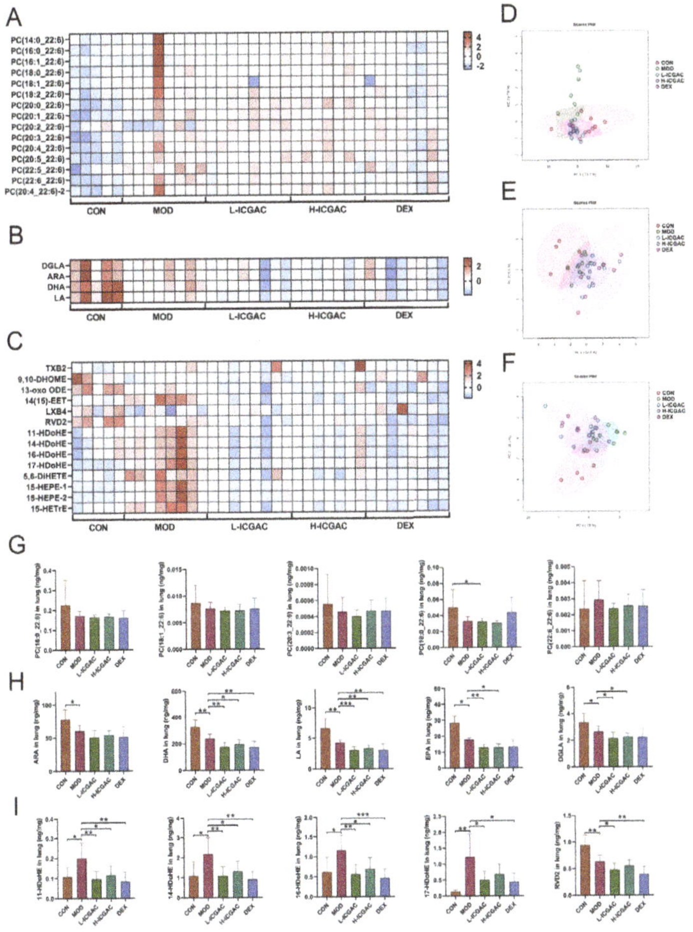

Figure 4. (**A–C**) Heatmap of PC (**A**), FFA (**B**) and LMs (**C**) in different groups. (**D–F**) PCA of PC (**D**), FFA (**E**) and LMs (**F**) in different groups. (**G–I**) PC (X_22:6) quantitation in lungs. (**H**) PUFA (ARA, DHA, LA, EPA, DGLA) quantitation in lungs. (**I**) LMs (11−HDoHE, 14−HDoHE, 16−HDoHE, 17−HDoHE, RVD2) quantitation in lungs. Data are shown as the mean ± SEM (n = 5–8), * $p < 0.05$, ** $p < 0.01$, *** $p < 0.001$.

To confirm the regulation of the polyunsaturated fatty acid (PUFA) metabolic network by ICGAC, we quantified arachidonic acid (ARA), docosahexaenoic acid (DHA), dihomo-γ-linolenic acid (DGLA), eicosapentaenoic acid (EPA), and linoleic acid (LA) in the lung using UPLC tandem mass spectrometry. In lung tissue, scatter plots based on PCA scores showed significant differences in polyunsaturated fatty acids levels between the control and model groups (Figure 4E). Further quantitative results revealed that OVA induction significantly reduced levels of ARA, DHA, DGLA, EPA, and LA. However, the concentration of ARA, DHA, DGLA, EPA, and LA were lower in the ICGAC group than in the model group (Figure 4H). Those trends were further supported by heatmap analysis, as shown in Figure 4B.

We quantified 105 ARA-, DHA-, DGLA-, EPA-, and LA-derived LMs in the lung using UPLC tandem mass spectrometry. In mouse lung tissue, 70 LMs were detected. Scatter plots based on PCA scores demonstrated significant differences in PUFA levels between the control and model groups (Figure 4F). PCA further revealed a significant difference in LM levels between the H-ICGAC and L-ICGAC groups compared to the model group. The ICGAC treatment induced a reversal of asthma-associated alterations in LM metabolism, suggesting a regulatory effect on lipid mediator pathways. Seven LMs (one LM was a metabolite of DGLA, one LM was a metabolite of ARA, one LM was a metabolite of EPA, and five LMs were metabolites of DHA) were determined to be significantly altered in the lungs based on further statistical analysis (Figure 4I). We focused on the DHA metabolites 11-Hydroxydocosahexaenoic acid (11-HDoHE), 14-HDoHE, 16-HDoHE, and resolvin D2 (RVD2), all of which are products of the arachidonate lipoxygenase (ALOX) enzyme. These metabolites were significantly up-regulated following asthmatic modeling. However, after ICGAC treatment, the levels of 11-HDoHE, 14-HDoHE, 16-HDoHE, and RVD2 were significantly down-regulated. These trends were further confirmed by the heatmap analysis, as shown in Figure 4C.

As shown in Figure 5, phospholipase A2 (PLA2) catalyzes the hydrolysis of the sn-2 position of PC (22:6/X) to release DHA. ALOX then catalyzes the conversion of DHA into LMs. Dexamethasone reduces LM production by inhibiting PLA2 activity, while ICGAC is metabolized by the gut microbiota to produce chlorogenic acid, which suppresses LM production by inhibiting ALOX12 activity.

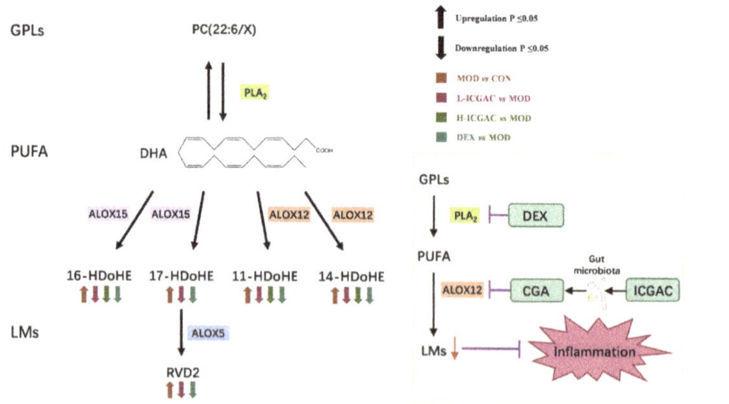

Figure 5. Metabolic pathway of PC and mechanism of anti-asthmatic action of DEX and ICGAC.

2.8. ICGAC Modulates ALOX12, SLC7A11, Nrf2, and GPX4 Expression to Enhance Antioxidant Defense in OVA Mice

We determined ALOX12, solute carrier family 7 member 11 (SLC7A11), nuclear factor erythroid 2-related factor 2 (Nrf2), and GPX4 expression in the lung tissues using real-time quantitative PCR. As shown in Figure 6A, H-ICGAC and L-ICGAC intervention significantly upregulated the expression of SLC7A11, GPX4 and Nrf2. H-ICGAC and L-ICGAC intervention significantly downregulated the expression of ALOX12. GSH levels in the lung tissue of mice were significantly upregulated after ICGAC treatment compared to the model group (Figure 6B). As shown in Figure 6D, ICGAC significantly reversed the OVA-induced downregulation of GPX4. As shown in Figure 6C, ICGAC increased the expression of SLC7A11, promoting the uptake of extracellular cystine into cells for conversion to cysteine. This led to enhanced GSH synthesis, increased GPX4 activity, and the subsequent inhibition of IgE production.

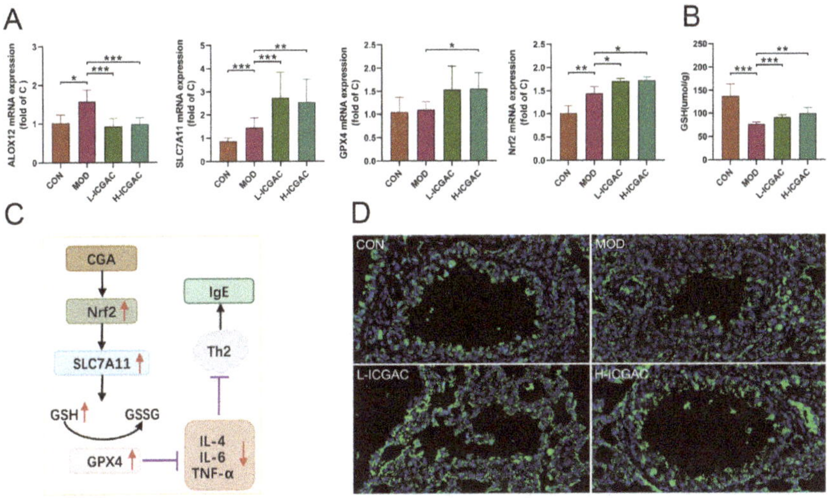

Figure 6. (**A**) The expression of lung ALXO12, SLC7A11, GPX4, Nfr2 using qRT-PCR. (**B**) The level of GSH in lung. Data are shown as the mean ± SEM (n = 3–8), * $p < 0.05$, ** $p < 0.01$, *** $p < 0.001$. (**C**) CGA inhibits the IgE production effect based on the Nrf2/GPx4 pathway. (**D**) Representative photomicrographs obtained using confocal microscopy after the immunofluorescence staining of GPX4. (GPX4 is displayed in green. Nuclei of cells are displayed in blue. scale bar: 100 μm).

3. Material and Method

3.1. Material

ELISA kits for the determination of tumor necrosis factor-α (TNF-α), interleukin (IL)-1β, IL-4, IL-5, IL-6, and immunoglobulin E (IgE) were purchased from Cusabio Biotech Co., Ltd. (Wuhan, China). Glutathione (GSH) assay kits were obtained from Nanjing Jiancheng Bioengineering Institute (Nanjing, China). Mass spectrometry (MS)-grade water, acetonitrile, formic acid, acetic acid, ethanol, isopropanol, methyl tert-butyl ether, and methanol were purchased from Thermo Fisher Scientific (Waltham, MA, USA). Glycerol, OVA, and butylated hydroxytoluene (BHT) were purchased from Sigma-Aldrich (St. Louis, MO, USA). Deuterated lipid mediator (LM) standards (IS) were purchased from Cayman (Ann Arbor, MI, USA). Deuterated phosphatidyl choline (PC) standards (IS) were purchased from Avanti (Alabaster, AL, USA). Deuterated propionic acid was purchased from Shanghai Zzbio Co., Ltd. (Shanghai, China). All other reagents were of analytical or high-performance liquid chromatography (HPLC) grade.

3.2. Establishment of the Asthma Mouse Model

All animal experiments were conducted in accordance with the guidelines established by the Animal Care and Use Committee (IACUC) of Xinjiang Medical University and the

Ethics Committee of the same institution (XJIMM-20230704). After purchase, the animals were kept in an animal house at a constant temperature (22 °C) and 50% relative humidity; they were fed ad libitum and acclimatized for 1 week during the rearing period.

BALB/C female mice (18–22 g) were divided into five groups: a naïve control group (A), a model group (B), a high-dose ICGAC group (C) (2 mg/mL, 20 mg/kg), a low-dose ICGAC group (D) (1 mg/mL, 10 mg/kg), and a dexamethasone (DEX) group (E) (0.16 mg/mL, 1.6 mg/kg). Group A consists of 6 mice, while Groups B, C, D, E, and F each comprise 8 mice. Mice in groups B, C, D, and E were intraperitoneally injected with 0.2 mL of OVA sensitization solution (aluminum hydroxide gel saline dilution 0.5 mg/mL), and mice in group A were injected with an equal amount of saline from days 1–8. On days 15–21, mice in group A were given saline stimulation, and mice in group B, C, D, E were given atomized OVA stimulation solution (OVA 10 mg/mL) for 30 min once a day for 7 days. Groups C, D, E were administered ICGAC and dexamethasone intragastrically. The dose was determined according to the previous study. After 1 h, atomized OVA was used for stimulation, once a day for 30 min per dose for 7 days. After gavage on the first day, three mice were randomly selected as single-dosage groups from the control and high-dose ICGAC groups. During the experiment, the mice were allowed to eat and drink freely.

3.3. Collection of Bio-Samples

Serum: Blood serum was collected from mouse eye samples and transferred into Eppendorf tubes. The tubes were then left to sit at room temperature for 2 h, followed by centrifugation at 3000 rpm for 15 min. The serum was subsequently stored at −20 °C for IgE detection and a serum components assay.

Bronchoalveolar lavage fluid (BALF): Mice were euthanized and skinned, followed by immediate isolation of the peritracheal tissues. A transverse incision was made in the upper trachea (between the third and fourth tracheal rings), and a 1 mL syringe needle was inserted into the trachea (approximately 2 mm) and secured. The right main bronchus was ligated, and the left lung was flushed three times with saline, 0.5 mL each time (with a recovery rate of 80–90%). The collected BALF was centrifuged twice at 2000 rpm for 10 min each, and the supernatant was aliquoted into 200 μL portions and stored at −80 °C for a subsequent cytokine assay.

Lung tissues: After serum collection, the mice were dislocated and executed. Lung tissues were immediately removed and rinsed clean with saline. The left lung was placed in a 5 mL Eppendorf tube containing 4% paraformaldehyde solution for morphological analysis. The right lung was immediately frozen in liquid nitrogen for 30 min and then transferred to a −80 °C refrigerator for the later analysis of phosphatidyl cholines (PCs), polyunsaturated fatty acids (PUFAs), and lipid mediators (LMs).

Fecal samples: Fecal samples were collected from the colon and immediately snap-frozen in liquid nitrogen. Subsequently, they were stored at −80 °C for later biochemical analysis, including the determination of short-chain fatty acids (SCFAs).

3.4. Biological Sample Preparation for PUFA, LM, and PC Assays

Preparation for the LM Assay: The preparation of lipid mediators begins with homogenizing 5 mg of finely ground lung tissue in 175 μL of acetonitrile through vigorous vortexing for 10 min. After centrifugation at 13,300 rpm for 10 min, the clarified supernatant is carefully collected and combined with stabilizing reagents: 35 μL of 10% glycerol, 1 μL of 10 mg/mL BHT to prevent oxidation, and 10 μL of 50 ng/mL internal standard. The mixture is diluted with 359 μL of 25% acetonitrile and vortexed again to ensure homogeneity. For purification, the sample is passed through a Waters MAX solid phase extraction (SPE) column (Milford, MA, USA) preactivated with acetonitrile and equilibrated with

25% acetonitrile. Sequential washes with 25% acetonitrile and pure acetonitrile eliminate interfering substances, while the target analytes are eluted using 1% formic acid in acetonitrile. The eluate is concentrated under vacuum and stored at −20 °C until analysis. Prior to ultra-high-performance liquid chromatography—multiple reaction monitoring (UHPLC-MRM) detection, the samples are reconstituted in a methanol/acetonitrile solution (50:50, *v*/*v*) to optimize chromatographic compatibility.

Preparation for the PC Assay: Building on the residual material from the LMs assay, phosphatidylcholine extraction involves supplementing the remaining supernatant with 50 μL of methanol and 1 mL of methyl tert-butyl ether (MTBE). The biphasic system is vortexed for 10 min to enhance lipid partitioning, followed by centrifugation at 13,300 rpm to isolate the organic phase. A 200 μL aliquot of the upper layer is transferred and dried under vacuum to remove solvents, with the resulting residue stored at −20 °C for stability. For downstream ultra-high-performance liquid chromatography—high-resolution mass spectrometry (UHPLC-HRMS) analysis, dried PC extracts are resolubilized in acetonitrile containing a 200 ng/mL internal standard.

This sequential methodology not only maximizes sample utilization but also minimizes cross-contamination risks between lipid classes, ensuring analytical specificity for both LMs and PCs within a unified workflow.

3.5. Biochemical Assays in Bronchoalveolar Lavage Fluid

The collected BALF was centrifuged at 3500 rpm for 15 min. The cell pellet was suspended in 1 mL phosphate-buffered saline (PBS) for inflammatory cell counting and classification. Levels of basophils (BAS), eosinophils (EOS) white blood cells (WBC), and lymphocytes (LYM) in BALF were measured using an automatic cell counter (Nihon Kohden, Shinjuku, Japan) according to the manufacturer's protocol. Levels of IL-4, IL-5, IL-6, IL-1β, TNF-α in BALF and the total IgE level in serum were measured using corresponding kits.

3.6. Biological Sample Preparation for Active Components Analysis

We mixed 10 μL of serum and 40 μL of methanol containing 0.1% formic acid. After centrifugation at 13,300 rpm for 10 min, 30 μL of the supernatant was taken and then subjected to mass spectrometry analysis.

3.7. Biological Sample Preparation for Fecal SCFAs Analysis

In brief, 10 mg of feces was mixed with 600 μL of water containing 0.5% concentrated sulfuric acid. The mixture was vortexed for 5 min and sonicated for 30 min and then centrifuged for 10 min at 13,300 rpm. Then, 500 μL of the supernatant was collected and extracted with 800 μL of MTBE. The mixture was vortexed for 5 min and centrifuged for 10 min at 13,300 rpm. Subsequently, 700 μL of the supernatant was collected and mixed with 0.15 g of anhydrous sodium sulfate. From this, another 500 μL aliquot was taken and combined with 500 μL of Propionic Acid-d3 (200 μM). After vortexing for another 5 min, the mixture was transferred to a chromatographic vial for analysis.

3.8. Real-Time Quantitative PCR Analysis

Total RNA was extracted from lung tissues using the RNAeasy™ Animal RNA Isolation Kit with Spin Column (Beyotime, Shanghai, China). The concentration of total RNA was quantified using a NanoDrop 2000 spectrophotometer (Thermo Fisher Scientific, Waltham, MA, USA). Then, 1 μg of purified RNA from each sample was reverse-transcribed into complementary DNA (cDNA) using the HiFi-Script cDNA Synthesis Kit (CWBIO, Taizhou, Jiangsu, China). Quantitative reverse transcription PCR (qRT-PCR) was performed using an ABI 7500 Fast PCR system (Applied Biosystems, San Francisco, CA, USA) using

UltraSYBR Mixture (Low ROX) (CWBIO, Taizhou, China). Glyceraldehyde-3-phosphate dehydrogenase (GAPDH) was employed as an internal control for normalization. The relative expression levels were calculated as fold changes in comparison to GAPDH, using the $2^{-\Delta\Delta CT}$ method.

3.9. The Determination of Glutathione (GSH)

First, we accurately weighed the tissue (approximately 0.005 g) and proceeded to homogenize it in physiological saline using a weight-to-volume ratio of 1:9 (g:mL). Subsequently, we centrifuged the homogenate at 2500 rpm for 10 min and then collected 0.1 mL of the supernatant. Following the manufacturer's instructions from the Reduced Glutathione (GSH) Assay Kit (Njjcbio, Nanjing, China), we added 0.1 mL of the reagent and mixed it thoroughly. We then centrifuged the mixture at 3500 rpm for 10 min and collected the supernatant for colorimetric analysis.

3.10. Histology and Immunofluorescence Staining

Lung tissues were fixed with 4% paraformaldehyde for 48 h, routinely dehydrated, paraffin embedded, cut into 4 μm thick sections, stained with hematoxylin-eosin (H&E), and observed under light microscope for inflammatory cell infiltration in lung tissues. The expression of glutathione peroxidase 4 (GPX4) in the lung tissue was detected via an immunofluorescence assay using anti-GPX4 antibody. After washing, the slides were incubated with Alexa Fluor 488-conjugated goat anti-mouse (Lot: A10680) and CY3-conjugated goat anti-rabbit secondary antibody (Lot: A10520, Thermo Fisher Scientific). Slides were washed with PBS and counterstained with 4′,6-Diamidino-2-Phenylindole (DAPI). Fluorescence images were obtained using a whole-slide imaging system.

As described in our previous study [14]. Microbiota analysis was conducted using the Majorbio Cloud Platform (https://cloud.majorbio.com/; access date: 7 October 2024). α- and β-diversity, along with bacterial community composition, were assessed using the Quantitative Insights Into Microbial Ecology (QIIME) pipeline. Additionally, the linear discriminant analysis (LDA) effect size (LEfSe) was employed with a threshold LDA score >3.0 to identify differentially abundant taxa.

3.11. UPLC-MRM Analysis

The analysis of active components in serum samples was carried out using Waters Synapt G2-Si High-Definition Mass Spectrometry (HDMS) with H-class ultra-performance liquid chromatography (UPLC) and an Acquity Bridged Ethylene Hybrid (BEH) C18 column. Mobile phase A consisted of water containing 0.1% acetic acid, while mobile phase B comprised acetonitrile with 0.1% acetic acid. The solvent procedure is shown in Table 1. The injection volume was 2 μL, and the column temperature was 30 °C. MS was operated in negative mode under MRM. Data acquisition was performed using MassLynx 4.2.

Table 1. Gradient elution program for active components analysis.

Time (min)	A%	B%
0	95	5
1	95	5
4	40	60
4.5	0	100
5	0	100
5.1	95	5

PUFAs and LMs analyses were performed on a SCIEX 5500plus mass spectrometer equipped with an Electrospray Ionization (ESI) source (Framingham, MA, USA) and a

Thermo Scientific Dionex ultimate 3000 HPLC (Waltham, MA, USA). The Waters ACQUITY UPLC BEH C18 (1.7 μm, 2.1 × 50 mm) column was used. Mobile phase A consisted of water with 0.1% acetic acid, whereas mobile phase B was a mixture of acetonitrile and isopropanol in a 9:1 volume ratio. The solvent procedure is shown in Table 2. The flow rate was set at 0.4 mL/min, with the column temperature maintained at 40 °C. An injection volume of 5 μL was used, and the MS was operated in negative mode under MRM. All data were acquired and processed using SCIEX Analyst software (version 1.7.1).

Table 2. Gradient elution program for the PUFA and LM analysis.

Time (min)	A%	B%
0	75	25
1	75	25
8	5	95
10	5	95
10.01	75	25
12	75	25

3.12. UPLC-MS Analysis

PCs analysis was performed on a SCIEX 5500plus mass spectrometer equipped with an ESI source (MA, USA) and a Thermo Scientific Dionex ultimate 3000 HPLC (MA, USA). The Waters ACQUITY UPLC C8 (1.7 μm, 2.1 × 100 mm) column was used. Mobile phase A consisted of a mixture of acetonitrile and water (3:2 *v*/*v*) containing 0.1% acetic acid and 10 mM ammonium acetate. Mobile phase B was a blend of acetonitrile and isopropanol (1:9 *v*/*v*) with 0.1% acetic acid and 10 mM ammonium acetate. The solvent procedure is shown in Table 3. The flow rate was set at 0.3 mL/min and the column temperature was maintained at 55 °C. The injection volume was 2 μL. MS was operated in positive mode under full scans. All data were acquired and processed using SCIEX Analyst software (version 1.7.1).

Table 3. Gradient elution program for the PC analysis.

Time (min)	A%	B%
0	80	20
2	75	25
2.1	70	30
12	65	35
12.1	30	70
18	1	99
18.1	80	20
20	80	20

3.13. GC-MS Analysis

The SCFA analysis was conducted on an Agilent 7890B (Santa Clara, CA, USA) equipped with an automatic sampler (7693A) and coupled to an Agilent 5973 mass selective detector. The column employed was a fused-silica capillary column featuring a free fatty acid phase (DB-FFAP 122-3232) with dimensions of 0.25 mm internal diameter, 30 m length, and 0.25 μm film thickness. The helium carrier gas flow rate was set at 1 mL/min. The oven temperature was initially set at 50 °C and held for 1.5 min. It was then ramped up to 90 °C at a rate of 20 °C/min and held for 2.1 min. Subsequently, it was increased to 120 °C at a rate of 5 °C/min and held for 1 min. Finally, it was raised to 180 °C at a rate of 25 °C/min and held for an additional 1.5 min. The injected sample volume for the GC analysis was 0.5 μL, and the run time for each analysis was 16.5 min. The detector was operated in the selection ionization mode (SIM). Ion selection of the SCFAs was based on the retention time of standard compounds. All data were acquired and processed using Agilent MassHunter (version B.07.01).

3.14. Data Processing and Statistical Analysis

The open-source tool MetaboAnalyst 6.0 (available at http://www.metaboanalyst.ca/; access date: 15 January 2025) was utilized for the statistical analysis. The threshold for selecting important predictive variables was based on the variable importance in projection (VIP) score. Multivariate statistical differences were identified using an independent sample *t*-test. Variables with a VIP score greater than 1 and a *p*-value less than 0.05 were considered significant and selected as differential variables.

4. Discussion

In this study, ICGAC, identified as the key active ingredient in SXCF (a traditional folk remedy historically used for asthma relief in China), was assessed in an allergic asthma mouse model. Both dexamethasone and ICGAC significantly reduced typical lung morphological changes in asthma, such as pronounced inflammatory cell infiltration, bronchial epithelial degeneration, and interstitial inflammation. Remarkably, ICGAC exhibited therapeutic efficacy comparable to or exceeding that of dexamethasone, while demonstrating no experimentally detectable adverse effects commonly associated with steroid use. This highlights ICGAC's potential as a safe and effective natural therapeutic option for managing allergic asthma.

4.1. Oral ICGAC Boosts Gut-Microbiota-Derived CGA and SCFAs as Bioactive Metabolites

We discovered that orally administered ICGAC is not the direct effective component; instead, it undergoes transformation into Chlorogenic acid. This transformation process is mediated by the gut microbiota, a complex community of microorganisms that play crucial physiological roles in food digestion, nutrient absorption, drug metabolism, and maintenance of gut barrier function. Upon entering the intestine, ICGAC may interact with these microorganisms, leading to its conversion into CGA.

Although the involvement of the gut microbiota in such transformation processes is well recognized, the specific bacterial species or strains responsible for converting ICGAC to CGA remain unclear. However, several studies have emphasized the significant role of the gut microbiota in the biotransformation of bioactive compounds [21]. For example, certain gut bacteria are capable of hydrolyzing, reducing, or methylating compounds, thereby altering their chemical structures and biological activities [22]. Our study detected a significant increase in the *Eubacterium_brachy_group*, suggesting its potential role in the conversion of ICGAC to CGA through analogous microbial metabolic pathways. To date, several studies have demonstrated that chlorogenic acid exhibits anti-inflammatory effects via the modulation of the intestinal microbiota [23].

Natural compounds frequently undergo structural modifications during absorption. Moreover, the effective form and pharmacological mechanisms of natural drugs can vary between single-dose and prolonged administrations [24]. Further research is essential to fully elucidate the intricate interactions between natural products, gut microbiota, and therapeutic efficacy.

Using UPLC-tandem mass spectrometry, we measured serum ICGAC and chlorogenic acid levels. Both H-ICGAC and L-ICGAC groups showed similar chlorogenic acid concentrations in the bloodstream (Figure 2A). This similarity can be attributed to the enhanced expression of P-glycoprotein (PGP) in the gut endothelial cells of the H-ICGAC group, facilitated by ICGAC.

PGP, an ATPase efflux pump that is a member of the ATP-binding cassette transporter family, is recognized for its ability to extrude a variety of lipophilic drugs. Contemporary studies have underscored a significant association between gut microbiota and PGP expression within the colon [25]. Specifically, *Clostridium* and *Bacillus* have been identified as key in stimulating PGP expression in the gut epithelium of mouse models, contributing to the observed correlation [25]. This suggests that ICGAC may upregulate PGP, leading to stable chlorogenic acid serum levels irrespective of the initial dose administered.

ICGAC also enhances the synthesis of SCFAs—notably, acetic acid and butyric acid—upon administration. SCFAs play a crucial role in modulating the immune system and reducing inflammation [26]. They also regulate various immune cells and influence T and B cell differentiation, impacting both innate and adaptive responses [27]. Furthermore, SCFAs reduce airway inflammation, a hallmark feature of allergic asthma [14].

Indeed, multiple studies emphasize the connection between SCFAs and allergic asthma. Changes in the gut microbiota, which produce SCFAs, can influence the risk of developing allergic asthma [14]. While, early-life antibiotic exposure disrupts the gut microbiota, increasing the risk of asthma and other allergic conditions [28]. Moreover, SCFAs modulate immune-related genes, potentially contributing to allergic asthma [29]. Intriguingly, our findings further show a positive correlation between butyrate production and PGP expression, highlighting its importance. A detailed analysis revealed that *Eubacterium_brachy_group* plays a key role in SCFA production following ICGAC administration (Figure 3).

4.2. ICGAC Modulates ALOX Activity and LMs to Reduce Asthmatic Inflammation

We explored the regulatory role of ICGAC in LMs metabolism and its anti-inflammatory potential through Lipidomics analysis. Our investigation focused on the activities of arachidonate lipoxygenases (ALOX-isoforms), particularly ALOX12 and ALOX15, which catalyze the oxidation of polyunsaturated fatty acids (PUFAs) and play crucial roles in inflammatory responses [30].We found that ICGAC treatment inhibited ALOX12 and ALOX15 activities, specifically reducing the enzyme-catalyzed oxidation of DHA (Figure 6A). DHA, an omega-3 PUFA, serves as a precursor for anti-inflammatory metabolites such as 11-HDoHE, 14-HDoHE, 16-HDoHE, 17-HDoHE, and RVD2, which are crucial in resolving inflammation in allergic asthma [31]. Disruption of their balance can result in heightened inflammatory reactions and asthma symptoms.

Consistently with this, chlorogenic acid, the active compound in ICGAC, has also been identified as a competitive inhibitor of ALOX, thereby inhibiting lipid peroxidation [32]. Taken together, our findings indicate that ICGAC modulates the metabolic networks of LMs and PUFAs, potentially exerting anti-inflammatory effects by regulating ALOX activity and LMs levels.

4.3. ICGAC Enhances GPX4, SLC7A11, and Nrf2 to Strengthen Antioxidant Defense

In this study, we explored how ICGAC suppresses asthma-related inflammation and oxidative stress, specifically by upregulating GPX4, SLC7A11, and Nrf2 in lung tissues.

Firstly, GPX4, an enzyme found on the outer surface of the inner mitochondrial membrane and within the mitochondrial matrix, is unique in its ability to directly reduce phospholipid peroxides (PL-OOH) in the mitochondrial membrane [33]. By using GSH as a co-substrate, GPX4 plays a vital role in regulating cellular lipid peroxides. Our findings revealed that the ICGAC treatment markedly increased GPX4 expression (Figure 6A), thereby enhancing the uptake and reduction in the peroxide load in PL-OOH, converting it into nontoxic lipid alcohol while oxidizing glutathione (reduced, GSH) to glutathione (oxidized, GSSG) [34].

Secondly, ICGAC significantly upregulated SLC7A11 expression (Figure 6A), which aids in the uptake of extracellular cystine for conversion to cysteine, a crucial precursor for glutathione synthesis [35]. Consistently with this upregulation, we observed a notable increase in GSH content in the lung tissue of ICGAC-treated mice compared to the model group.

Additionally, Nrf2, a member of the basic leucine zipper transcription factor family, is a central regulator of the antioxidant response. GPX4, a well-known transcriptional target of Nrf2, is upregulated by Nrf2 activation, creating a positive feedback loop that enhances cellular antioxidant defenses [36]. Our results indicated that ICGAC also increased Nrf2 and GPX4 expression (Figure 6A), further strengthening the antioxidant response.

Moreover, dendritic cells (DCs) facilitate the differentiation of naive CD4^{+} T cells to Th2 cells, which further stimulate B cells to produce IgE [37]. IL-4 amplifies Th2 activity, but elevated levels of Nrf2 inhibit its release [38]. ICGAC suppressed IL-4 and other asthma-related cytokines, thus reducing IgE production by B cells and alleviating asthma symptoms by lowering airway responsiveness (Figure 7).

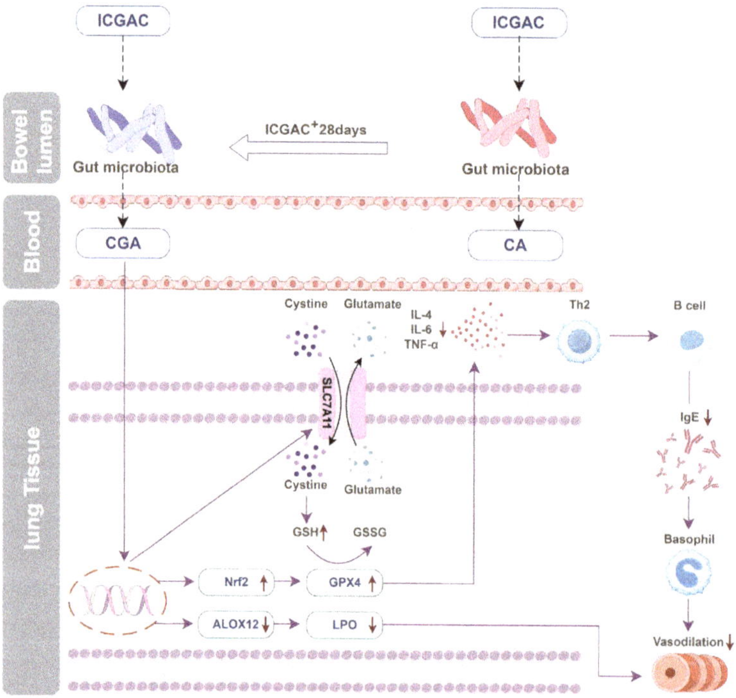

Figure 7. Illustration of the anti-asthmatic mechanism of ICGAC through the gut–lung axis. (The process of the change in the effective substances is related to the gut microbiota in mice. When ICGAC is administered for a long time, ICGAC in mice will be converted into CGA; when administered once, ICGAC will be converted into CA. ICGAC significantly upregulated the expression of SLC7A11 in mouse lung tissue. SLC7A11 helps to absorb cystine outside the cell and convert it into cysteine. ICGAC also upregulated the expression of GPX4 in mouse lung tissue. It reduced the release of inflammatory factor IL-4, thereby reducing the release of IgE by B cells. ICGAC also increased the expression of Nrf2, further strengthening the antioxidant response. CGA can inhibit the expression of ALOX12, thereby reducing the generation of inflammatory mediators such as 11-HDoHE, 14-HDoHE, 16-HDoHE, 17-HDoHE and RvD2, and alleviating asthma inflammation.).

5. Conclusions

In conclusion, our study integrates lipid metabolomics and network pharmacology to elucidate the mechanisms underlying ICGAC's efficacy in treating allergic asthma. We demonstrate that, through gut microbiota-derived chlorogenic acid and SCFAs, ICGAC inhibits lipid peroxidation and inflammatory mediators. Our findings also highlight the critical roles of the GPX4 and Nrf2 pathways in ICGAC's therapeutic effects. This research provides a robust foundation for the potential clinical application of ICGAC in asthma management and paves the way for future research and therapeutic innovations.

Supplementary Materials: The following supporting information can be downloaded at: https://www.mdpi.com/article/10.3390/ijms26104864/s1.

Author Contributions: J.-Y.X.: writing—original draft, investigation, methodology. X.-J.R.: methodology, funding acquisition. Z.S.: investigation. Y.-D.G.: investigation. Y.-X.Z.: investigation. C.-C.D.: investigation. Y.W.: investigation, formal analysis. Y.-X.H.: validation, supervision. T.-L.G.: writing—review and editing, formal analysis, supervision, funding acquisition. C.T.: writing—review and editing, supervision, project administration, funding acquisition, conceptualization, data curation. All authors have read and agreed to the published version of the manuscript.

Funding: This work was supported by the Natural Science Foundation of Xinjiang Uygur Autonomous Region (2022D01E65), the CAMS Innovation Fund for Medical Sciences (2022-I2M-1-016, 2024-I2M-ZH-012), the Tianshan Talent Training Program (2023TSYCCX0065), and the Fundamental Research Funds for the Central Universities (2024ZKPYHH01).

Institutional Review Board Statement: The animal study was reviewed and approved by the Animal Care and Used Committee of XinJiang Medical University (protocol code: XJIMM-20230704 and date of approval: 2023-07-04).

Informed Consent Statement: Not applicable.

Data Availability Statement: Data is unavailable due to privacy or ethical restrictions.

Acknowledgments: Support for Figure 7 was provided by Figdraw. Technical validation was conducted with the assistance of QuantSpec (Sciprecis, Beijing).

Conflicts of Interest: The authors declare that they have no known competing financial interests or personal relationships that could have appeared to influence the work reported in this study.

List of Abbreviations

DEX, Dexamethasone; ICGAC, Isochlorogenic acid C; LMs, Lipid mediators; IgE, Immunoglobulin E; qRT-PCR, Quantitative reverse transcription PCR; OVA, Ovalbumins; TNF-α, Tumor necrosis factor-α; IL-1β, Interleukin-1β; IL-4, Interleukin-4; IL-5, Interleukin-5; IL-6, Interleukin-6; PCs, phosphatidyl cholines; PUFAs, polyunsaturated fatty acids; BALF, Bronchoalveolar lavage fluid; SCFAs, Bronchoalveolar lavage fluid; UHPLC, Ultra High Performance Liquid Chromatography; MRM, Multiple Reaction Monitoring; HRMS, High Resolution Mass Spectrometry; MTBE, methyl tert-butyl ether; BAS, Basophil; EOS, Eosinophil; WBC, White blood cell; LYM, Lymphocyte; MON, Monocytes; GSH, Glutathione; H&E, hematoxylin-eosin; GPX4, Glutathione Peroxidase 4; SPE, Solid-Phase Extraction; GC-MS, Gas Chromatography-Mass Spectrometry; BHT, Butylated hydroxytoluene; LDA, Linear Discriminant Analysis; LEfSe, LDA Effect Size; VIP, Variable Importance in Projection; Nrf2, Nuclear factor erythroid 2-related factor 2; ALOX12, Arachidonate 12-Lipoxygenase; DC, Dendritic cells; PCoA, Principal coordinate analysis; NMDS, Non-metric multidimensional scaling; ARA, Arachidonic acid; DHA, Docosahexaenoic Acid; DGLA, Dihomo-γ-linolenic acideicosapentaenoic Acid; EPA, Eicosapentaenoic Acid; LA, Linoleic Acid; PCA, Principal Component Analysis; RVD2, Resolvin D2; PLA2, Phospholipase A2 enzymes; SLC7A11, Solute Carrier Family 7 Member 11; SXCF, Hyssopus cuspidatus Boriss extract; ELASA, Enzyme linked immunosorbent assay; CGA, Chlorogenic acid; CA, Caffeic acid; PGP, P-glycoprotein.

References

1. Wang, Z.; Li, Y.; Gao, Y.; Fu, Y.; Lin, J.; Lei, X.; Zheng, J.; Jiang, M. Global, regional, and national burden of asthma and its attributable risk factors from 1990 to 2019: A systematic analysis for the Global Burden of Disease Study 2019. *Respir. Res.* **2023**, *24*, 169. [CrossRef] [PubMed]
2. Huang, K.; Yang, T.; Xu, J.; Yang, L.; Zhao, J.; Zhang, X.; Bai, C.; Kang, J.; Ran, P.; Shen, H. Prevalence, risk factors, and management of asthma in China: A national cross-sectional study. *Lancet* **2019**, *394*, 407–418. [CrossRef] [PubMed]
3. Kostakou, E.; Kaniaris, E.; Filiou, E.; Vasileiadis, I.; Katsaounou, P.; Tzortzaki, E.; Koulouris, N.; Koutsoukou, A.; Rovina, N. Acute severe asthma in adolescent and adult patients: Current perspectives on assessment and management. *J. Clin. Med.* **2019**, *8*, 1283. [CrossRef] [PubMed]
4. Hough, K.P.; Curtiss, M.L.; Blain, T.J.; Liu, R.-M.; Trevor, J.; Deshane, J.S.; Thannickal, V.J. Airway remodeling in asthma. *Front. Med.* **2020**, *7*, 191. [CrossRef]
5. Lemanske, R.F., Jr. Viruses and asthma: Inception, exacerbation, and possible prevention. *J. Pediatr.* **2003**, *142*, S3–S8. [CrossRef]
6. Kwah, J.H.; Peters, A.T. Asthma in adults: Principles of treatment. *Allergy Asthma Proc.* **2019**, *40*, 396–402. [CrossRef]
7. Papi, A.; Ryan, D.; Soriano, J.B.; Chrystyn, H.; Bjermer, L.; Rodríguez-Roisin, R.; Dolovich, M.B.; Harris, M.; Wood, L.; Batsiou, M. Relationship of inhaled corticosteroid adherence to asthma exacerbations in patients with moderate-to-severe asthma. *J. Allergy Clin. Immunol. Pract.* **2018**, *6*, 1989–1998.e3. [CrossRef]
8. Pizzichini, M.M.; Kerstjens, H.A.; Pizzichini, E. Current role of anticholinergic drugs in the treatment of asthma. *Pol. Arch. Med. Wewnętrznej* **2015**, *125*, 859–866. [CrossRef]
9. Gosens, R.; Gross, N. The mode of action of anticholinergics in asthma. *Eur. Respir. J.* **2018**, *52*, 1701247. [CrossRef]
10. Wenzel, S. Severe asthma in adults. *Am. J. Respir. Crit. Care Med.* **2005**, *172*, 149–160. [CrossRef]
11. Heffler, E.; Madeira, L.N.G.; Ferrando, M.; Puggioni, F.; Racca, F.; Malvezzi, L.; Passalacqua, G.; Canonica, G.W. Inhaled Corticosteroids Safety and Adverse Effects in Patients with Asthma. *J. Allergy Clin. Immunol. Pract.* **2018**, *6*, 776–781. [CrossRef]
12. Gao, T.-L.; Guo, H.-H.; Jiang, J.-D. Neg-Entropy Mechanism as a Target for Natural Medicines. *Engineering* **2024**, *38*, 11–12. [CrossRef]
13. Gao, T.; Hao, J.; Wiesenfeld-Hallin, Z.; Wang, D.-Q.; Xu, X.-J. Analgesic effect of sinomenine in rodents after inflammation and nerve injury. *Eur. J. Pharmacol.* **2013**, *721*, 5–11. [CrossRef] [PubMed]
14. Guo, H.-H.; Han, Y.-X.; Rong, X.-J.; Shen, Z.; Shen, H.-R.; Kong, L.-F.; Guo, Y.-D.; Li, J.-Z.; Xu, B.; Gao, T.-L. Alleviation of allergic asthma by rosmarinic acid via gut-lung axis. *Phytomedicine* **2024**, *126*, 155470. [CrossRef]
15. Tanno, L.K.; Gonzalez-Estrada, A.; Olivieri, B.; Caminati, M. Asthma and anaphylaxis. *Curr. Opin. Allergy Clin. Immunol.* **2019**, *19*, 447–455. [CrossRef]
16. Lee, Y.-C. Synergistic effect of various regulatory factors in TH1/TH2 balance; immunotherapeutic approaches in asthma. *Int. J. Biomed. Sci. IJBS* **2008**, *4*, 8. [CrossRef]
17. Jiang, T.; Dai, L.; Li, P.; Zhao, J.; Wang, X.; An, L.; Liu, M.; Wu, S.; Wang, Y.; Peng, Y. Lipid metabolism and identification of biomarkers in asthma by lipidomic analysis. *Biochim. Et Biophys. Acta (BBA)-Mol. Cell Biol. Lipids* **2021**, *1866*, 158853. [CrossRef]
18. Wang, S.; Tang, K.; Lu, Y.; Tian, Z.; Huang, Z.; Wang, M.; Zhao, J.; Xie, J. Revealing the role of glycerophospholipid metabolism in asthma through plasma lipidomics. *Clin. Chim. Acta* **2021**, *513*, 34–42. [CrossRef]
19. Kim, S.H.; Kim, B.K.; Lee, Y.C. Antiasthmatic effects of hesperidin, a potential Th2 cytokine antagonist, in a mouse model of allergic asthma. *Mediat. Inflamm.* **2011**, *2011*, 485402. [CrossRef]
20. Kim, H.-R.; Lee, D.-M.; Lee, S.-H.; Seong, A.-R.; Gin, D.-W.; Hwang, J.-A.; Park, J.-H. Chlorogenic acid suppresses pulmonary eosinophilia, IgE production, and Th2-type cytokine production in an ovalbumin-induced allergic asthma: Activation of STAT-6 and JNK is inhibited by chlorogenic acid. *Int. Immunopharmacol.* **2010**, *10*, 1242–1248. [CrossRef]
21. Dey, P. Gut microbiota in phytopharmacology: A comprehensive overview of concepts, reciprocal interactions, biotransformations and mode of actions. *Pharmacol. Res.* **2019**, *147*, 104367. [CrossRef] [PubMed]
22. Selma, M.V.; Espin, J.C.; Tomas-Barberan, F.A. Interaction between phenolics and gut microbiota: Role in human health. *J. Agric. Food Chem.* **2009**, *57*, 6485–6501. [CrossRef] [PubMed]
23. Shi, A.; Li, T.; Zheng, Y.; Song, Y.; Wang, H.; Wang, N.; Dong, L.; Shi, H. Chlorogenic Acid Improves NAFLD by Regulating gut Microbiota and GLP-1. *Front. Pharmacol.* **2021**, *12*, 693048. [CrossRef] [PubMed]
24. Satoskar, R.; Bhandarkar, S. *Pharmacology and Pharmacotherapeutics*; Elsevier: Bengaluru, India, 2020.
25. Foley, S.E.; Tuohy, C.; Dunford, M.; Grey, M.J.; De Luca, H.; Cawley, C.; Szabady, R.L.; Maldonado-Contreras, A.; Houghton, J.M.; Ward, D.V. Gut microbiota regulation of P-glycoprotein in the intestinal epithelium in maintenance of homeostasis. *Microbiome* **2021**, *9*, 1–17. [CrossRef]
26. Parada Venegas, D.; De la Fuente, M.K.; Landskron, G.; González, M.J.; Quera, R.; Dijkstra, G.; Harmsen, H.J.; Faber, K.N.; Hermoso, M.A. Short chain fatty acids (SCFAs)-mediated gut epithelial and immune regulation and its relevance for inflammatory bowel diseases. *Front. Immunol.* **2019**, *10*, 277.

27. Gonçalves, P.; Araújo, J.R.; Di Santo, J.P. A cross-talk between microbiota-derived short-chain fatty acids and the host mucosal immune system regulates intestinal homeostasis and inflammatory bowel disease. *Inflamm. Bowel Dis.* **2018**, *24*, 558–572. [CrossRef]
28. Russell, S.L.; Gold, M.J.; Hartmann, M.; Willing, B.P.; Thorson, L.; Wlodarska, M.; Gill, N.; Blanchet, M.R.; Mohn, W.W.; McNagny, K.M. Early life antibiotic-driven changes in microbiota enhance susceptibility to allergic asthma. *EMBO Rep.* **2012**, *13*, 440–447. [CrossRef]
29. Sbihi, H.; Boutin, R.C.; Cutler, C.; Suen, M.; Finlay, B.B.; Turvey, S.E. Thinking bigger: How early-life environmental exposures shape the gut microbiome and influence the development of asthma and allergic disease. *Allergy* **2019**, *74*, 2103–2115. [CrossRef]
30. Kulkarni, A.; Nadler, J.L.; Mirmira, R.G.; Casimiro, I. Regulation of tissue inflammation by 12-lipoxygenases. *Biomolecules* **2021**, *11*, 717. [CrossRef]
31. Zhang, X.; Yang, N.; Ai, D.; Zhu, Y. Systematic metabolomic analysis of eicosanoids after omega-3 polyunsaturated fatty acid supplementation by a highly specific liquid chromatography–tandem mass spectrometry-based method. *J. Proteome Res.* **2015**, *14*, 1843–1853. [CrossRef]
32. Zhai, L.; Ning, Z.-W.; Huang, T.; Wen, B.; Liao, C.-H.; Lin, C.-Y.; Zhao, L.; Xiao, H.-T.; Bian, Z.-X. Cyclocarya paliurus leaves tea improves dyslipidemia in diabetic mice: A lipidomics-based network pharmacology study. *Front. Pharmacol.* **2018**, *9*, 973. [CrossRef] [PubMed]
33. Reichert, C.O.; de Freitas, F.A.; Sampaio-Silva, J.; Rokita-Rosa, L.; Barros, P.d.L.; Levy, D.; Bydlowski, S.P. Ferroptosis mechanisms involved in neurodegenerative diseases. *Int. J. Mol. Sci.* **2020**, *21*, 8765. [CrossRef] [PubMed]
34. Averill-Bates, D.A. The antioxidant glutathione. In *Vitamins and Hormones*; Elsevier: Amsterdam, The Netherlands, 2023; Volume 121, pp. 109–141.
35. Wu, G.; Lupton, J.R.; Turner, N.D.; Fang, Y.-Z.; Yang, S. Glutathione metabolism and its implications for health. *J. Nutr.* **2004**, *134*, 489–492. [CrossRef]
36. Brigelius-Flohé, R.; Flohé, L. Regulatory phenomena in the glutathione peroxidase superfamily. *Antioxid. Redox Signal.* **2020**, *33*, 498–516. [CrossRef]
37. Kadowaki, N. Dendritic Cells—A Conductor of T Cell Differentiation. *Allergol. Int.* **2007**, *56*, 193–199. [CrossRef]
38. Kikuchi, N.; Ishii, Y.; Morishima, Y.; Yageta, Y.; Haraguchi, N.; Itoh, K.; Yamamoto, M.; Hizawa, N. Nrf2 protects against pulmonary fibrosis by regulating the lung oxidant level and Th1/Th2 balance. *Respir. Res.* **2010**, *11*, 31. [CrossRef]

International Journal of
Molecular Sciences

MDPI

Article

Exploring the Multi-Faceted Effects of Berberine in Ameliorating Diastolic Dysfunction in Rats with Heart Failure with Preserved Ejection Fraction

Yu Mu [1], Jing Geng [2], Chilu Liu [1], Shuang Jiang [1], Yanxing Han [1], Jiandong Jiang [1,*] and Yuhong Wang [1,*]

1 State Key Laboratory of Bioactive Substances and Function of Natural Medicine, Institute of Materia Medica, Chinese Academy of Medical Sciences and Peking Union Medical College, Beijing 100050, China
2 Institute of Medicinal Biotechnology, Chinese Academy of Medical Sciences and Peking Union Medical College, Beijing 100050, China
* Correspondence: jiangjd@imm.ac.cn (J.J.); yuhongwang78@hotmail.com or wangyh@imm.ac.cn (Y.W.)

Abstract: Heart failure with preserved ejection fraction (HFpEF), marked by cardiac diastolic dysfunction, contributes to half of all heart failure cases globally and poses a significant public health challenge. Effective therapies for HFpEF are rare, largely due to its complex and heterogeneous pathophysiology, which often involves multiple comorbidities. Berberine (BBR), an isoquinoline alkaloid, has demonstrated beneficial effects on multiple metabolic and cardiovascular disorders; however, its impact on cardiac diastolic dysfunction in HFpEF remains poorly understood. In this study, we utilized a rat model of HFpEF induced by a sustained high-fat/high-sucrose (HFHS) diet to explore the impact and mechanisms of BBR on diastolic dysfunction. The results revealed that BBR administration effectively alleviated cardiac diastolic dysfunction and alleviated extracardiac comorbidities, including increased weight, impaired glucose tolerance, hypercholesterolemia and hypertension, in rats fed an HFHS diet. Furthermore, BBR mitigated myocardial inflammation, oxidative stress, microvascular endothelial dysfunction, and notably restored the disturbed NO-cGMP-PKG pathway. Additionally, BBR reduced myocardial fibrosis and inhibited the abnormally activated TGF-β/Smads signaling. Moreover, BBR attenuated the systemic inflammation and corrected immune dysregulation in an HFHS diet-fed rats. Our study suggests that BBR exhibits multi-beneficial effects in the prevention and management of HFpEF, demonstrating its potential as a holistic therapeutic candidate for HFpEF.

Keywords: berberine; heart failure; diastolic dysfunction; inflammation; oxidative stress; cardiovascular fibrosis; endothelial dysfunction

1. Introduction

Heart failure (HF) poses a substantial clinical and public health challenge, with nearly half of the cases attributed to HF with preserved ejection fraction (HFpEF), which is marked by cardiac diastolic dysfunction with normal ejection fraction (EF) values [1]. Despite its high prevalence and poor prognosis with five-year mortality rate exceeding 50%, conventional pharmacological therapies for HF with reduced ejection fraction (HFrEF) have demonstrated limited efficacy in improving clinical outcomes for patients with HFpEF [2,3]. This highlights an urgent medical need for effective treatments to HFpEF.

Cardiac diastolic dysfunction constitutes a core pathological feature in the development of HFpEF, with low-grade inflammation being a key mechanism leading to this dysfunction. Patients with HFpEF often present with numerous extracardiac inflammatory comorbidities, notably obesity, hypertension, type 2 diabetes and aging [4]. In recent years, novel therapeutic agents, including glucagon-like peptide-1 (GLP-1) receptor agonists and sodium-glucose cotransporter 2 (SGLT2) inhibitors, have shown promising results in improving symptoms and prognosis in patients with HFpEF [5,6]. Rising evidence indicates that, beside hypoglycemic effect, SLTG2 inhibitors and GLP-1 receptor agonists have a systemic impact via indirectly targeting heart by regulating inflammation, oxidative stress, nitric oxide (NO) bioavailability, cardiac fibrosis, and other mechanisms [7–9]. Furthermore, the anti-inflammatory properties of SGLT2 inhibitors and GLP-1 receptor agonists may account for their cardiovascular benefits that extend beyond glucose-lowering effects [9,10]. However, their application may be constrained by a limited target population, high drug costs, and potential adverse effects, such as urinary tract infections. In spite of this, these findings suggest a novel and intriguing therapeutic strategy for HFpEF.

Plant-based medicines have a long history in the treatment of chronic HF. Berberine (BBR), a bioactive isoquinoline alkaloid derived from the medicinal plant *Coptis chinensis*, is well known for its antibacterial and antidiarrheal properties [11]. It has also demonstrated significant benefits for cardiovascular disease due to its pleiotropic effects, including the regulation of glucose homeostasis and lipid profile, attenuation of inflammatory responses and oxidative damage, and inhibition of fibrotic processes [12,13]. Clinical investigations have established the safety and clinical effectiveness of BBR in the treatment of HF patients [14]. Additionally, BBR holds promise as a therapeutic agent for systemic inflammation, as it has shown therapeutic potential in both prophylactic and therapeutic interventions for chronic inflammation associated disease, including atherosclerosis and/or non-alcoholic fatty liver disease by controlling inflammation [15]. These findings underscore its promise as a novel therapeutic candidate for HFpEF. Although a recent study has shown that BBR mitigated diastolic dysfunction in a murine HFpEF model possibly by inhibiting mitochondrial fragmentation [16], the mechanism by which BBR improves cardiac diastolic dysfunction still needs to be further explored.

Previous investigations have established that chronic exposure to calorie-dense diets can impair cardiac diastolic function and contribute to the progression of HFpEF [17,18]. This study aims to comprehensively assess the impact of BBR on diastolic dysfunction in rats with HFpEF caused by a sustained high-fat, high-sucrose (HFHS) diet, hoping to establish a theoretical foundation for its potential translation into clinical practice in the future.

2. Results

2.1. BBR Ameliorates Cardiac Diastolic Dysfunction, Hypertrophy, and Extracardiac Comorbidities in HFHS Diet-Fed Rats

Echocardiography results showed that rats fed an HFHS diet had similar ejection fraction (EF) and fractional shortening (FS) values to normal control (NC) rats. However, they exhibited evident cardiac diastolic dysfunction, evidenced by a decreased mitral E/A ratio (Figure 1B–E), and cardiac hypertrophy, indicated by increases in left ventricular anterior wall thickness (LVAW), LV mass, and heart weight to body weight ratio (HW/BW) (Figure 1F–H). Furthermore, myocardial mRNA expressions levels of ANP and BNP were significantly increased in HFHS diet-fed rats (Figure 1I,J). Histological analysis using H&E staining showed disordered cardiomyocytes, infiltration of inflammatory cells, and increased cardiomyocyte cross-sectional area (CSA) (Figure 1K,L). Additionally, serum levels of BNP, lactate dehydrogenase (LDH) and creatine kinase (CK) were elevated (Figure 1M–O). Importantly, BBR administration significantly mitigated these cardiac abnormalities, indicating a protective effect against cardiac injury and improvement in diastolic function. Moreover, BBR reduced body weight, improved glucose tolerance, regulated blood lipid levels, and decreased blood pressure in HFHS diet rats (Figure 1P–Y), suggesting the mitigation of associated extracardiac comorbidities.

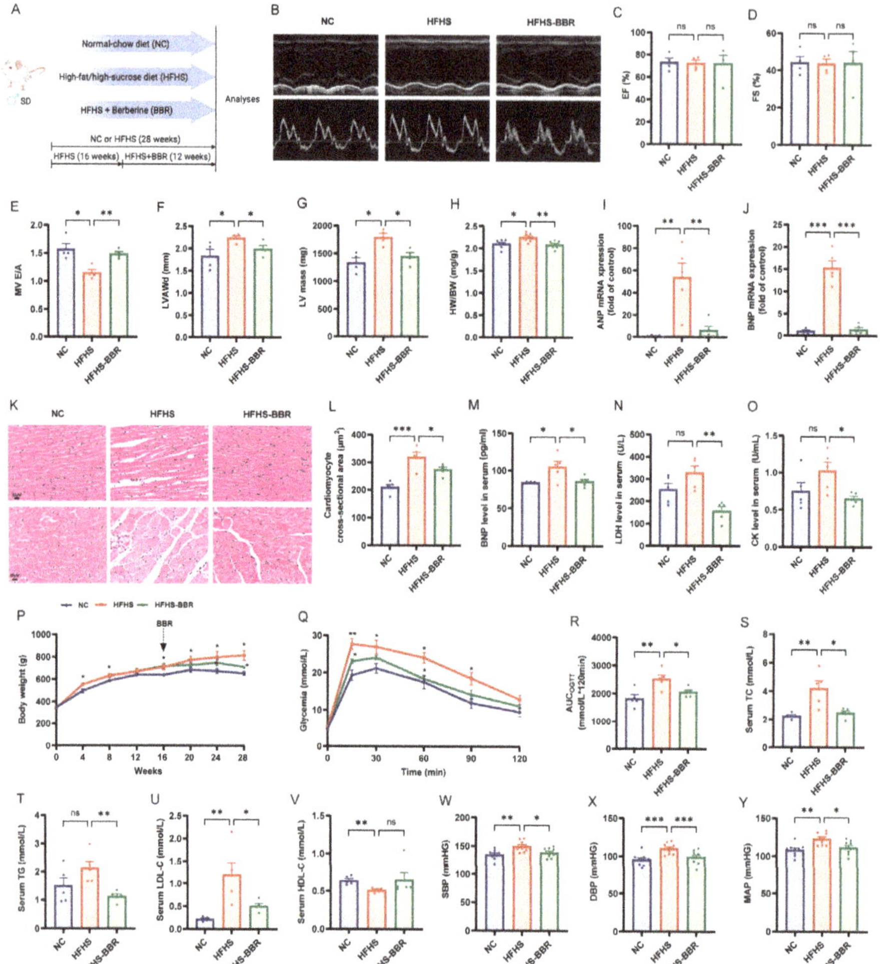

Figure 1. BBR ameliorates cardiac diastolic dysfunction, hypertrophy and metabolic disorders in HFHS diet rats. (**A**) Schematic diagram of experiments in rats. (**B**) Representative M-mode and pulse-wave Doppler echocardiograms. (**C–G**) Statistical analyses of left ventricular ejection fraction (EF), left ventricular fractional shortening (FS), mitral inflow E and A wave ratio (MV E/A), LV anterior wall thickness (LVAW) and LV mass. (**H**) Statistical analysis of the heart weight (HW) to body weight (BW) ratio. (**I**) ANP and (**J**) BNP mRNA expression levels in myocardium. (**K**) HE staining and (**L**) quantitative analysis of cardiomyocyte cross-sectional area (CSA). Scale bars: 20 μm. (**M–O**) Serum BNP, lactate dehydrogenase (LDH), and creatine kinase (CK) levels. (**P**) Body weight. (**Q**) Oral glucose tolerance test (OGTT) and (**R**) OGTT area under the curve (AUC). (**S–V**) Serum levels of total cholesterol (TC), triglyceride (TG), low-density lipoprotein cholesterol (LDL-C) and high-density lipoprotein cholesterol (HDL-C). (**W–Y**) Systolic blood pressure (SBP), diastolic blood pressure (DBP), and mean arterial pressure (MAP). Data are expressed as the mean ± SEM. * $p < 0.05$, ** $p < 0.01$, *** $p < 0.001$; ns, not significant.

2.2. Network Pharmacology Predicates the Potential Mechanisms of BBR in the Treatment of HFpEF

Given BBR's observed therapeutic benefits in diastolic dysfunction, we conducted network pharmacology to investigate the potential mechanisms of BBR in HFpEF treatment. From various databases, we identified 49 common targets for BBR and HFpEF (Figure 2A). To obtain key regulators among these genes, the protein–protein interaction (PPI) network was established using the STRING database and graphically represented using Cytoscape visualization software (https://cytoscape.org). Based on the betweenness centrality scores of PPI network (Figure 2B and Table S3), we selected 20 key genes: TNF, PPARGC1A, MMP9, NOS3, ACHE, CDKN1A, BCL2L1, FOXO3, PPARG, EDN1, IL1B, MMP2, DRD1, DPP4, NAMPT, PARP1, FABP4, MAPK1, ICAM1, and H2AX. Subsequently, we performed GO and KEGG analysis on these targets, which suggest that the therapeutic effects of BBR in HFpEF may be closely linked to the positive regulation of nitric oxide biosynthetic process, collagen catabolic process, T cell selection, regulation of inflammatory response, and regulation of endothelial cell development (Figure 2C, Table S4). Furthermore, our finding implicated the TNF, NF-κB and TGF-β pathways as underlying signaling mechanisms (Figure 2D, Table S5). Based on these results and literature reviews [1,18–20], we identified several pathological processes associated with HFpEF that may be modulated by BBR, including inflammation, oxidative stress, myocardial interstitial fibrosis, microvascular endothelial dysfunction and immune dysregulation. Subsequently, we conducted further investigations to elucidate the potential mechanisms through which BBR may exert its therapeutic effects against HFpEF.

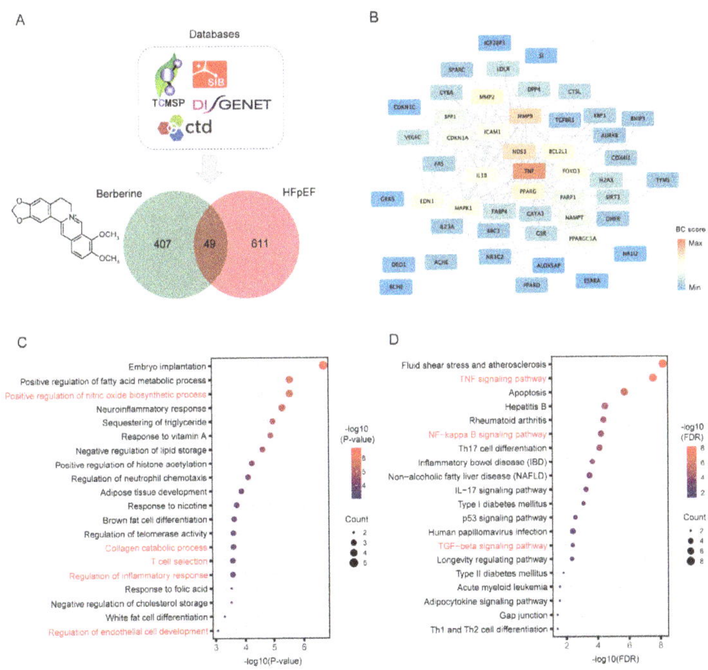

Figure 2. Network pharmacology predicates the potential mechanisms of BBR in the treatment of HFpEF. (**A**) Venn diagram of 49 common target genes of BBR and HFpEF. (**B**) The protein–protein interaction (PPI) network of 49 common targets of BBR and HFpEF. (**C**) GO enrichment bubble plots of BBR and HFpEF related pathways (**D**) KEGG enrichment bubble plots of BBR and HFpEF related pathways.

2.3. BBR Alleviates Myocardial Inflammation and Microvascular Inflammation

Compared to NC rats, HFHS diet-fed rats exhibited elevated myocardium inflammation, as evidenced by increased levels of IL-1β, IL-6, MCP-1 and TNF-α (Figure 3A–C,E). NF-κB, a key regulator of gene involved in inflammation, cellular transformation, and various chronic diseases, was found to be upregulated in HFHS diet-fed rats (Figure 3D,F). Microvascular inflammation was assessed by analyzing the expression of VCAM-1 and ICAM-1, both of which were increased in HFHS diet-fed rats compared to NC rats (Figure 3G–J). Notably, BBR application significantly reduced this increase in inflammatory marker in rats fed an HFHS diet. Overall, these findings suggest that BBR inhibits both myocardial inflammation and microvascular inflammation.

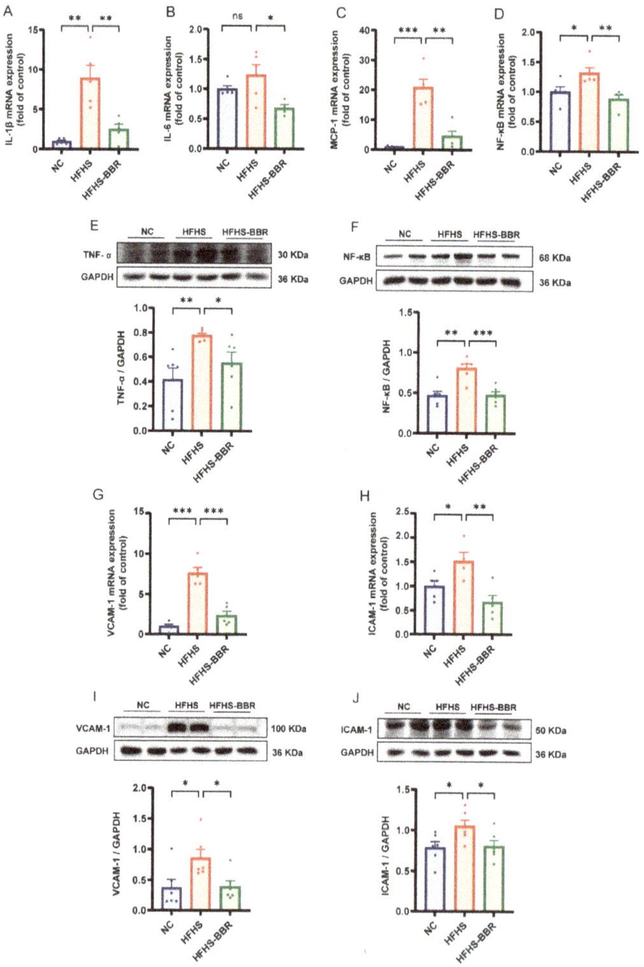

Figure 3. BBR alleviates myocardial inflammation and microvascular inflammation in HFHS diet rats. (**A–D**) Statistical analyses of IL-1β, IL-6, MCP-1 and NF-κB mRNA expression levels in myocardium. (**E,F**) Representative immunoblots and quantitative analysis of TNF-α and NF-κB protein expression levels. (**G,H**) Statistical analyses of VCAM-1 and ICAM-1 mRNA expression levels in myocardium. (**I,J**) Representative immunoblots and quantitative analysis of VCAM-1 and ICAM-1 protein expression levels. Data are expressed as the mean ± SEM. * $p < 0.05$, ** $p < 0.01$, *** $p < 0.001$; ns, not significant.

2.4. BBR Suppresses Myocardial Oxidative Stress

To assess myocardial oxidative stress, the levels of malondialdehyde (MDA) content and the activities of superoxide dismutase (SOD), catalase (CAT) and glutathione peroxidase (GSH-Px) were tested (Figure 4A–D). Rats fed an HFHS diet exhibited significantly elevated MDA levels along with markedly reduced activities of SOD, CAT, and GSH-Px when compared with NC rats. Notably, BBR treatment effectively counteracted these HFHS-induced oxidative disturbances. Furthermore, the mRNA and protein expression levels of NOX-2 and NOX-4 were upregulated in HFHS diet-fed rats (Figure 4E–H). Importantly, BBR administration significantly reversed these changes. Collectively, these results suggest that BBR inhibits myocardial oxidative stress and enhances myocardial antioxidant capacity in HFHS diet-fed rats.

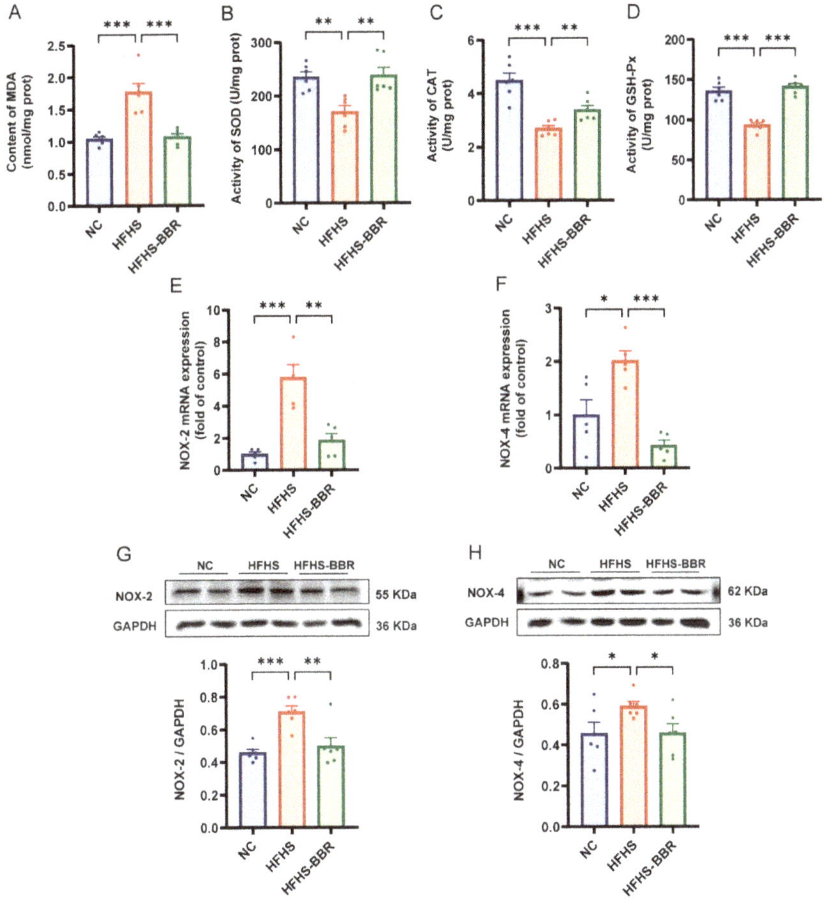

Figure 4. BBR suppresses myocardial oxidative stress in HFHS diet rats. (**A**) Quantitative analysis of the content of malondialdehyde (MDA) in myocardium. (**B–D**) Quantitative analysis of the activities of superoxide dismutase (SOD), catalase (CAT) and glutathione peroxidase (GSH-Px) in myocardium. (**E,F**) Statistical analyses of NOX-2 and NOX-4 mRNA expression levels in myocardium. (**G,H**) Representative immunoblots and quantitative analysis of NOX-2 and NOX-4 protein expression levels. Data are expressed as the mean ± SEM. * $p < 0.05$, ** $p < 0.01$, *** $p < 0.001$.

2.5. BBR Diminishes Myocardial Fibrosis and Inhibits TGF-β/Smads Pathway

Masson's trichrome staining revealed an increased collagen volume fraction in the myocardium of HFHS diet-fed rats compared to NC rats (Figure 5A,B). Additionally, the expression of cardiac fibrosis markers, namely Collagen I, Collagen III, α-SMA and CTGF, was elevated in these rats (Figure 5C–F,J–L). Notably, BBR administration significantly reduced these changes, indicating its anti-fibrotic effect. Furthermore, mRNA expression of TGF-β and Smad2/3 was upregulated in the HFHS diet rats to NC rats (Figure 5G–I). Similarly, protein expression levels of TGF-β and phosphorylation levels of Smad2 and Smad3 were elevated in these rats (Figure 5M–O). Importantly, BBR significantly decreased these alterations, suggesting that BBR attenuates myocardial fibrosis by inhibiting the TGF-β/Smads pathway.

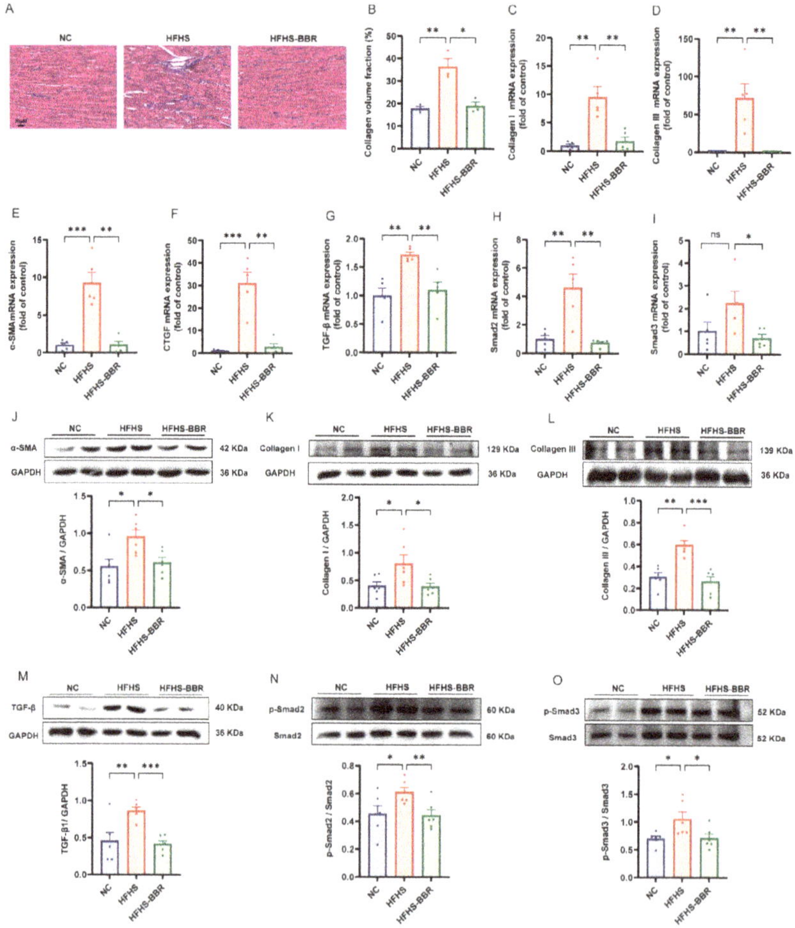

Figure 5. BBR diminishes myocardial interstitial fibrosis and inhibited TGF-β/Smads pathway in HFHS diet rats. (**A**,**B**) Masson staining and quantitative analysis of collagen volume fraction; Scale bar: 20 μm. (**C**–**I**) Statistical analyses of Collagen I, Collagen III, α-SMA, CTGF, TGF-β, Smad2 and Smad3 mRNA expression levels in myocardium. (**J**–**O**) Representative immunoblots and quantitative analysis of α-SMA, Collagen I, Collagen III, TGF-β, p-Smad2/Smad2 and p-Smad3/Smad3 protein expression. Data are expressed as the mean ± SEM. * $p < 0.05$, ** $p < 0.01$, *** $p < 0.001$; ns, not significant.

2.6. BBR Restores the Impaired NO/cGMP/PKG Pathway and Reduces Cardiomyocytes Stiffness

Elevated oxidative stress levels resulted in decreased nitric oxide (NO) bioavailability, which subsequently reduced soluble guanylate cyclase (sGC) activity and led to a decrease in cGMP production, the key regulator of protein kinase G (PKG) activity. To assess NO bioavailability, myocardial nitrite and nitrate concentrations were measured. HFHS diet-fed rats diet exhibited significantly lower NO production compared to NC rats (Figure 6A). Furthermore, these rats showed reduced levels of phosphorylated endothelial nitric oxide synthase (eNOS), the catalytic enzyme mediating NO biosynthesis (Figure 6B). These effects were restored upon BBR administration. Additionally, HFHS diet-fed rats had decreased expression of sGC and reduce cGMP content compared to NC rats, and BBR reversed these changes (Figure 6B–C). Moreover, PKG expression were lower in the myocardium of HFHS diet-fed rats compared to NC rats (Figure 6D), and BBR restored the PKG expression. Increase arginase (ARG) activity is known to decrease in NO production, while activation of phosphodiesterases (PDEs) promotes cGMP degradation [21]. HFHS-fed rats exhibited increased expression of ARG and PDE5A compared to NC rats, and BBR significantly decreased these elevated protein levels (Figure 6D). Overall, BBR restored the impaired NO/cGMP/PKG pathway. PKG reduces cardiomyocyte stiffness by phosphorylating titin, an essential sarcomere structural protein that regulates passive tension in cardiac myocytes [22]. Immunofluorescence analysis of the myocardium revealed the increased titin expression in HFHS diet-fed rats compared to NC rats, which decreased following BBR administration (Figure 6E). The lower titin expression indirectly reflected an increase in titin phosphorylation, resulting in reduced cardiomyocyte stiffness [23]. Thereby, BBR may reduce the cardiomyocytes stiffness by restoring the impaired NO/cGMP/PKG pathway.

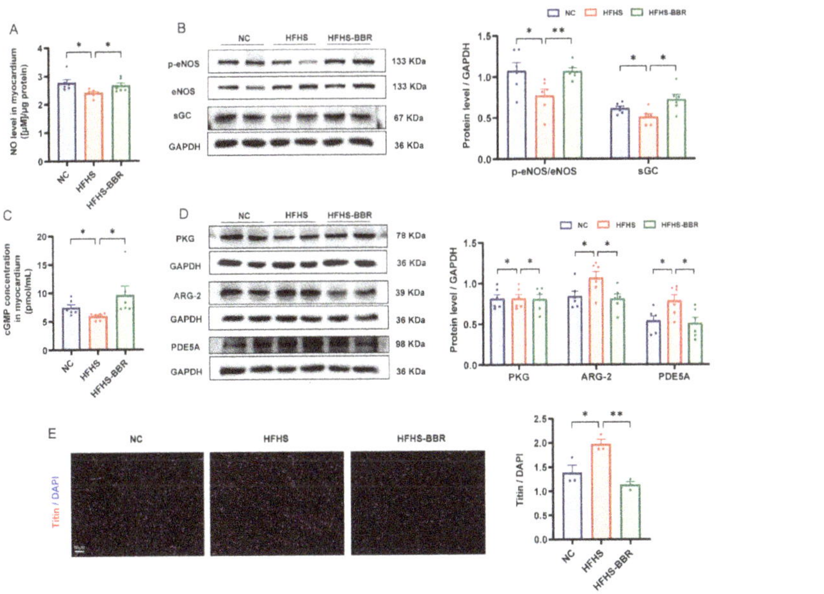

Figure 6. BBR activates NO/cGMP/PKG pathway in HFHS diet rats. (**A**) Quantitative analysis of total nitric oxide (NO) production in myocardium. (**B**) Representative immunoblots and quantitative analysis of p-eNOS/eNOS and sGC protein expression levels. (**C**) Quantitative analysis of cGMP concentration in myocardium. (**D**) Representative immunoblots and quantitative analysis of PKG, ARG-2 and PDE5A protein expression levels. (**E**) Immunofluorescence staining and quantitative analysis of titin (red) and DAPI (blue), Scale bar: 50 μm. Data are expressed as mean ± SEM. * $p < 0.05$, ** $p < 0.01$.

2.7. BBR Mitigates Systemic Inflammation in HFHS Diet-Fed Rats

Calorie-dense diets trigger systemic low-grade chronic inflammation through up-regulation of pro-inflammatory cytokine expression and secretion [24]. In the current study, serum levels of CRP, IL-1β, IL-6 and TNF-α were significantly elevated in HFHS diet-fed rats compared to NC rats (Figure 7A–D). Similarly, the mRNA expression of pro-inflammatory cytokines, including IL-1β, IL-6, TNF-α and MCP-1, was increased in the visceral adipose tissue of HFHS diet-fed rats (Figure 7E–H). Furthermore, hemogram analysis revealed higher levels of WBCs and NEUs, as well as elevated systemic inflammatory indices such as neutrophil to lymphocyte ratio (NLR) and systemic immune inflammation index (SII), in HFHS diet-fed rats compared to NC rats (Figure 7I–L, Table S6). Collectively, these results indicate a systemic pro-inflammatory state in rats fed an HFHS diet. Importantly, BBR administration significantly reduced the levels of inflammatory cytokines in both serum and adipose tissue, as well as the systemic inflammatory indices, indicating that BBR effectively mitigates the systemic inflammation in HFHS diet-fed rats.

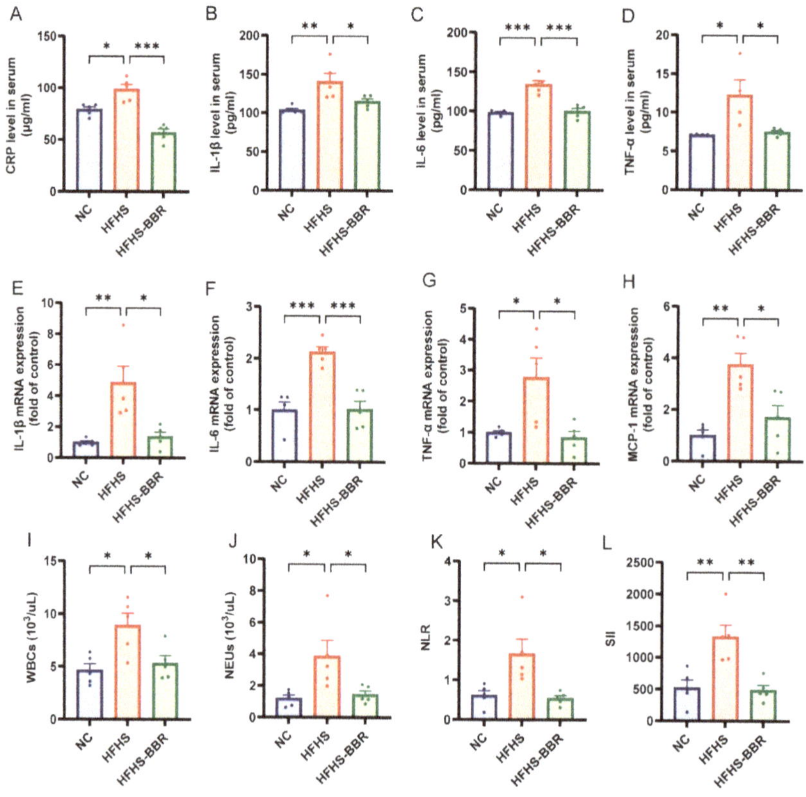

Figure 7. BBR ameliorates systemic inflammation in HFHS diet rats. (**A–D**) Quantitative analysis of serum CRP, IL-1β, IL-6 and TNF-α levels. (**E–H**) Statistical analyses of IL-1β, IL-6, TNF-α and MCP-1 mRNA expression levels in adipose tissue. (**I–L**) Statistical analyses of white blood cells (WBCs) counts, neutrophils (NEUs) counts, neutrophil to lymphocyte ratio (NLR) and systemic immune inflammation index (SII) in peripheral blood. Data are expressed as the mean ± SEM. * $p < 0.05$, ** $p < 0.01$, *** $p < 0.001$.

2.8. BBR Correctes Immune Dysregulation in HFHS Diet-Fed Rats

Calorie-dense diets have been shown to induce dysregulation of immune cells [25]. T-lymphocyte subsets proportion, particularly the $CD4^+/CD8^+$, which is traditionally used as an immune-stimulatory marker, are essential for evaluating the immune function. We then compared the proportion of $CD4^+$ and $CD8^+$ T cells in the $CD3^+$ population of PBMC and spleen in different groups. In both PBMC and spleen, HFHS diet-fed rats exhibited a decrease in $CD4^+$ T cells and an increase in $CD8^+$ T cells compared to NC rats (Figure 8A–D). Furthermore, the $CD4^+/CD8^+$ ratio was significantly reduced in these tissues of HFHS diet-fed rats. Importantly, BBR administration altered the composition of T-lymphocyte subsets and significantly restored the $CD4^+/CD8^+$ ratio to a near normal levels in rats fed an HFHS diet, suggesting a potential immunoregulatory effect of BBR.

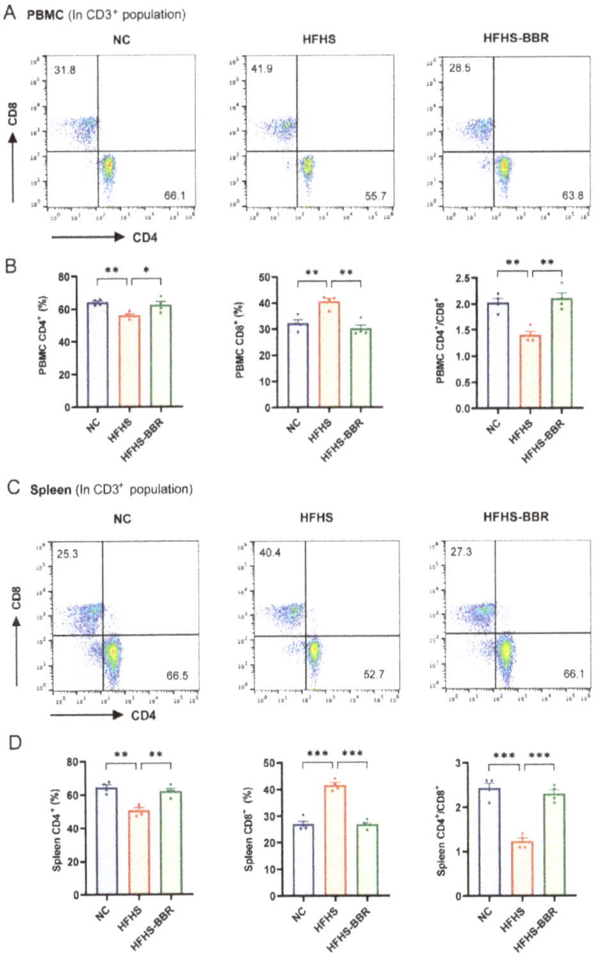

Figure 8. BBR regulates lymphocyte subsets disorder in HFHS diet rats. (**A**) The flow cytometric plots of $CD4^+$ and $CD8^+$ T cell subsets in PBMC. (**B**) Statistical analysis of the proportion of $CD4^+$ T cell and $CD8^+$ T cell and the $CD4^+/CD8^+$ ratio in PBMC. (**C**) The flow cytometric plots of $CD4^+$ and $CD8^+$ T cell subsets in spleen. (**D**) Statistical analysis of the proportion of $CD4^+$ T cell and $CD8^+$ T cell and the $CD4^+/CD8^+$ ratio in spleen. Data are expressed as mean ± SEM. * $p < 0.05$, ** $p < 0.01$, *** $p < 0.001$.

3. Discussion

HFpEF is a complex, multi-factorial syndrome with clinical heterogeneity and multiple comorbidities contributing to its overall presentation. The prevailing HFpEF paradigm, first proposed by Paulus W.J. and Tschöpe C in 2013 [26], posits that systemic pathological factors trigger sustained inflammatory responses and impair endothelial homeostasis, leading to microvascular impairment and subsequent disease advancement, ultimately resulting in cardiac diastolic dysfunction. This study has demonstrated that BBR exerts multi-beneficial effects on rats with HFpEF, which are relevant to the pathophysiology of HFpEF.

Cardiac inflammation and oxidative stress are known contributors to diastolic dysfunction [18,27]. BBR has demonstrated significant efficacy in attenuating inflammation and oxidative stress in various pathological conditions. Notably, BBR alleviates diabetic cardiomyopathy by inhibiting inflammasome activation [28,29], protects against doxorubicin-induced cardiomyopathy by diminishing cardiac oxidative stress [30], and alleviates myocardial ischemia-reperfusion injury by suppressing the inflammatory response and oxidative stress [31]. Our study exhibited a significant decrease in inflammation and oxidative stress in rats fed an HFHS diet following BBR treatment. The linkage between ROS and myocardial inflammation may involves the stimulation of NF-κB signaling pathways, triggering transcriptional upregulation of pro-inflammatory mediators, leading to the generation of ROS [27,32,33]. Inflammation also induces ROS production primarily through NADPH oxidase enzymes [34], resulting in elevated oxidative stress. Our results further demonstrated that BBR suppressed NF-κB signaling pathway and significantly downregulated NADPH oxidase isozyme expression, particularly NOX-2 and NOX-4, in rats fed an HFHS diet. Collectively, our findings indicate that BBR improves cardiac diastolic dysfunction by suppressing cardiac inflammation and oxidative stress.

In HFpEF, cardiac remodeling is initially driven by chronic microvascular inflammation, as supported by compelling preclinical and clinical evidence [35]. Excessive ROS generation upregulates the expression levels of vascular adhesion molecules, particularly VCAM-1 and ICAM-1 [36], facilitating myocardial infiltration by activated leukocytes, transformation of fibroblast into myofibroblasts, and the progression of myocardial fibrosis. The TGF-β/Smads signaling cascade represents one of the most crucial regulatory pathways for cardiac fibrosis formation [37]. Our results showed that BBR application inhibited microvascular inflammation, inhibited the TGF-β/Smads signaling pathway, and alleviated myocardial fibrosis in HFHS diet-fed rats. Myocardial fibrosis is a crucial contributor of cardiac diastolic dysfunction. Compelling evidence has demonstrated that reducing myocardial fibrosis would be beneficial for HFpEF patients [38]. Therefore, BBR presents profound potential for alleviating myocardial fibrosis, thereby improving diastolic dysfunction in HFpEF.

Furthermore, microvascular inflammation triggers oxidative stress and diminishes NO bioavailability, resulting in microvascular endothelial dysfunction, a pivotal event in the progression of HFpEF [36]. Additionally, sGC enzymatic activity, intracellular cGMP concentration, and PKG activation status in adjacent cardiomyocytes leads to titin hypophosphorylation, ultimately contributing to increased passive stiffness and diastolic dysfunction in HFpEF [39]. Substantial evidence has demonstrated the disruption of NO/cGMP/PKG pathway in the occurrence and development of HFpEF [40]. Importantly, our findings reveal that BBR application effectively mitigates microvascular endothelial dysfunction, rescues the disturbed NO-cGMP-PKG pathway in rats fed an HFHS diet. A previous study indicated that decreased titin expression served as an indirect indicator of elevated titin phosphorylation, leading to a decrease in cardiomyocyte rigidity [23]. BBR also effectively decreases cardiac titin expression. Thereby, BBR may reduce cardiomyocyte stiffness by restoring the disturbed NO-cGMP-PKG pathway, contributing to improve diastolic dysfunction in HFpEF.

Sustained low-grade systemic inflammation represents a key driver in the initiation and advancement of HFpEF pathophysiology [26]. Elevated levels of inflammatory biomarkers, including CRP, IL-6, IL-1β and TNF-α, have been strongly associated with disease progression and unfavorable outcomes in HFpEF [41]. A dose–response meta-analysis of randomized controlled trials has demonstrated that BBR interventions significantly reduce IL-6, TNF-α, and CRP levels in adults, indicating improved systemic inflammation status [42]. In our study, BBR administration significantly improved extracardiac pro-inflammatory comorbidities in rats with HFpEF. Furthermore, BBR significantly decreased the level of serum CRP, IL-1, IL-6, and TNF-α. Chronic inflammation is associated with adipose tissue inflammation [43]. BBR also reduced the mRNA expression of TNF-α, IL-6, IL-1β and MCP-1 in visceral adipose tissue. In addition to traditional inflammatory markers, the NLR and SII derived from peripheral blood serve as novel indices of systemic inflammation for assessing the prognosis of patients with HFpEF [41,44]. Our study also revealed that BBR significantly lowered NLR and SII in HFHS diet-fed rats. These findings show that BBR suppresses systemic low-grade inflammation in rats with HFpEF.

Multiple lines of evidence suggest that immune dysregulation is linked to the pathology and progression of HFpEF. Alteration of T-lymphocyte subsets, particularly the $CD4^+/CD8^+$ ratio, has shown predictive potential for HF in clinical and experimental studies [45]. A recent study by Cao et al. reported a decreased peripheral $CD4^+/CD8^+$ in elderly patients with chronic HF, and immunomodulatory agents administration significantly altered the T-cell subsets and restored the $CD4^+/CD8^+$ ratio, improving cardiac function and quality of life [46]. Our study demonstrated a significant decline in $CD4^+/CD8^+$ in peripheral blood and spleen of HFHS diet-fed rats. However, BBR administration altered peripheral T-cell subsets and restored $CD4^+/CD8^+$ ratio to a near-normal status. This aligns with previous studies highlighting BBR as an effective immunomodulator [47,48]. Furthermore, $CD4^+$ T cells play pivotal roles in mediating chronic inflammatory responses in obese mouse models [49]. Thereby, our results suggest that BBR may attenuate chronic inflammation state through regulating immune cell balance.

4. Materials and Methods

4.1. Animals

Male Sprague Dawley rats (8–10 weeks old, 300–400 g body weight) were procured from Beijing Vital River Laboratory Animal Technology Co., Ltd. (Beijing, China). The animals were maintained under standard laboratory conditions (2–3 rats per cage) with ad libitum access to food and water, in a temperature- and humidity-controlled environment. Following acclimatization, rats were randomly allocated into three experimental groups ($n = 9$ per group): NC group, HFHS group, and HFHS plus berberine (HFHS-BBR) group. The NC group was fed a standard chow diet, while both HFHS and HFHS-BBR groups received a specially formulated high-fat/high-sucrose diet (composition: 10% fat, 20% sucrose, 2.5% cholesterol, and 0.5% sodium cholate by weight) supplied by Beijing Keao Xieli Feed Co., Ltd. (Beijing, China) for a duration of 28 weeks. Starting from week 16 through week 28, the HFHS-BBR group was supplemented with 100 mg/kg body weight/day of berberine (BBR; Macklin, Shanghai, China), based on established protocols [19]. The experimental design timeline is presented in Figure 1A.

Body weight measurements were recorded at 4-week intervals throughout the study period. Upon completion of the experimental protocol, comprehensive assessments were performed, including evaluation of cardiac function, glucose tolerance, and blood pressure measurements. Following functional assessments, rats were anesthetized using 3–5%

isoflurane (RWD, Shenzhen, China), and blood samples were obtained via abdominal aorta puncture. Immediately after sacrifice, heart, visceral adipose tissue and spleen were rapidly excised for subsequent biochemical analyses. Cardiac mass was precisely measured, and the heart index was determined by calculating the heart weight to body weight ratio (HW/BW).

All experimental procedures involving animals were performed in strict compliance with the ethical guidelines and regulatory standards established by the Chinese Council on Animal Care. The study protocol was approved by the Animal Care and Use Committee of the Chinese Academy of Medical Sciences (No. 00003917).

4.2. Echocardiographic Analysis

Cardiac functional assessment was performed at study termination utilizing transthoracic echocardiography with the Vevo2100 Imaging System (FUJIFILM VisualSonics, Toronto, ON, Canada). Rats were maintained under anesthesia with 2–3% isoflurane (RWD, Shenzhen, China) during the procedure. Comprehensive LV evaluation included measurement of LV mass, LVAW, EF, and FS using conventional M-mode echocardiography. Mitral valve inflow patterns were analyzed through pulse wave Doppler imaging to determine the E/A wave ratio. All parameter calculations were performed using VevoLAB software (Version 2.2.3) according to standard protocols.

4.3. Oral Glucose Tolerance Test (OGTT)

Following an overnight fasting period, rats received an oral glucose load of 2 g/kg body weight. Serial blood samples were obtained via tail vein puncture at predetermined time intervals (0, 15, 30, 60, 90, and 120 min) following glucose administration. Blood glucose concentrations were determined using a commercial glucose meter (Yuwell, Danyang, China). Glucose tolerance was quantitatively assessed by calculating the area under the curve (AUC) using GraphPad Prism 8.0 software for statistical analysis.

4.4. Blood Pressure Measurements

Systolic and diastolic blood pressure measurements were obtained non-invasively from conscious rats using the tail-cuff method with a CODA monitoring system (ADInstruments, Shanghai, China). Animals were individually restrained in temperature-regulated holders maintained at 37 °C, and hemodynamic recordings were performed under stable physiological conditions. For each measurement session, a minimum of ten readings were acquired and subsequently averaged to ensure data reliability.

4.5. Network Pharmacology

The collection of 407 targets associated with BBR were obtained from CTD (https://ctdbase.org, accessed on 4 March 2023), SwissTargetPrediction (www.SwissTargetPrediction.ch, accessed on 4 March 2023), and TCMSP (https://www.tcmsp-e.com/#/home, accessed on 4 March 2023) databases. The collection of 611 disease genes related to HFpEF was retrieved from the CTD and DiseGeNET (https://www.disgenet.org/, accessed on 4 March 2023) databases. The co-targeted genes were identified through taking intersection of BBR and HFpEF targeting genes. The protein–protein interaction (PPI) information of co-targeted genes was obtained from STRING (https://cn.string-db.org, accessed on 5 March 2023) database and visualized by Cytoscape software (Version 3.8.0). Functional and pathway enrichment analyses of 49 targets were performed using R programming language (Version 4.0.3) based on GO (http://geneontology.org/, accessed on 5 March 2023) and KEGG (https://www.kegg.jp/kegg/pathway.html, accessed on 5 March 2023) databases, presenting the results in bubble plots.

4.6. Biochemical Analyses

Creatine kinase (CK) activity and concentrations of brain natriuretic peptide (BNP), C-reactive protein (CRP), interleukin 1 beta (IL-1β), interleukin 6 (IL-6), and tumor necrosis factor alpha (TNF-α) were quantified using commercial ELISA kits (Elabscience, Wuhan, China) following the manufacturer's protocols. Serum metabolic profiles, including lactate dehydrogenase (LDH) activity and levels of total cholesterol (TC), triglycerides (TG), low-density lipoprotein cholesterol (LDL-C), and high-density lipoprotein cholesterol (HDL-C), were determined using a Hitachi biochemical analyzer (Tokyo, Japan) in conjunction with commercial assay kits (BioSin Bio-Technology and Science Inc., Beijing, China). Myocardial tissue analyses included measurement of superoxide dismutase (SOD) activity and quantification of malondialdehyde (MDA), catalase (CAT), glutathione peroxidase (GSH-Px), and cyclic guanosine monophosphate (cGMP) using ELISA kits (Elabscience, Wuhan, China). Total nitric oxide (NO) production in myocardial tissue was assessed by determining nitrate and nitrite concentrations through a modified Griess reaction assay (Beyotime, Shanghai, China).

4.7. Hemogram Analysis

EDTA-anticoagulated blood samples were processed using the automated CELL-DYN® CD3700 hematology analyzer (Abbott, Alameda, CA, USA) for complete blood count analysis. The system quantified white blood cell (WBC) populations, including neutrophil (NEU) and lymphocyte counts, along with platelet enumeration. The neutrophil-to-lymphocyte ratio (NLR) and systemic immune-inflammation index (SII) were subsequently calculated according to the formula provided in reference [20]:

$$\text{NLR} = [\text{absolute neutrophils count (ANC)}]/[\text{absolute lymphocytes count (ALC)}]$$

$$\text{SII} = [\text{ANC} \times \text{absolute platelets count (APC)}]/\text{ALC}$$

4.8. Histopathological Analysis

Cardiac tissue samples were processed for histological analysis through fixation in 4% paraformaldehyde (PFA), paraffin embedding, and sectioning at 4 μm thickness. Tissue sections were subjected to H&E staining and Masson's trichrome staining for morphological evaluation. Histopathological examination was performed using a Zeiss microscope (Shanghai, China), with digital image acquisition for quantitative analysis. Cardiomyocyte cross-sectional area (CSA) was determined from H&E-stained sections, while collagen volume fraction was assessed from Masson-stained sections, both quantified using ImageJ software (Version 1.53).

4.9. Immunofluorescence Analysis

LV tissue sections underwent sequential processing including deparaffinization, rehydration, and permeabilization. Following blocking with goat serum, sections were incubated with primary antibodies (1:200 dilution of rabbit anti-Titin polyclonal antibody, Bioss, Beijing, China) at 4 °C overnight. After PBS washing, sections were treated with appropriate secondary antibodies. Nuclear staining was performed using DAPI solution, followed by application of anti-fade mounting medium. Fluorescence imaging was conducted using a Zeiss microscope (Shanghai, China), and titin-positive areas were quantitatively analyzed using ImageJ software (Version 1.53).

4.10. RNA Isolation and qRT-PCR

Total RNA isolation from LV and visceral adipose tissues was performed using the PureLinkTM RNA Mini Kit (Thermo Fisher Scientific, Waltham, MA, USA), with subsequent quantification using NanoDrop spectrophotometry (Thermo Fisher Scientific, Waltham, MA, USA). Complementary DNA (cDNA) synthesis was carried out with the HiFiScript cDNA Synthesis Kit (CoWin, Taizhou, China). Quantitative reverse transcription polymerase chain reaction (qRT-PCR) analysis was executed on the CFX ConnectTM Real-Time PCR Detection System (Bio-Rad, Hercules, CA, USA) employing the Ultra-SYBR Mixture (CoWin, Taizhou, China). The specific primer sequences for target gene amplification are detailed in Table S1.

4.11. Western Blotting

Protein extraction from LV tissue was performed using RIPA lysis buffer (Thermo Fisher Scientific, Waltham, MA, USA). Protein samples (equal loading amounts) were separated on 8% SDS-PAGE gels and subsequently transferred onto PVDF membranes. Following 1 h blocking with 10% skimmed milk at room temperature, membranes were incubated with primary antibodies at 4 °C overnight, then with corresponding secondary antibodies for 2 h at room temperature. Protein band visualization was achieved using the BLT GelView6000 ProII imaging system (Biolight Biotechnology, Guangzhou, China), with quantitative analysis performed using ImageJ software (Version 1.53). Comprehensive antibody information is provided in Table S2.

4.12. Flow Cytometry

After mechanical homogenization of spleen, splenic cells were filtered through a 300-mesh sieve. T lymphocytes were collected from blood and splenic cells using a rat lymphocyte separation kit (Solarbio, Beijing, China). Cells were washed, stained with antibodies specific to FITC-CD3, PerCP-eF710-CD4, and PE-CD8, and analyzed using a flow cytometer (Thermo Fisher Scientific, Waltham, MA, USA). Data analysis was performed using FlowJo software (Version 10.8.1). All antibodies were from Invitrogen (Thermo Fisher Scientific, Shanghai, China).

4.13. Statistical Analysis

All data were presented as mean ± SEM. Normality of data was assessed using the Shapiro–Wilk and Kolmogorov–Smirnov tests. The difference between two groups were examined with two-tailed paired and unpaired t tests. One-way ANOVA was used for multiple comparisons followed by Tukey's post hoc test. All statistical analysis was conducted with GraphPad Prism 8.0. A p value < 0.05 was considered statistically significant.

5. Conclusions

In conclusion, this research has demonstrated that BBR markedly ameliorated cardiac diastolic dysfunction, decreased myocardial inflammation and oxidative stress, improved microvascular endothelial dysfunction, and reversed the pathological suppression of the NO/cGMP/PKG pathway in rats fed an HFHS diet. BBR also attenuated cardiac fibrosis and inhibited the TGF-β/Smads signaling pathway. Furthermore, BBR exhibited a broad spectrum of extracardiac benefits by ameliorating pro-inflammatory comorbidities, reducing systemic inflammation, and correcting immune dysregulation in rats with HFpEF. These findings collectively indicate that BBR exerts multi-beneficial effects on rats with HFpEF. Future research and clinical trials are essential to fully explain the molecular mechanisms mediating BBR's therapeutic effects and to establish its clinical efficacy in HFpEF treatment strategies.

Supplementary Materials: The following supporting information can be downloaded at: https://www.mdpi.com/article/10.3390/ijms26104847/s1.

Author Contributions: Conceptualization, Y.W.; Data curation, Y.M., C.L. and S.J.; Formal analysis, Y.M.; Funding acquisition, J.J. and Y.W.; Investigation, Y.M., J.G., C.L. and S.J.; Methodology, Y.M., J.G. and Y.W.; Resources, Y.H.; Software, Y.M. and J.G.; Supervision, Y.H., J.J. and Y.W.; Writing—original draft, Y.M.; Writing—review and editing, Y.W. All authors have read and agreed to the published version of the manuscript.

Funding: This research was funded by the Chinese Academy of Medical Sciences (CAMS) Innovation Fund for Medical Sciences [grant numbers 2024-I2M-ZH-012; 2021-I2M-1-016; 2023-I2M-2-006].

Institutional Review Board Statement: The animal study was reviewed and approved by the Animal Care and Use Committee of the Chinese Academy of Medical Sciences (No. 00003917).

Informed Consent Statement: Not applicable.

Data Availability Statement: The original contributions presented in the study are included in the article/Supplementary Materials.

Conflicts of Interest: The authors declare no conflicts of interest.

References

1. Borlaug, B.A.; Sharma, K.; Shah, S.J.; Ho, J.E. Heart Failure with Preserved Ejection Fraction: JACC Scientific Statement. *J. Am. Coll. Cardiol.* **2023**, *81*, 1810–1834. [CrossRef] [PubMed]
2. Wintrich, J.; Kindermann, I.; Ukena, C.; Selejan, S.; Werner, C.; Maack, C.; Laufs, U.; Tschöpe, C.; Anker, S.D.; Lam, C.S.P.; et al. Therapeutic Approaches in Heart Failure with Preserved Ejection Fraction: Past, Present, and Future. *Clin. Res. Cardiol.* **2020**, *109*, 1079–1098. [CrossRef] [PubMed]
3. Shah, S.J.; Borlaug, B.A.; Kitzman, D.W.; McCulloch, A.D.; Blaxall, B.C.; Agarwal, R.; Chirinos, J.A.; Collins, S.; Deo, R.C.; Gladwin, M.T.; et al. Research Priorities for Heart Failure with Preserved Ejection Fraction: National Heart, Lung, and Blood Institute Working Group Summary. *Circulation* **2020**, *141*, 1001–1026. [CrossRef]
4. Dunlay, S.M.; Roger, V.L.; Redfield, M.M. Epidemiology of Heart Failure with Preserved Ejection Fraction. *Nat. Rev. Cardiol.* **2017**, *14*, 591–602. [CrossRef]
5. Solomon, S.D.; McMurray, J.J.V.; Claggett, B.; de Boer, R.A.; DeMets, D.; Hernandez, A.F.; Inzucchi, S.E.; Kosiborod, M.N.; Lam, C.S.P.; Martinez, F.; et al. Dapagliflozin in Heart Failure with Mildly Reduced or Preserved Ejection Fraction. *N. Engl. J. Med.* **2022**, *387*, 1089–1098. [CrossRef] [PubMed]
6. Borlaug, B.A.; Kitzman, D.W.; Davies, M.J.; Rasmussen, S.; Barros, E.; Butler, J.; Einfeldt, M.N.; Hovingh, G.K.; Møller, D.V.; Petrie, M.C.; et al. Semaglutide in HFpEF across Obesity Class and by Body Weight Reduction: A Prespecified Analysis of the STEP-HFpEF Trial. *Nat. Med.* **2023**, *29*, 2358–2365. [CrossRef]
7. Gallo, G.; Volpe, M. Potential Mechanisms of the Protective Effects of the Cardiometabolic Drugs Type-2 Sodium-Glucose Transporter Inhibitors and Glucagon-like Peptide-1 Receptor Agonists in Heart Failure. *Int. J. Mol. Sci.* **2024**, *25*, 2484. [CrossRef] [PubMed]
8. Belli, M.; Barone, L.; Bellia, A.; Sergi, D.; Lecis, D.; Prandi, F.R.; Milite, M.; Galluccio, C.; Muscoli, S.; Romeo, F.; et al. Treatment of HFpEF beyond the SGLT2-Is: Does the Addition of GLP-1 RA Improve Cardiometabolic Risk and Outcomes in Diabetic Patients? *Int. J. Mol. Sci.* **2022**, *23*, 14598. [CrossRef]
9. Kolijn, D.; Pabel, S.; Tian, Y.; Lódi, M.; Herwig, M.; Carrizzo, A.; Zhazykbayeva, S.; Kovács, Á.; Fülöp, G.Á.; Falcão-Pires, I.; et al. Empagliflozin Improves Endothelial and Cardiomyocyte Function in Human Heart Failure with Preserved Ejection Fraction via Reduced Pro-Inflammatory-Oxidative Pathways and Protein Kinase Gα Oxidation. *Cardiovasc. Res.* **2021**, *117*, 495–507. [CrossRef]
10. Bonfioli, G.B.; Rodella, L.; Metra, M.; Vizzardi, E. GLP-1 Receptor Agonists as Promising Anti-Inflammatory Agents in Heart Failure with Preserved Ejection Fraction. *Heart Fail. Rev.* **2025**, *30*, 131–136. [CrossRef]
11. Habtemariam, S. Berberine Pharmacology and the Gut Microbiota: A Hidden Therapeutic Link. *Pharmacol. Res.* **2020**, *155*, 104722. [CrossRef] [PubMed]
12. Kong, W.-J.; Vernieri, C.; Foiani, M.; Jiang, J.-D. Berberine in the Treatment of Metabolism-Related Chronic Diseases: A Drug Cloud (dCloud) Effect to Target Multifactorial Disorders. *Pharmacol. Ther.* **2020**, *209*, 107496. [CrossRef] [PubMed]
13. Feng, X.; Sureda, A.; Jafari, S.; Memariani, Z.; Tewari, D.; Barrea, L.; Hassan, S.T.S.; Šmejkal, K.; Malaník, M.; Sychrová, A.; et al. Berberine in Cardiovascular and Metabolic Diseases: From Mechanisms to Therapeutics. *Theranostics* **2019**, *9*, 1923–1951. [CrossRef]

14. Zeng, X.-H.; Zeng, X.-J.; Li, Y.-Y. Efficacy and Safety of Berberine for Congestive Heart Failure Secondary to Ischemic or Idiopathic Dilated Cardiomyopathy. *Am. J. Cardiol.* **2003**, *92*, 173–176. [CrossRef]
15. Wang, S.; Lu, K.; Lin, L.; Li, G.; Han, Y.; Lin, Z.; Chu, Q.; Wu, K.; Liu, P.; Zhou, G.; et al. Exploring the Mechanism of Berberine Treatment for Atherosclerosis Combined with Non-Alcoholic Fatty Liver Disease Based on Bioinformatic and Experimental Study. *PLoS ONE* **2024**, *19*, e0314961. [CrossRef]
16. Abudureyimu, M.; Yang, M.; Wang, X.; Luo, X.; Ge, J.; Peng, H.; Zhang, Y.; Ren, J. Berberine Alleviates Myocardial Diastolic Dysfunction by Modulating Drp1-Mediated Mitochondrial Fission and Ca2+ Homeostasis in a Murine Model of HFpEF. *Front. Med.* **2023**, *17*, 1219–1235. [CrossRef] [PubMed]
17. Mu, Y.; Yang, Y.; Jiang, S.; Liu, C.; Han, Y.; Jiang, J.; Wang, Y. Benefits of Puerarin on Metabolic Syndrome and Its Associated Cardiovascular Diseases in Rats Fed a High-Fat/High-Sucrose Diet. *Nutrients* **2024**, *16*, 1273. [CrossRef]
18. Liu, H.; Huang, Y.; Zhao, Y.; Kang, G.-J.; Feng, F.; Wang, X.; Liu, M.; Shi, G.; Revelo, X.; Bernlohr, D.; et al. Inflammatory Macrophage Interleukin-1β Mediates High-Fat Diet-Induced Heart Failure with Preserved Ejection Fraction. *JACC Basic Transl. Sci.* **2023**, *8*, 174–185. [CrossRef]
19. Li, C.; Jiang, S.; Wang, H.; Wang, Y.; Han, Y.; Jiang, J. Berberine Exerts Protective Effects on Cardiac Senescence by Regulating the Klotho/SIRT1 Signaling Pathway. *Biomed. Pharmacother.* **2022**, *151*, 113097. [CrossRef]
20. Zhao, Y.; Shao, W.; Zhu, Q.; Zhang, R.; Sun, T.; Wang, B.; Hu, X. Association between Systemic Immune-Inflammation Index and Metabolic Syndrome and Its Components: Results from the National Health and Nutrition Examination Survey 2011–2016. *J. Transl. Med.* **2023**, *21*, 691. [CrossRef]
21. James, A.S.; Eteng, O.E.; Dosumu, O.A.; Moses, C.A.; Ogbonna, C.U.; Adeleye, O.A.; Ugwor, E.I.; Omilo, B.C.; Fabunmi, R.F.; Olakitan, A.M.; et al. Morin Augmented Myocardial eNOS/cGMP/PKG Signaling Pathway and Abated Oxidative and Inflammo-Apoptotic Responses in Diethyl Phthalate and Bisphenol-S Co-Exposed Male Albino Rats. *Inflammation* **2023**, *46*, 175–189. [CrossRef] [PubMed]
22. Van Heerebeek, L.; Hamdani, N.; Falcão-Pires, I.; Leite-Moreira, A.F.; Begieneman, M.P.V.; Bronzwaer, J.G.F.; Van Der Velden, J.; Stienen, G.J.M.; Laarman, G.J.; Somsen, A.; et al. Low Myocardial Protein Kinase G Activity in Heart Failure With Preserved Ejection Fraction. *Circulation* **2012**, *126*, 830–839. [CrossRef]
23. Huang, Y.; Zhang, K.; Liu, M.; Su, J.; Qin, X.; Wang, X.; Zhang, J.; Li, S.; Fan, G. An Herbal Preparation Ameliorates Heart Failure with Preserved Ejection Fraction by Alleviating Microvascular Endothelial Inflammation and Activating NO-cGMP-PKG Pathway. *Phytomedicine* **2021**, *91*, 153633. [CrossRef] [PubMed]
24. Duan, Y.; Zeng, L.; Zheng, C.; Song, B.; Li, F.; Kong, X.; Xu, K. Inflammatory Links Between High Fat Diets and Diseases. *Front. Immunol.* **2018**, *9*, 2649. [CrossRef]
25. Kiran, S.; Rakib, A.; Kodidela, S.; Kumar, S.; Singh, U.P. High-Fat Diet-Induced Dysregulation of Immune Cells Correlates with Macrophage Phenotypes and Chronic Inflammation in Adipose Tissue. *Cells* **2022**, *11*, 1327. [CrossRef]
26. Paulus, W.J.; Tschöpe, C. A Novel Paradigm for Heart Failure with Preserved Ejection Fraction: Comorbidities Drive Myocardial Dysfunction and Remodeling through Coronary Microvascular Endothelial Inflammation. *J. Am. Coll. Cardiol.* **2013**, *62*, 263–271. [CrossRef]
27. Jeong, E.-M.; Chung, J.; Liu, H.; Go, Y.; Gladstein, S.; Farzaneh-Far, A.; Lewandowski, E.D.; Dudley, S.C. Role of Mitochondrial Oxidative Stress in Glucose Tolerance, Insulin Resistance, and Cardiac Diastolic Dysfunction. *J. Am. Heart Assoc.* **2016**, *5*, e003046. [CrossRef] [PubMed]
28. Yang, L.; Cheng, C.-F.; Li, Z.-F.; Huang, X.-J.; Cai, S.-Q.; Ye, S.-Y.; Zhao, L.-J.; Xiong, Y.; Chen, D.-F.; Liu, H.-L.; et al. Berberine Blocks Inflammasome Activation and Alleviates Diabetic Cardiomyopathy via the miR-18a-3p/Gsdmd Pathway. *Int. J. Mol. Med.* **2023**, *51*, 49. [CrossRef]
29. Zhong, C.; Xie, Y.; Wang, H.; Chen, W.; Yang, Z.; Zhang, L.; Deng, Q.; Cheng, T.; Li, M.; Ju, J.; et al. Berberine Inhibits NLRP3 Inflammasome Activation by Regulating mTOR/mtROS Axis to Alleviate Diabetic Cardiomyopathy. *Eur. J. Pharmacol.* **2024**, *964*, 176253. [CrossRef]
30. Wang, Y.; Liao, J.; Luo, Y.; Li, M.; Su, X.; Yu, B.; Teng, J.; Wang, H.; Lv, X. Berberine Alleviates Doxorubicin-Induced Myocardial Injury and Fibrosis by Eliminating Oxidative Stress and Mitochondrial Damage via Promoting Nrf-2 Pathway Activation. *Int. J. Mol. Sci.* **2023**, *24*, 3257. [CrossRef]
31. Jia, X.; Shao, W.; Tian, S. Berberine Alleviates Myocardial Ischemia-Reperfusion Injury by Inhibiting Inflammatory Response and Oxidative Stress: The Key Function of miR-26b-5p-Mediated PTGS2/MAPK Signal Transduction. *Pharm. Biol.* **2022**, *60*, 652–663. [CrossRef] [PubMed]
32. Xia, Z.; Sun, G. Blocking RIPK2 Function Alleviates Myocardial Ischemia/Reperfusion Injury by Regulating the AKT and NF-κB Pathways. *Immunol. Investig.* **2023**, *52*, 529–545. [CrossRef] [PubMed]
33. Chen, X.; Qian, J.; Wang, L.; Li, J.; Zhao, Y.; Han, J.; Khan, Z.; Chen, X.; Wang, J.; Liang, G. Kaempferol Attenuates Hyperglycemia-Induced Cardiac Injuries by Inhibiting Inflammatory Responses and Oxidative Stress. *Endocrine* **2018**, *60*, 83–94. [CrossRef] [PubMed]

34. Teuber, J.P.; Essandoh, K.; Hummel, S.L.; Madamanchi, N.R.; Brody, M.J. NADPH Oxidases in Diastolic Dysfunction and Heart Failure with Preserved Ejection Fraction. *Antioxidants* **2022**, *11*, 1822. [CrossRef]
35. Simmonds, S.J.; Cuijpers, I.; Heymans, S.; Jones, E.A.V. Cellular and Molecular Differences between HFpEF and HFrEF: A Step Ahead in an Improved Pathological Understanding. *Cells* **2020**, *9*, 242. [CrossRef]
36. Incalza, M.A.; D'Oria, R.; Natalicchio, A.; Perrini, S.; Laviola, L.; Giorgino, F. Oxidative Stress and Reactive Oxygen Species in Endothelial Dysfunction Associated with Cardiovascular and Metabolic Diseases. *Vasc. Pharmacol.* **2018**, *100*, 1–19. [CrossRef]
37. Wu, M.; Guo, Y.; Wu, Y.; Xu, K.; Lin, L. Protective Effects of Sacubitril/Valsartan on Cardiac Fibrosis and Function in Rats with Experimental Myocardial Infarction Involves Inhibition of Collagen Synthesis by Myocardial Fibroblasts Through Downregulating TGF-B1/Smads Pathway. *Front. Pharmacol.* **2021**, *12*, 696472. [CrossRef]
38. Duca, F.; Kammerlander, A.A.; Zotter-Tufaro, C.; Aschauer, S.; Schwaiger, M.L.; Marzluf, B.A.; Bonderman, D.; Mascherbauer, J. Interstitial Fibrosis, Functional Status, and Outcomes in Heart Failure with Preserved Ejection Fraction: Insights From a Prospective Cardiac Magnetic Resonance Imaging Study. *Circ. Cardiovasc. Imaging* **2016**, *9*, e005277. [CrossRef]
39. Franssen, C.; Chen, S.; Unger, A.; Korkmaz, H.I.; De Keulenaer, G.W.; Tschöpe, C.; Leite-Moreira, A.F.; Musters, R.; Niessen, H.W.M.; Linke, W.A.; et al. Myocardial Microvascular Inflammatory Endothelial Activation in Heart Failure with Preserved Ejection Fraction. *JACC Heart Fail.* **2016**, *4*, 312–324. [CrossRef]
40. Cai, Z.; Wu, C.; Xu, Y.; Cai, J.; Zhao, M.; Zu, L. The NO-cGMP-PKG Axis in HFpEF: From Pathological Mechanisms to Potential Therapies. *Aging Dis.* **2023**, *14*, 46. [CrossRef]
41. Sanders-van Wijk, S.; Tromp, J.; Beussink-Nelson, L.; Hage, C.; Svedlund, S.; Saraste, A.; Swat, S.A.; Sanchez, C.; Njoroge, J.; Tan, R.-S.; et al. Proteomic Evaluation of the Comorbidity-Inflammation Paradigm in Heart Failure with Preserved Ejection Fraction: Results From the PROMIS-HFpEF Study. *Circulation* **2020**, *142*, 2029–2044. [CrossRef] [PubMed]
42. Vahedi-Mazdabadi, Y.; Shahinfar, H.; Toushih, M.; Shidfar, F. Effects of Berberine and Barberry on Selected Inflammatory Biomarkers in Adults: A Systematic Review and Dose-Response Meta-Analysis of Randomized Clinical Trials. *Phytother. Res.* **2023**, *37*, 5541–5557. [CrossRef]
43. Engin, A. Reappraisal of Adipose Tissue Inflammation in Obesity. *Adv. Exp. Med. Biol.* **2024**, *1460*, 297–327. [CrossRef] [PubMed]
44. Tamaki, S.; Nagai, Y.; Shutta, R.; Masuda, D.; Yamashita, S.; Seo, M.; Yamada, T.; Nakagawa, A.; Yasumura, Y.; Nakagawa, Y.; et al. Combination of Neutrophil-to-Lymphocyte and Platelet-to-Lymphocyte Ratios as a Novel Predictor of Cardiac Death in Patients with Acute Decompensated Heart Failure with Preserved Left Ventricular Ejection Fraction: A Multicenter Study. *J. Am. Heart Assoc.* **2023**, *12*, e026326. [CrossRef] [PubMed]
45. Rai, A.; Narisawa, M.; Li, P.; Piao, L.; Li, Y.; Yang, G.; Cheng, X.W. Adaptive Immune Disorders in Hypertension and Heart Failure: Focusing on T-Cell Subset Activation and Clinical Implications. *J. Hypertens.* **2020**, *38*, 1878–1889. [CrossRef]
46. Xiaojing, C.; Yanfang, L.; Yanqing, G.; Fangfang, C. Thymopentin Improves Cardiac Function in Older Patients with Chronic Heart Failure. *Anatol. J. Cardiol.* **2017**, *17*, 24–30. [CrossRef]
47. Wang, Y.; Liu, Y.; Du, X.; Ma, H.; Yao, J. The Anti-Cancer Mechanisms of Berberine: A Review. *Cancer Manag. Res.* **2020**, *12*, 695–702. [CrossRef]
48. Ehteshamfar, S.-M.; Akhbari, M.; Afshari, J.T.; Seyedi, M.; Nikfar, B.; Shapouri-Moghaddam, A.; Ghanbarzadeh, E.; Momtazi-Borojeni, A.A. Anti-Inflammatory and Immune-Modulatory Impacts of Berberine on Activation of Autoreactive T Cells in Autoimmune Inflammation. *J. Cell Mol. Med.* **2020**, *24*, 13573–13588. [CrossRef]
49. Surendar, J.; Frohberger, S.J.; Karunakaran, I.; Schmitt, V.; Stamminger, W.; Neumann, A.-L.; Wilhelm, C.; Hoerauf, A.; Hübner, M.P. Adiponectin Limits IFN-γ and IL-17 Producing CD4 T Cells in Obesity by Restraining Cell Intrinsic Glycolysis. *Front. Immunol.* **2019**, *10*, 2555. [CrossRef]

International Journal of
Molecular Sciences

MDPI

Article

Liver-Targeted Scutellarin Nanoemulsion Alleviates Fibrosis with Ancillary Modulation of the Gut–Liver Microbiota

Haoyang Yu [1], Xia Niu [1], Bingyu Niu [1], Peng Lei [2], Ning Xu [1], Sitong Yang [1], Quanyong Yu [3], Guiling Li [1,*] and Lulu Wang [1,*]

1 State Key Laboratory of Bioactive Substance and Function of Natural Medicines, Institute of Medicinal Biotechnology, Chinese Academy of Medical Sciences & Peking Union Medical College, Beijing 100050, China; yuhy1014@163.com (H.Y.); niuxia307@163.com (X.N.); balala482@163.com (B.N.); xuning0624@163.com (N.X.); yangst0323@163.com (S.Y.)

2 Institute of Materia Medica, Chinese Academy of Medical Sciences & Peking Union Medical College, Beijing 100050, China; leipeng156@163.com

3 School of Pharmacy, Jiangning Campus, China Pharmaceutical University, Nanjing 210009, China; yuquanyong04@163.com

* Correspondence: liguiling@imb.pumc.edu.cn (G.L.); wanglulu@imb.cams.cn (L.W.)

Abstract

Liver fibrosis, a progressive condition with limited pharmacotherapies, poses a global health challenge. Scutellarin (SCU), a flavonoid derived from *Erigeron breviscapus*, has demonstrated anti-fibrotic activity and modulates gut microbiota. Emerging evidence suggests that SCU may also influence the hepatic microbiome. However, its clinical utility is constrained by poor water solubility and low oral bioavailability. Here, we developed an SCU-loaded nanoemulsion (SCE) to enhance solubility and liver-targeted delivery. In vitro, SCE increased SCU uptake in hepatic stellate cells (HSCs) and significantly inhibited TGF-β1-induced fibrogenesis. In a bile duct ligation (BDL) mouse model, oral administration of SCE improved hepatic SCU accumulation and produced superior anti-fibrotic efficacy. SCE treatment attenuated fibrosis and collagen deposition in the liver and improved liver function markers. Mechanistic investigations using 16S rRNA sequencing revealed that SCU treatment was associated with beneficial microbiota changes, although its main therapeutic effects were achieved through enhanced hepatic targeting. Notably, the SCE formulation was well-tolerated, showing no significant toxicity in vitro or in vivo. In conclusion, the SCU-loaded nanoemulsion achieved enhanced hepatic delivery of SCU and exerted potent anti-fibrotic effects via multiple mechanisms, including direct suppression of fibrogenesis and ancillary modulation of the gut–liver microbiome, offering a promising therapeutic strategy for liver fibrosis.

Keywords: liver fibrosis; scutellarin; gut microbiota; hepatic microbiota; liver-targeted delivery

1. Introduction

Liver fibrosis is a chronic condition characterized by the excessive accumulation of extracellular matrix proteins, primarily due to the differentiation of hepatic stellate cells (HSCs) into myofibroblasts during persistent liver injury. This pathological process disrupts the normal hepatic architecture and function, ultimately leading to cirrhosis and liver failure if untreated [1]. Managing liver fibrosis remains a major clinical challenge because effective anti-fibrotic therapies are scarce. Current treatments mainly address the underlying cause of injury—for example, antiviral therapy for hepatitis or alcohol abstinence in alcoholic liver disease—but they achieve only limited reversal of established fibrosis [2]. To date, no universally approved anti-fibrotic drug exists for broad clinical use. For instance, Rezdiffra (resmetirom) has demonstrated therapeutic benefit in patients with noncirrhotic NASH (nonalcoholic steatohepatitis) and fibrosis [3]; however, its efficacy across other etiologies of liver fibrosis remains unproven. These limitations underscore the urgent need for new therapeutic strategies that directly target established fibrosis, irrespective of etiology.

The gut–liver axis has been increasingly recognized as a critical determinant of liver health and disease. The gut microbiota, through its metabolic activities and cellular components, can influence the progression of liver fibrosis [4]. Dysbiosis—an imbalance in the gut microbial community—can increase intestinal permeability, allowing bacterial products such as endotoxins to translocate into the portal circulation. These products trigger hepatic inflammation and exacerbate fibrogenesis. Consequently, modulating the gut microbiota has emerged as a promising therapeutic strategy for liver disease. Previous studies have shown that scutellarin (SCU), a natural flavonoid derived from *Erigeron breviscapus* (Compositae), regulates intestinal microbiota composition [5,6]. This finding suggests that SCU's anti-fibrotic activity may partly depend on its ability to improve gut microbial balance. In line with the evolving paradigm of multi-target therapies for complex diseases [7], we hypothesized that SCU might also influence the liver's local microbiome. Emerging evidence, including our recent work, supports the existence of a resident "hepatic microbiota" that may contribute to liver fibrosis progression [8]. Thus, SCU's therapeutic effects may extend to modulating hepatic microbiota and, in turn, fibrogenesis. Nonetheless, research on the liver microbiome remains at an early stage, and further studies are needed to identify microbial changes that are beneficial or harmful in liver fibrosis.

SCU, a well-studied bioactive ingredient in traditional Chinese medicine, exhibits a broad spectrum of pharmacological activities. These include cardioprotective effects, such as the attenuation of cardiac hypertrophy [9] and the amelioration of ischemia–reperfusion injury [10,11], as well as antimicrobial [12] and antiviral properties [13]. SCU also exhibits neuroprotective effects, for example, in glaucoma [14], and antitumor activities through modulation of immune responses [15]. Notably, SCU has demonstrated anti-fibrotic efficacy in other organ systems, including the amelioration of cardiac interstitial fibrosis post-infarction [16] and the attenuation of pulmonary fibrosis [17]. Collectively, these findings suggest that SCU has potential as an anti-fibrotic agent in hepatic disease. Given its diverse pharmacological profile and anti-fibrotic activity in the heart and lung, it is reasonable to propose that SCU could also play a beneficial role in ameliorating liver fibrosis, especially if delivered effectively to hepatic targets.

Despite its therapeutic potential, the clinical use of SCU is limited by its unfavorable biopharmaceutical properties. Specifically, SCU's poor water solubility markedly limits its oral absorption and bioavailability, thereby reducing its efficacy. Overcoming these challenges requires advanced drug delivery strategies that improve solubility, protect SCU from degradation during gastrointestinal transit, and enable targeted delivery to the liver. Recent innovations, including phospholipid complexation and nanoformulations, have enhanced the solubility and bioavailability of poorly water-soluble drugs and, in some cases, conferred organ-targeting capabilities [18].

In this study, we developed a novel nanoemulsion-based delivery system for SCU to improve its solubility and liver-targeting capability. Initially, we prepared a scutellarin–phospholipid complex (SPC) to increase the lipophilicity of SCU and facilitate its incorporation into the nanoemulsion. Using this complex, we formulated a scutellarin-loaded nanoemulsion (SCE) stabilized with chitosan oligosaccharide. We then conducted a comprehensive physicochemical characterization of both SPC and SCE and evaluated their performance in vitro and in vivo. Specifically, we examined the ability of SCE to enhance cellular uptake and hepatic accumulation of SCU, along with its pharmacokinetic and pharmacodynamic profiles in a bile duct ligation (BDL)-induced liver fibrosis mouse model. We further assessed the therapeutic efficacy of SCE against liver fibrosis in vivo and in vitro, using a TGF-β1-activated HSC (LX-2 cell) model. In addition, we investigated whether free and nanoformulated SCU could modulate gut and liver microbiota in fibrotic mice, exploring the gut–liver axis as a potential mechanism of action. Finally, we confirmed the safety and biocompatibility of SCE through both in vitro assays (normal liver cells and LX-2 cells) and in vivo assessments (histopathology and blood chemistry in mice). Overall, our findings demonstrate that SCE significantly alleviates liver fibrosis by enhancing hepatic delivery and multi-target mechanisms, offering a promising therapeutic approach that combines microbiome modulation with direct anti-fibrotic activity.

2. Results

2.1. SCU Modulates Gut and Liver Microbiota Composition in BDL-Induced Fibrosis

2.1.1. Gut Microbiota

We first examined the impact of SCU on gut microbiota dysbiosis caused by BDL-induced liver fibrosis. High-throughput 16S rRNA sequencing of fecal samples yielded 2,164,399 high-quality sequences (average > 240,000 reads per sample), which clustered into 2394 OTUs (at 97% similarity) at 97% similarity across all groups. Sequencing depth was sufficient to capture the majority of microbial diversity, as indicated by the plateauing rarefaction curves (Supplementary Figure S1A,B). Alpha diversity indices, including Chao1 richness and ACE, revealed no significant differences in overall bacterial richness across the Sham, BDL, and SCU-treated groups at the phylum, genus, or OTU levels (Figures S1C and S2A). However, beta diversity analysis revealed substantial alterations in microbial community structure in the BDL group. PCoA and PLS-DA demonstrated distinct clustering of the BDL group's microbiota composition compared to the Sham controls, indicating fibrosis-induced changes in gut microbial composition (Figure 1A).

Although overall diversity did not change significantly, BDL induced notable taxonomic alterations in the gut microbiota. At the phylum level, BDL slightly increased the relative abundance of several potentially harmful bacterial phyla, including *Actinobacteriota*, *Cyanobacteria*, and *Desulfobacterota*, compared to the Sham group (Figure 1B,C). These changes, while subtle, are consistent with disease-associated dysbiosis reported in the literature. For instance, elevated *Actinobacteriota* abundance has been associated with steatosis severity in mouse models [19] and has been observed in both gut and liver microbiota in hepatocellular carcinoma [20]. Similarly, the elevated abundance of *Desulfobacterota*—a group comprising sulfate-reducing bacteria—has been linked to hepatic inflammation and injury in other studies [21], while overgrowth of *Cyanobacteria* and *Desulfobacterota* has been implicated in gut dysbiosis and various diseases [22–25]. At the family level, BDL increased the abundance of *Staphylococcaceae* and *Desulfovibrionaceae* (Figure S2B), both of which include pathogenic members. Specifically, increased abundance of *Staphylococcaceae* has been reported in pediatric nonalcoholic fatty liver disease [26] and cirrhosis-associated gut microbiota profiles [27], whereas *Desulfovibrionaceae* (sulfate-reducing, endotoxin-producing bacteria) have been implicated in steatohepatitis and liver cancer [28,29]. Conversely, BDL suppressed the abundance of beneficial taxa such as the genus *Clostridia_UCG-014* (family *Clostridiaceae*), which includes butyrate-producing bacteria linked to improved metabolic and inflammatory profiles [30]. BDL also increased the abundance of several genera considered harmful or pro-inflammatory, including *Bacteroides*, *Dubosiella*, and *Lacticigenium*. Notably, *Bacteroides* spp. are known to overgrow in gut dysbiosis and can produce metabolites that contribute to inflammation [31,32]. Similarly, increased *Dubosiella* abundance (family *Erysipelotrichaceae*) has been reported in certain inflammatory conditions [33], whereas *Lacticigenium* (a lactic acid bacterium) has been linked to gut metabolic disturbances [34].

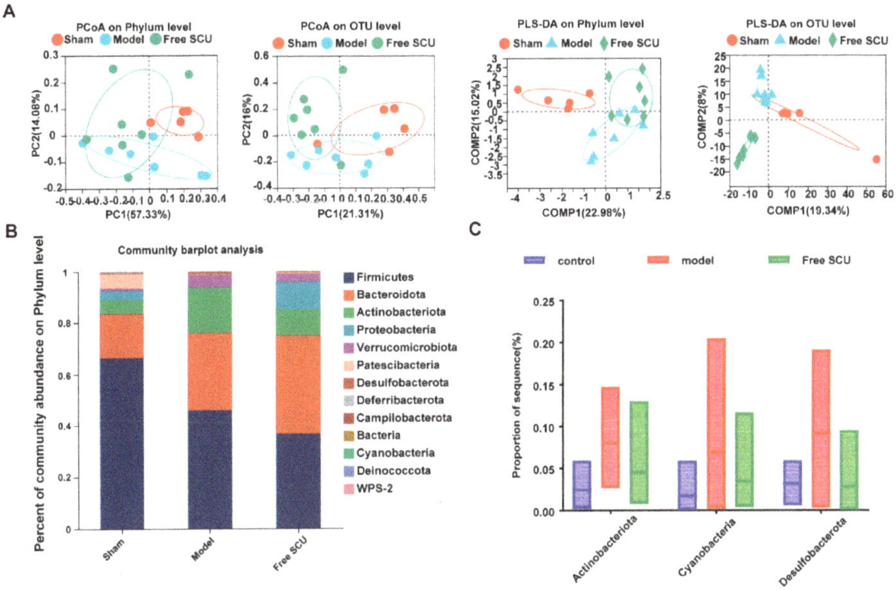

Figure 1. Scutellarin nanoemulsion (SCE) ameliorated BDL-induced gut microbiota dysbiosis. (**A**) Principal Coordinates Analysis (PCoA) plot based on Bray–Curtis distances at the OTU level. (**B**,**C**) Relative abundance of major bacterial phyla in the gut.

Importantly, SCU treatment corrected many of these BDL-induced microbial changes. Although some differences did not reach statistical significance due to variability, the microbiota composition of SCU-treated mice trended toward the Sham group profile. SCU administration reduced the relative abundance of *Actinobacteriota*, *Cyanobacteria*, and *Desulfobacterota* compared to untreated BDL mice (Figure 1C). At the family level, SCU reduced the abundance of *Staphylococcaceae* and *Desulfovibrionaceae*, while partially restoring *Clostridiaceae* populations, like *Clostridia_UCG-014* (Figure S2B). Similarly, the overgrowth of *Bacteroides*, *Dubosiella*, and *Lacticigenium* observed in BDL mice was moderated in the SCU group. These results suggest that SCU can beneficially modulate the gut microbiota in liver fibrosis, reducing potentially deleterious bacteria and supporting the growth of beneficial taxa. Overall, SCU appears to mitigate the dysbiosis induced by BDL, shifting the microbial community toward a healthier composition.

2.1.2. Liver Microbiota

We subsequently examined the liver microbiome composition in the same cohort of mice. Although liver tissue contains low microbial biomass, 16S rRNA gene sequencing was successful, yielding 1,491,466 high-quality sequences, with an average of approximately 165,000 reads per sample. Clustering at 97% sequence similarity identified 8512 OTUs. As expected, liver tissue exhibited a broader diversity of taxa than fecal samples, likely reflecting blood-borne or translocated bacteria. In total, sequences were classified into 34 phyla, 436 families, and 891 genera. Rarefaction analysis confirmed sufficient sequencing depth (Supplementary Figure S3A,B). Alpha diversity analyses revealed subtle differences among groups (Figure S3C–F). BDL decreased the liver microbial richness and diversity compared to Sham, whereas SCU treatment appeared to prevent this decrease, although the changes in ACE and Chao1 indices were not statistically significant. Beta diversity, assessed by PLS-DA, revealed distinct clustering among the three experimental groups, indicating that both fibrotic injury and SCU intervention impacted liver microbial composition (Figure 2A).

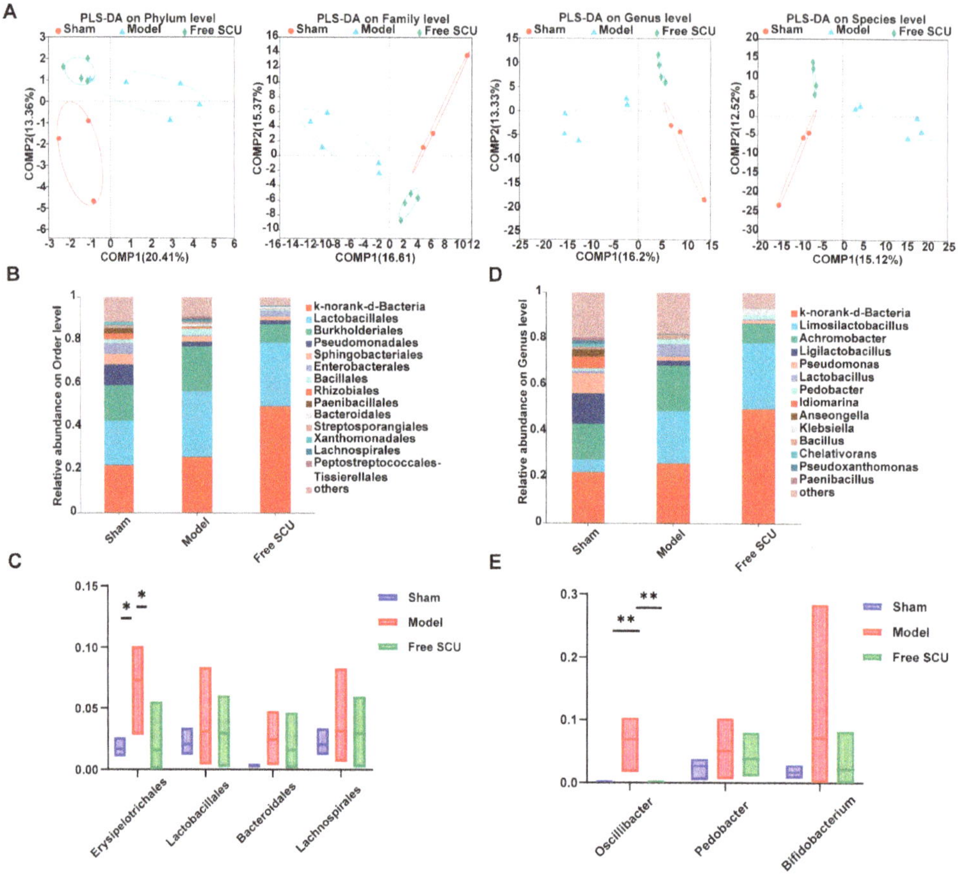

Figure 2. SCU treatment influenced the diversity and composition of the liver microbiome in BDL mice. (**A**) Partial Least Squares Discriminant Analysis (PLS-DA) of liver microbiota profiles at multiple taxonomic levels. (**B**,**C**) Relative abundance of selected bacterial orders in liver tissue. (**D**,**E**) Relative abundance of selected genera in liver microbiota. Liver microbiota data should be interpreted as exploratory due to low biomass. ** $p < 0.01$, * $p < 0.05$.

Taxonomic comparisons revealed several notable changes in liver microbiota composition due to BDL and its modulation by SCU. At the order level, the relative abundance of *Erysipelotrichales* was significantly elevated in BDL mice compared to Sham controls (Figure 2B,C), consistent with previous studies linking this order to high-fat diet-induced nonalcoholic steatohepatitis (NASH) and disease progression [35]. Notably, SCU administration significantly reduced the abundance of *Erysipelotrichales*, restoring levels close to those of the Sham group (Figure 2C), suggesting a corrective effect on fibrosis-associated microbial dysregulation. Additionally, SCU had a mild regulatory effect on other taxa perturbed by BDL. At the genus level, *Oscillibacter* (family *Ruminococcaceae*) emerged as another taxon affected by BDL. Its abundance was significantly reduced in the livers of BDL mice relative to Sham controls but was partially restored following SCU treatment

(Figure 2E). Although the role of *Oscillibacter* in liver disease remains unclear, with some studies associating it with anti-inflammatory properties and others linking it to disease states [36,37], its partial normalization in response to SCU suggests a potentially beneficial modulation. In addition to these major changes, BDL induced several minor alterations in the abundance of families within the *Proteobacteria* and *Firmicutes* phyla, which were also moderated by SCU treatment. Collectively, these data indicate that SCU regulates the liver microbial community in fibrotic mice, paralleling its influence on the gut microbiota. Given the relatively limited research on the hepatic microbiome, these findings should be interpreted with caution. Nonetheless, they suggest that SCU can influence microbial populations in the liver. Given that the oral route initially exposes the drug to gut microbiota before systemic absorption, part of the therapeutic effect may arise from modulation of the gut–liver axis. However, the enhanced hepatic accumulation achieved by the nanoemulsion suggests that these microbial alterations are likely secondary and supportive to SCU's primary liver-targeted antifibrotic action.

These findings suggest that SCU modulates the liver-resident microbiota disturbed by cholestatic injury, although the functional consequences of these changes remain unclear. The therapeutic benefits of SCE in BDL-induced fibrosis involve a multifaceted mechanism. Our pharmacokinetic and tissue distribution data confirm significantly enhanced hepatic accumulation of SCU, which correlates with the observed reduction in the expression of liver fibrosis markers, suggesting direct pharmacodynamic action at the target site. Simultaneously, oral administration exposes SCU transiently to the gut microbiota, partially correcting fibrosis-associated dysbiosis. Although microbial modulation may be limited by the nanoemulsion's efficient hepatic uptake, this early-phase interaction with the gut epithelium and microbiota could influence immune-metabolic pathways via the gut-liver axis. While these microbiota-mediated effects likely do not constitute the primary mechanism, they may complement and reinforce SCU's antifibrotic action. Overall, this integrated perspective highlights the dual advantage of oral nanoformulations: targeted hepatic delivery coupled with supportive modulation of gut microbiota. Given the relatively low bacterial biomass in liver tissue, the results from liver 16S sequencing should be interpreted as exploratory with caution.

2.2. *Preparation and Physicochemical Characterization of SPC and SCE*

2.2.1. SCU–Phospholipid Complex (SPC) Formation

SPC was successfully prepared using the solvent evaporation method, and its formation was confirmed through spectroscopic and thermal analyses. The FTIR spectra (Figure 3C) revealed that pure SCU has characteristic O–H stretching vibration peaks around 3500–3300 cm^{-1} due to phenolic hydroxyl groups and distinct absorption bands in the 1500–1700 cm^{-1} region due to aromatic ring vibrations and carbonyl groups. In the physical mixture of SCU and phospholipid, peaks from both components were present without significant shifts, indicating just a simple superposition of spectra. However, the FTIR spectrum of SPC exhibited notable differences: the broad O–H stretch band was greatly diminished or shifted, and several peaks in the 1500–1700 cm^{-1} region were attenuated or disappeared. These changes suggest strong intermolecular interactions in SPC, likely due to hydrogen bonding between SCU and phospholipid molecules, which can alter the vibrational frequencies of functional groups.

Thermal analysis by DSC provided further evidence of complex formation. The thermogram of pure SCU displayed endothermic peaks at 163.6 °C and 353.3 °C, which are attributed to melting or decomposition processes of SCU, as well as an exothermic peak at 206.7 °C, likely due to a phase transition or crystallization event. These thermal transitions confirm the crystalline nature of SCU. The physical mixture exhibited endothermic peaks around 234 °C, corresponding to the melting of the phospholipid (which has a broad transition) and perhaps a shifted SCU melt due to some mixing. In contrast, the DSC thermogram of SPC (Figure 3D) showed no sharp endothermic or exothermic peaks over the range up to 400 °C, indicating the absence of the characteristic thermal transitions associated with crystalline SCU. This absence suggests that SCU exists in an amorphous or molecularly dispersed state within the phospholipid matrix. The phospholipid likely stabilizes SCU in this amorphous state, which is typically associated with enhanced solubility.

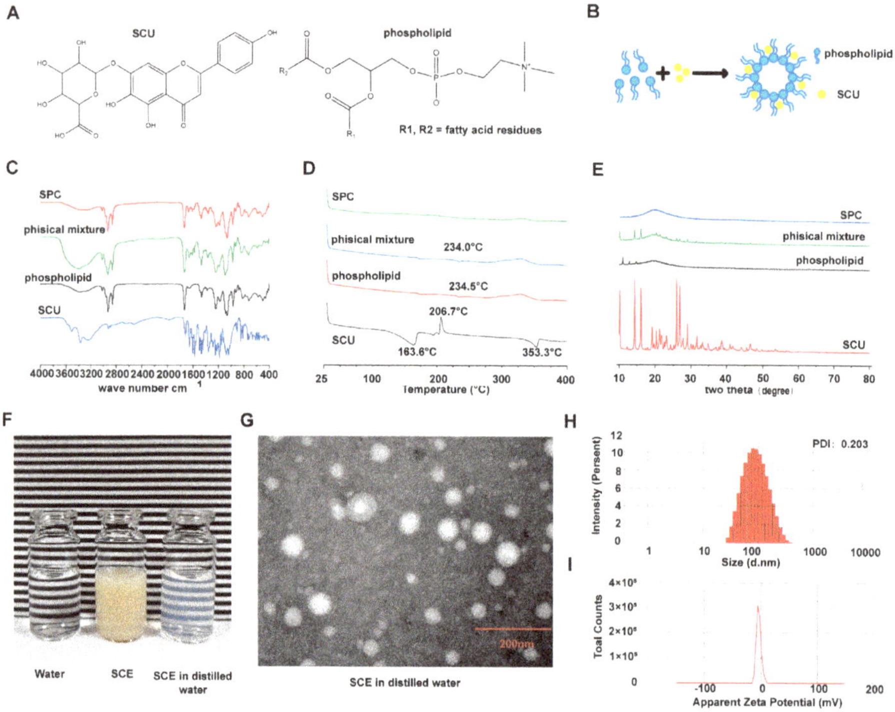

Figure 3. Preparation and physicochemical characterization of the SCU-phospholipid complex (SPC) and the SCU nanoemulsion (SCE). (**A**) Chemical structures of scutellarin (SCU, left) and the phospholipid (soy lecithin, primarily phosphatidylcholine, right) used to form the SPC. (**B**) Schematic illustration of the SPC formation process. (**C**) FTIR spectra of SCU, phospholipid, their physical mixture, and SPC. (**D**) Differential Scanning Calorimetry thermograms of SCU, phospholipid, physical mixture, and SPC. (**E**) X-ray diffraction patterns of the same four samples. (**F**) Photograph of vials containing distilled water, undiluted SCE, and SCE after dilution in water. (**G**) Transmission electron microscopy (TEM) image of SCE. (**H**,**I**) Particle size distribution and zeta potential of SCE measured by DLS.

XRD analysis further substantiated the structural transformation of SCU in the complex. The XRD patterns (Figure 3E) of pure SCU displayed numerous distinct peaks, confirming its crystalline nature. The phospholipid showed a few broad peaks, while the physical mixture retained the distinct diffraction patterns of both components, suggesting no significant interaction upon simple grinding. Notably, SPC exhibited a diffractogram with no pronounced diffraction peaks. Instead, only a diffuse halo was observed, characteristic of amorphous substances. The disappearance of SCU's crystalline peaks in SPC confirms that SCU was successfully converted into a non-crystalline form through complexation with the phospholipid. Together, the FTIR, DSC, and XRD results demonstrate that SCU and the phospholipid formed a complex, likely via hydrogen bonding and hydrophobic interactions, resulting in an amorphous solid. This complexation is expected to enhance the apparent solubility of SCU in the oil phase, providing a basis for the subsequent nanoemulsion formulation.

2.2.2. Nanoemulsion Properties and Stability

Incorporating SPC into the nanoemulsion produced a milky white formulation, which became transparent with a slight bluish opalescence upon sufficient dilution in water (Figure 3F). This Tyndall effect qualitatively indicates the presence of nanoscale particles. DLS confirmed that the SCE droplets were indeed in the nanometer range. The average particle size of the freshly prepared nanoemulsion was 78.5 ± 2.5 nm (Figure 3H), with a PDI of approximately 0.20, suggesting a relatively narrow size distribution. The zeta potential was slightly negative at −3.8 mV, likely due to the combined effects of the negatively charged phospholipid and the partially cationic COS coating. TEM imaging confirmed the droplet morphology and size, revealing roughly spherical and uniform particles ranging from 70 to 100 nm (Figure 3G), consistent with the DLS results.

The SCE formulation also exhibited satisfactory short-term stability at 4 °C. During a 7-day storage period, its characteristics changed minimally. The mean particle size increased slightly, from approximately 78 nm to nearly 89 nm by day 7, while the PDI remained between 0.19 and 0.23, indicating no significant aggregation or broadening of size distribution. The zeta potential stayed within −3 to −4 mV throughout the period. Notably, the SCU content in the nanoemulsion remained above 90% of the initial concentration, with fluctuations below 10% over the study period (Table 1). These results confirm that SCE maintains both physical and chemical stability under refrigerated conditions for at least one week, sufficient for experimental use and suggesting potential for practical shelf-life upon further optimization. Taken together, the characterization results show that the SPC-based nanoemulsion was successfully prepared, yielding nanosized SCU carriers with suitable stability. This short-term stability was sufficient for the experimental duration of this study. However, long-term stability over periods of months, which is critical for clinical translation, was not assessed and will be an important focus of future formulation optimization studies.

Table 1. Physical stability of SCE at 4 °C.

Day	Mean PS ± SD	Mean PDI ± SD	Mean ZP ± SD	Mean Concentration ± SD
0	78.51 ± 2.45	0.23 ± 0.01	−3.6 ± 0.23	106.11 ± 3.49
1	81.93 ± 3.08	0.21 ± 0	−3.72 ± 0.35	98.39 ± 10.75
3	84.37 ± 3.2	0.20 ± 0.01	−3.86 ± 0.31	90.99 ± 18.58
5	86.62 ± 2.77	0.19 ± 0.01	−3.99 ± 0.38	105.22 ± 5.29
7	89.22 ± 2.94	0.19 ± 0.01	−3.77 ± 0.36	102.87 ± 5.56

2.3. *SCE Formulation Enhances Cellular Uptake of SCU*

We next investigated whether nanoemulsion-based delivery could enhance SCU uptake by SCU by HSCs. To this end, LX-2 cells were employed to compare the cellular uptake kinetics of SCU delivered via SCE versus free SCU. Because SCU exhibits limited intrinsic fluorescence, we used NR, a lipophilic fluorescent probe, as a surrogate tracer. Confocal microscopy revealed pronounced differences in uptake dynamics between the two formulations. In cells treated with free NR (dissolved in PBS), virtually no red fluorescence was detectable in the cytoplasm during the first hour of incubation. A faint signal appeared by 2 h and became more apparent by 4 h, indicating slow diffusion of NR into cells (Figure 4B). In contrast, cells treated with NR-loaded SCE showed detectable red fluorescence as early as 5 min after exposure, and the intensity increased rapidly over time. At 15 min, 30 min, and 1 h, the NR signal in the SCE group was already substantial, whereas it remained negligible in the free NR group at those same time points. Even at later time points (2 h and 4 h), the fluorescence intensity in SCE-treated cells significantly exceeded that of cells treated with free NR. This qualitative observation implies that the nanoemulsion facilitated a much faster and higher uptake of the hydrophobic compound into HSCs.

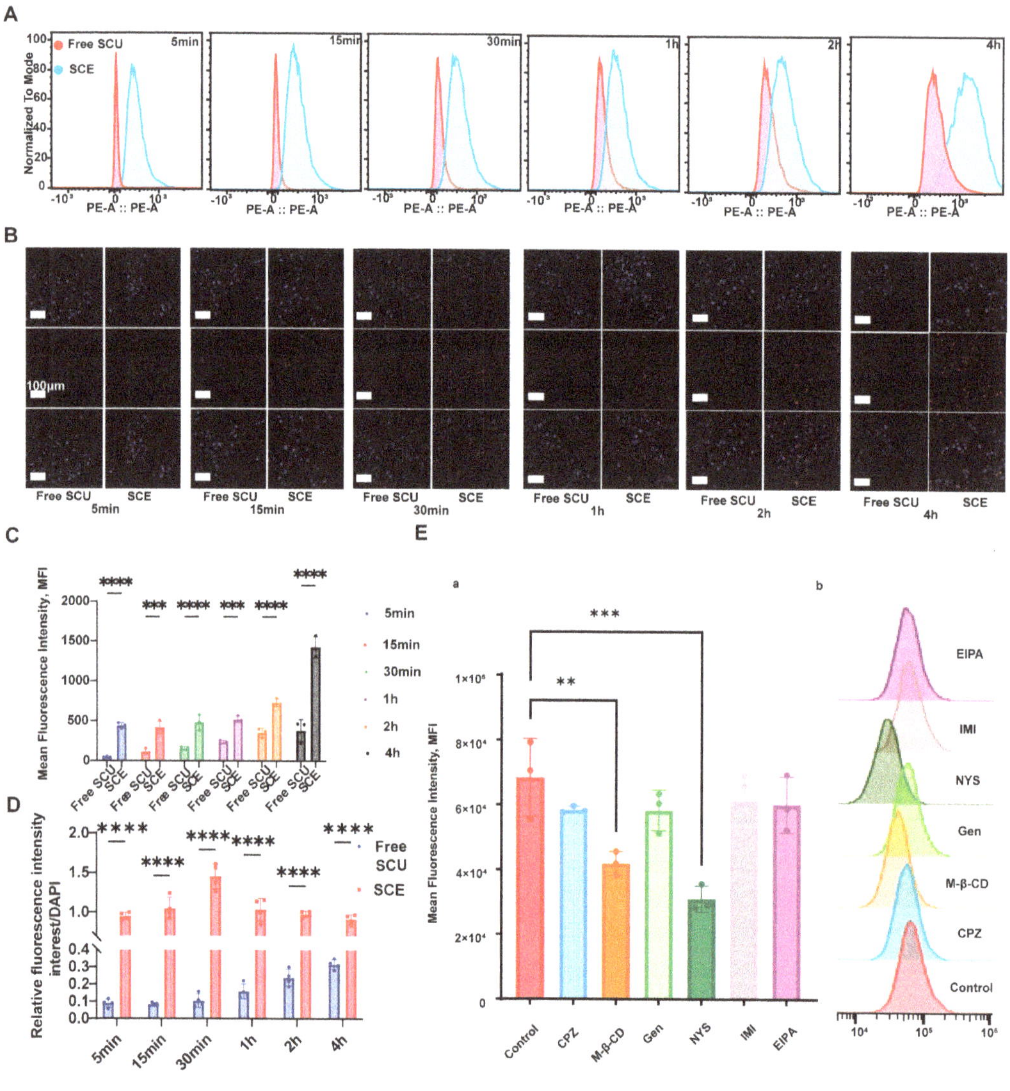

Figure 4. Enhanced cellular uptake of SCU by LX-2 hepatic stellate cells using the SCU nanoemulsion (SCE). (**A**) Flow cytometry representative histograms comparing Nile Red (NR) uptake over time. (**B**) Confocal microscopy images of LX-2 cells at 5 min, 1 h, and 4 h after treatment with free NR vs. NR-SCE. (**C**) Quantification of cellular NR uptake by flow cytometry (mean fluorescence intensity ± SD, n = 3). (**D**) Quantification of cellular NR uptake by Confocal microscopy. (Mean ± SD, n = 3). (**E**) Uptake of NR-SCE in LX-2 cells after pretreatment with endocytosis inhibitors. (**a**) Quantification of cellular NR uptake by flow cytometry (Mean ± SD, n = 3). (**b**) representative histograms. **** $p < 0.0001$, *** $p < 0.001$, ** $p < 0.01$.

Quantitative assessment by flow cytometry confirmed the confocal microscopy observations. The mean fluorescence intensity of NR in LX-2 cells treated with SCE was consistently and significantly higher than in cells treated with free NR at all measured time points (Figure 4C). For example, at 1 h, NR fluorescence in the SCE group was several-fold greater than in the free NR group, consistent with the confocal imaging results. These results collectively demonstrate that nanoemulsion-based delivery markedly enhances the intracellular uptake of hydrophobic compounds into HSCs, likely due to the efficient endocytosis of nano-sized droplets.

To elucidate the mechanisms underlying the enhanced cellular uptake of SCE, we employed a panel of endocytosis inhibitors. As shown in Figure 4D,E, caveolae-mediated endocytosis emerged as the primary pathway for SCE internalization. Specifically, treatment with genistein (a caveolin-dependent endocytosis inhibitor) or methyl-β-cyclodextrin (which depletes membrane cholesterol and disrupts lipid raft/caveolae structures) significantly reduced NR uptake from SCE, suggesting these pathways mediate the majority of the nanoemulsion's uptake. Conversely, EIPA (an inhibitor of macropinocytosis) had no significant effect on SCE uptake, indicating macropinocytosis was not involved. Chlorpromazine, an inhibitor of clathrin-mediated endocytosis, caused only a slight decrease in uptake, implying that clathrin pathways play a secondary role. Similarly, nystatin, another caveolae pathway inhibitor, reduced uptake. These findings suggest that SCE enters LX-2 cells primarily through caveolae/caveolin-1-mediated endocytosis, a pathway commonly utilized by nanoparticles for efficient intracellular delivery. The nanoemulsion's small size and composition may likely promote its association with lipid rafts and caveolar pits on the cell membrane, driving rapid internalization. In summary, formulating SCU into SCE significantly increases its cellular uptake by HSCs via energy-dependent, caveolae-mediated endocytosis, which is expected to improve intracellular bioavailability and therapeutic efficacy.

2.4. SCE Alleviates Liver Fibrosis In Vitro

Having established that SCE improves SCU uptake, we next assessed whether this improvement translates into greater anti-fibrotic efficacy in vitro. LX-2 cells activated with TGF-β1 were used as a model of fibrogenic HSCs. We evaluated two hallmark features of activated HSCs in the presence or absence of SCU treatments: increased migratory capacity and overexpression of fibrosis-related genes.

In the scratch wound healing assay, untreated activated LX-2 cells (model group) migrated rapidly to close the scratch, reflecting their high motility upon activation. After 12 h, the wound area in the model group was partly filled by migrating cells (Figure 5A). In contrast, LX-2 cells treated with free SCU exhibited a modest reduction in wound closure; the scratch area remained slightly more open at 12 h compared to the model group, suggesting that SCU can attenuate HSC migration to some extent. Notably, SCE-treated cells exhibited a substantially larger remaining scratch area at 12 h, demonstrating significant impairment in migration. By 24 h, the differences between groups were even more pronounced, indicating a strong anti-migratory effect of the nanoemulsion. These results indicate that while free SCU can suppress HSC migration to some degree, SCE substantially enhanced this activity, likely due to improved intracellular delivery of SCU.

We next examined fibrogenic gene expression in LX-2 cells. Real-time PCR analysis (Figure 5B) demonstrated that TGF-β1 stimulation markedly upregulated key fibrogenesis-associated genes such as *ACTA2* (encoding α-SMA) and *COL1A1* (collagen type I) in LX-2 cells (model vs. quiescent control). Treatment with free SCU produced only a modest reduction in gene expression, which did not reach statistical significance relative to the model group. In contrast, SCE treatment significantly suppressed both *ACTA2* and *COL1A1* expression. Specifically, α-SMA mRNA in SCE-treated cells decreased to nearly baseline levels, and collagen I mRNA was greatly reduced, indicating an effective suppression of HSC activation at the transcriptional level. These results suggest that the intracellular SCU delivered via SCE effectively interfered with TGF-β1–driven fibrogenic signaling.

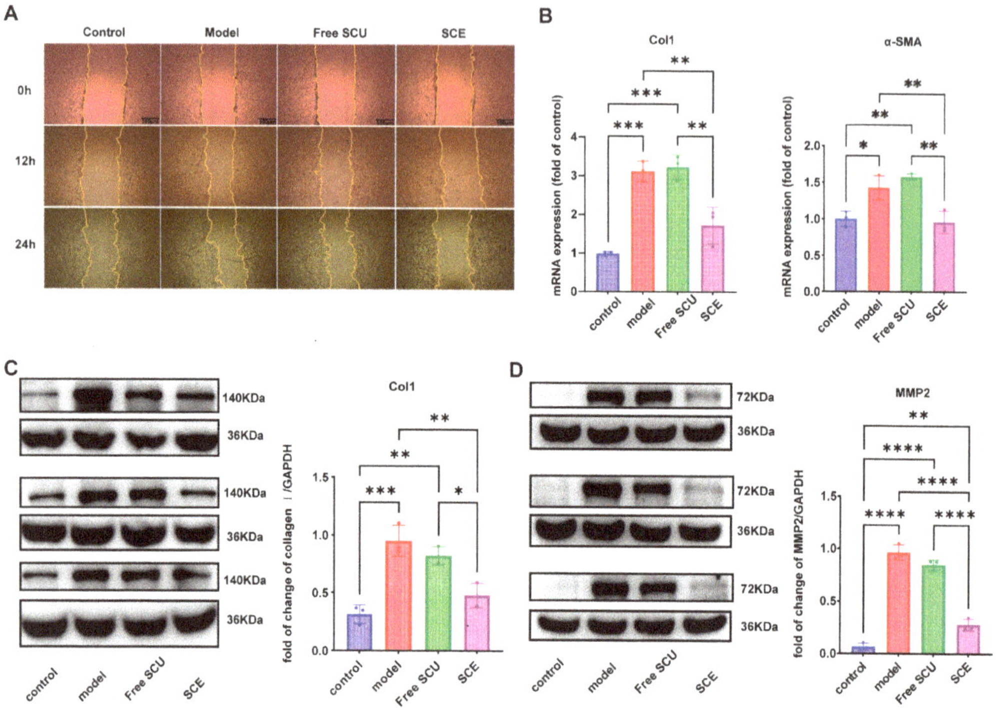

Figure 5. Anti-fibrotic effects of SCU (free vs. nanoemulsion) in TGF-β1-activated LX-2 hepatic stellate cells. (**A**) Wound healing (scratch) assay images at 0, 12, and 24 h for activated LX-2 cells under different treatments (no treatment, free SCU, SCE). (**B**) Relative mRNA expression of fibrosis-related genes (α-SMA and collagen I) in LX-2 cells after 12 h of treatment, as determined by qPCR (normalized to β-actin) (n = 3). (**C**,**D**) Western blot analysis of fibrotic proteins in LX-2 cells. (**C**) Collagen I and (**D**) MMP2 protein levels are shown (bands and quantification) (n = 3). **** $p < 0.0001$, *** $p < 0.001$, ** $p < 0.01$, * $p < 0.05$.

Consistent with the mRNA data, Western blot analysis confirmed that SCE more effectively suppressed fibrotic protein production in LX-2 cells compared to free SCU (Figure 5C,D). Untreated activated HSCs exhibited high collagen I expression, as expected for myofibroblastic cells, and elevated MMP2 levels, reflecting matrix remodeling activity despite excessive collagen accumulation. Free SCU treatment marginally reduced *collagen I* and *MMP2* levels relative to the model group. In contrast, SCE markedly downregulated *ACTA2*, *COL1A1*, and *MMP2* expression, indicating a shift toward a less fibrogenic, more quiescent phenotype. Densitometric analysis indicated that *collagen I* and *MMP2* in the SCE group were decreased by nearly 40–50% compared to the model, whereas free SCU achieved only approximately 10–20% reductions. These findings demonstrate that SCU can mitigate HSC activation and fibrogenesis, and that delivering SCU via the nanoemulsion significantly amplifies these anti-fibrotic effects. By enhancing cellular uptake and retention of SCU in HSCs, SCE achieves better suppression of the fibrotic phenotype (reduced motility, collagen production, and myofibroblastic markers) than the free drug.

2.5. SCE Exhibits Enhanced Liver-Targeting In Vivo

We then evaluated whether the nanoemulsion formulation could improve the delivery of SCU to the liver in vivo, a key factor for maximizing therapeutic efficacy against liver fibrosis. To visualize and quantify tissue distribution, we used the near-infrared fluorescent dye DIR as a surrogate for SCU, incorporated either into the nanoemulsion or administered as a free compound. Following oral administration in mice, we monitored real-time in vivo fluorescence and performed ex vivo imaging of major organs to compare biodistribution profiles. In contrast, mice treated with DIR-loaded SCE showed a notably stronger fluorescence signal in the liver region at early time points, suggesting that more of the administered dose reached the liver, likely via uptake of nanoemulsion through intestinal lymphatics and subsequent accumulation in the liver.

Ex vivo fluorescence imaging of major organs revealed marked enhancement of liver targeting by SCE (Figure 6A). At 1 h post-treatment, livers from SCE-treated mice fluoresced more intensely than those from mice treated with free DIR. This difference became increasingly evident over time: by 6 h, the fluorescence signal in the free DIR group had declined, whereas SCE-treated livers retained high intensity. At 12 h, the liver signal in the free DIR group had largely dissipated, while SCE-treated livers remained brightly fluorescent. Even at 24 h, livers from the SCE group emitted a detectable signal, indicating prolonged retention of the nanoemulsion or its payload in hepatic tissue, whereas the free DIR was almost undetectable in the liver by that time. Quantitative analysis of liver fluorescence confirmed these observations (Figure 6B). The area-normalized radiant efficiency (or total photon count) in the livers of SCE-treated mice was significantly higher at all measured time points compared to the free DIR group. For instance, the peak liver fluorescence (observed around 3–6 h) in the SCE group was several-fold greater than the peak in the free group. Moreover, while the free DIR liver signal decreased by approximately 52% between 6 h and 24 h, the SCE liver signal only declined by nearly 8% over the same period, indicating that SCE significantly prolongs the residence time of the compound in the liver.

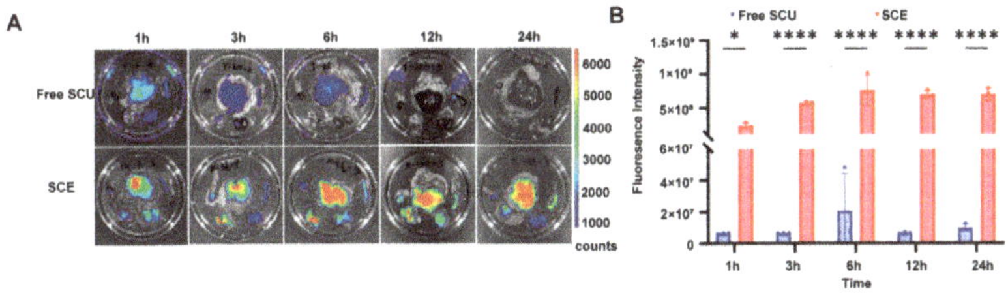

Figure 6. Enhanced hepatic accumulation and retention of scutellarin achieved by the nanoemulsion (SCE) in vivo. (**A**) Ex vivo fluorescence imaging of major organs (heart, liver, spleen, lung, and kidneys) collected from mice at 1, 3, 6, 12, and 24 h after oral administration of free DIR (a near-IR dye, surrogate for free SCU) vs. DIR-loaded SCE. (**B**) Quantification of liver fluorescence intensity over time (mean ± SD, n = 3). **** $p < 0.0001$, * $p < 0.05$.

In addition to enhancing liver targeting, the nanoemulsion formulation reduced or delayed distribution to non-target organs. In the SCE group, fluorescence in organs other than the liver remained low beyond the first hour post-administration. In contrast, mice treated with free DIR exhibited a more diffuse fluorescence pattern, with substantial signal detected in the lungs and kidneys shortly after dosing, suggesting rapid systemic distribution and clearance. This difference is likely attributable to the distinct absorption pathway of the nanoemulsion. With a particle size of approximately 80 nm, SCE is efficiently taken up via Peyer's patches and intestinal lymphatics, bypassing immediate hepatic (first-pass) metabolism as intact particles. The nanoemulsion then accumulates in the liver, either within the fenestrated endothelium or through uptake by Kupffer cells and hepatocytes. This pathway reduces plasma spikes and renal excretion, enabling sustained delivery of the payload to the liver over time.

In summary, the in vivo imaging results highlight the pronounced liver-targeting capacity of the nanoemulsion formulation. SCE delivers a greater proportion of SCU to the liver and prolongs its retention compared to an equivalent dose of free SCU. This enhanced hepatic delivery is expected to translate into improved anti-fibrotic efficacy, as more drug is available at the site of pathology for an extended period. Consistent with these findings, SCE treatment produced markedly superior therapeutic outcomes in the BDL model relative to free SCU, consistent with the observed improvements in pharmacokinetics.

2.6. Therapeutic Efficacy of SCE in BDL-Induced Liver Fibrosis

The ultimate goal of developing SCE was to achieve superior therapeutic effects against liver fibrosis. To evaluate this, we employed the BDL mouse model and compared pathological and biochemical outcomes among groups treated with SCE, free SCU, or vehicle. Sham-operated mice served as healthy controls.

Gross examination of the livers at sacrifice revealed visible signs of treatment efficacy. Livers from untreated BDL mice were enlarged, cholestatic (yellow-brown), and had an irregular surface with nodularity and bile accumulation, indicative of severe injury and fibrosis (Figure 7A). Livers from the free SCU group exhibited modest improvements, with reduced swelling and less discoloration; however, they remained morphologically abnormal. In contrast, livers from the SCE-treated group displayed near-normal morphology: they were smaller in size (less hepatomegaly), with a smoother surface more akin to Sham livers, and fewer visible lesions. This visual improvement suggested that SCE substantially mitigated the BDL-induced liver damage.

Histological analyses confirmed that SCE was more effective in reducing liver injury and fibrosis (Figure 7B–D). H&E staining of livers from the BDL group revealed extensive architectural disruption, with large areas of hepatocyte necrosis, inflammation, ductular proliferation, and fibrosis. Free SCU treatment led to a modest improvement: H&E sections from this group still exhibited significant damage, but with somewhat reduced necrotic areas and inflammation compared to untreated BDL. However, l SCE treatment markedly improved tissue architecture: necrotic foci were rare, inflammatory cell infiltration was reduced, and normal hepatocyte cords were more apparent. Although fibrotic septa remained, their extent was significantly diminished.

We performed Masson's trichrome and Sirius Red staining to visualize collagen deposition. Untreated livers from the BDL group showed extensive bridging fibrosis, indicated by dense blue (Masson) and red (Sirius Red) staining between portal areas. Free SCU treatment slightly reduced the thickness and extent of fibrotic bands, suggesting a partial reduction in collagen deposition. Remarkably, SCE treatment markedly attenuated fibrosis: collagen-positive septa became thinner, more fragmented, and localized, while large collagen-free parenchymal regions resembled an earlier stage of fibrosis. Quantitative image analysis confirmed these findings (Figure 7E,F). SCE significantly reduced the collagen-positive area compared to both the model and free SCU groups ($p < 0.01$), whereas the free SCU group showed no significant difference from the untreated model group in Sirius Red quantification. Thus, SCE curtailed the progression of fibrosis, halting it at a much lower level of collagen deposition.

Biochemical analyses of liver function further corroborated the therapeutic benefit of SCE (Figure 7G–J). The BDL model group exhibited very high serum levels of ALT and AST, reflecting substantial hepatocellular injury, along with elevated ALP and total bile acids (TBA) due to cholestasis from bile duct ligation. Free SCU treatment modestly decreased the levels of these markers, but the reductions were not statistically significant. In free SCU-treated mice, these liver injury markers were slightly lower on average, but the differences were not statistically significant for most markers, indicating limited hepatoprotection by free SCU. In contrast, SCE significantly improved all parameters: it reduced ALT and AST by approximately 50% and decreased ALP and TBA levels similarly, reflecting both hepatoprotection and enhanced bile flow. Across all groups, SCE restored liver function closest to Sham levels—for example, ALT in SCE-treated mice remained only slightly above Sham, whereas BDL mice exhibited several-fold higher levels. Combined with the histological findings, these results demonstrate that SCE provided the greatest therapeutic benefit, effectively protecting hepatocytes, reducing fibrosis, and improving overall liver function, far outperforming the free SCU treatment.

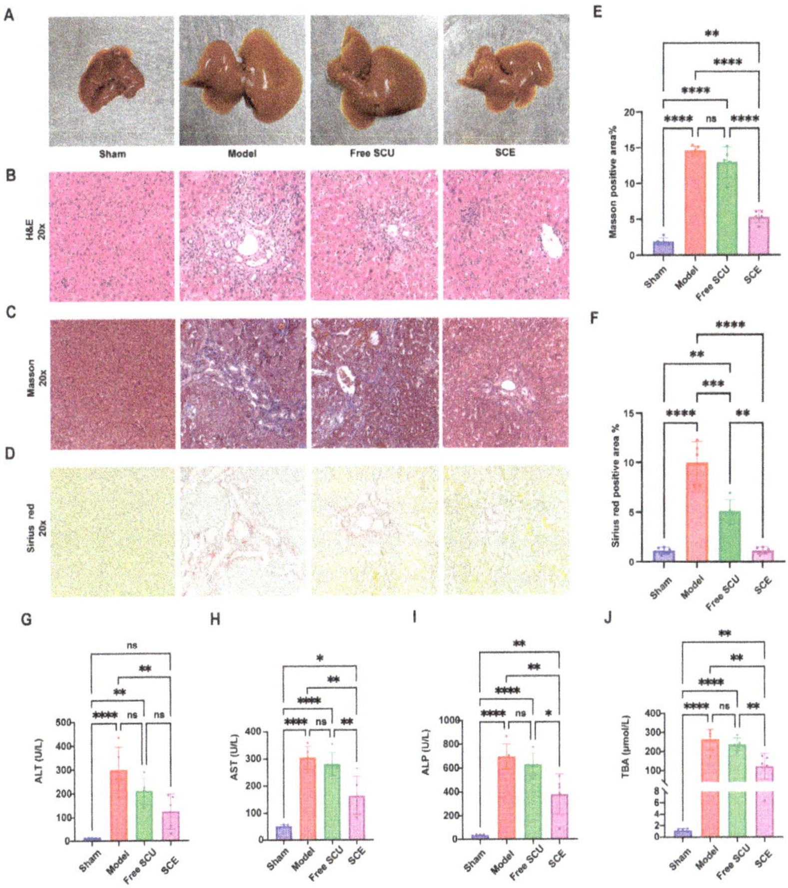

Figure 7. Therapeutic effects of SCE vs. free SCU in the BDL-induced mouse liver fibrosis model. (**A**) Representative gross morphology of livers from each group at day 14. (**B**) H&E-stained liver sections (20× magnification). (**C**) Masson's trichrome staining for collagen (blue). (**D**) Sirius Red staining for collagen (red) corroborates Masson's results. (**E**,**F**) Quantification of fibrotic area from Masson (**E**) and Sirius Red (**F**) stains (percentage of stained area ± SD, n = 5). (**G**–**J**) Serum levels of ALT, AST, ALP, and total bile acids (TBA) across different groups (mean ± SD, n = 5). Statistical analysis was conducted using the complete dataset. Q-values were calculated from *p*-values using the FDR method. **** $p < 0.0001$, *** $p < 0.001$, ** $p < 0.01$, * $p < 0.05$, ns, not significant ($p \geq 0.05$).

To further validate the anti-fibrotic efficacy of SCE, we analyzed molecular markers of fibrosis and inflammation in the liver tissues. Western blot results (Figure 8A–C) showed that BDL strongly upregulated α-SMA, collagen I, and the pro-inflammatory cytokine IL-6, reflecting HSC activation and an inflammatory fibrotic response. Free SCU treatment slightly attenuated the expression of these proteins (e.g., weaker α-SMA and IL-6 bands), but these reductions were not statistically significant. In contrast, SCE markedly decreased the expression of all three markers. Quantification revealed that SCE lowered α-SMA and collagen I protein levels by 47% and 63%, respectively, compared to BDL ($p < 0.01$), indicating substantial inhibition of stellate cell activation and extracellular matrix production. SCE also reduced IL-6 to near-Sham levels, highlighting its potent anti-inflammatory effect. The differences between SCE and free SCU were significant ($p < 0.05$), demonstrating that the nanoemulsion achieved excellent molecular-level therapeutic impact.

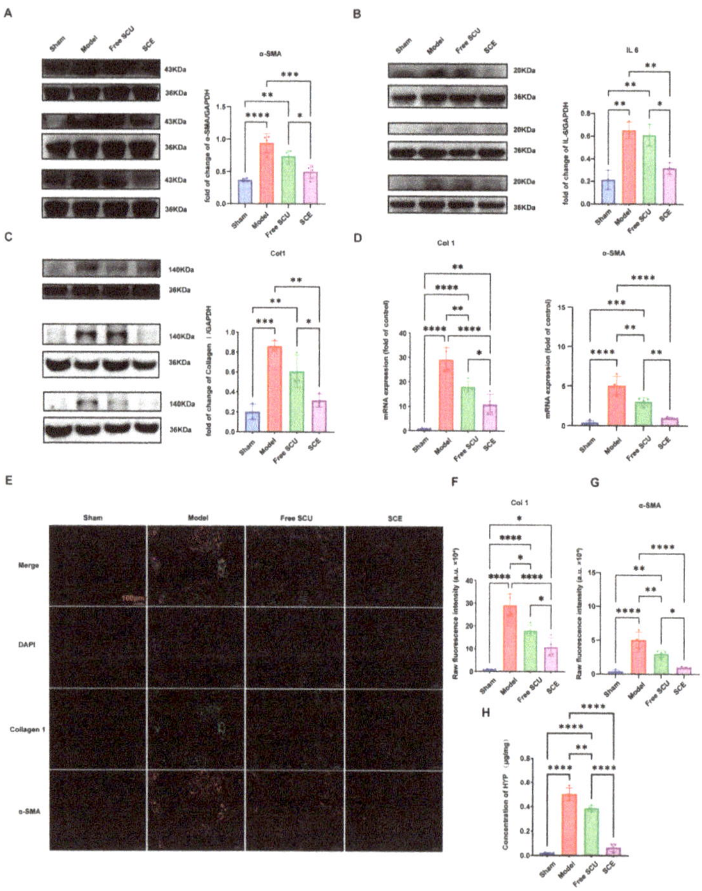

Figure 8. SCE treatment suppresses fibrogenic and inflammatory markers in fibrotic livers more effectively than free SCU. (**A–C**) Western blot analysis of liver tissue proteins (with GAPDH as loading control). (**A**) α-SMA, (**B**) IL-6, and (**C**) collagen I protein levels in Sham, BDL, free SCU, and SCE groups. (**D**) Quantitative PCR analysis of hepatic mRNA expression of α-SMA and collagen I. (**E–G**) Immunofluorescence staining of liver sections for collagen I (green) and α-SMA (red), with nuclei in blue (DAPI). (**H**) Hepatic hydroxyproline content (μg per gram of liver). Statistical analysis was conducted using the complete dataset. Q-values were calculated from p-values using the FDR method. **** $p < 0.0001$, *** $p < 0.001$, ** $p < 0.01$, * $p < 0.05$.

Hepatic mRNA levels mirrored the protein results. BDL strongly increased collagen I and α-SMA transcripts (Figure 8D). Free SCU modestly decreased these mRNA levels, with some reductions reaching statistical significance, suggesting a moderate effect on fibrogenic gene expression. SCE, however, produced the strongest suppression: it significantly downregulated collagen I and α-SMA mRNA compared to both BDL and free SCU groups, confirming potent inhibition of fibrogenesis at the transcriptional level. These findings indicate that SCE not only reduces existing fibrotic protein deposition but also actively suppresses ongoing fibrogenic signaling in the liver.

Immunofluorescence staining provided spatial context to these molecular findings (Figure 8E–G). In the BDL model liver sections, α-SMA-positive activated HSCs (green fluorescence) densely lined the fibrotic septa, and collagen I (red fluorescence) was extensively deposited, co-localizing in fibrous strands. Free SCU treatment slightly reduced the number of α-SMA-positive cells and modestly decreased collagen I intensity, but fibrotic streaks remained prominent. In contrast, SCE-treated liver sections had markedly weaker fluorescence for both α-SMA and collagen I. Only a few scattered α-SMA-positive cells were detected, and collagen I staining appeared faint and confined to periportal zones, signifying a markedly reduced fibrosis. Image analysis of the fluorescent areas indicated SCE significantly cut down α-SMA and collagen I positivity compared to BDL ($p < 0.01$), whereas free SCU had a smaller effect. These observations visually confirm that SCE effectively inactivates HSCs and reduces scar matrix in vivo.

Hydroxyproline quantification further supported these results (Figure 8H). BDL significantly elevated hepatic hydroxyproline content, reflecting excessive collagen accumulation. Free SCU slightly reduced hydroxyproline levels by approximately 16%, but the change was not statistically significant, consistent with the modest histological improvements. In contrast, SCE reduced hepatic hydroxyproline content by roughly 71% compared to BDL ($p < 0.01$), approaching the levels observed in Sham mice. These data indicate that SCE substantially curbed collagen accumulation and may have even promoted partial resorption of existing scar tissue, highlighting the formulation's superior efficacy.

Collectively, these evaluations demonstrate that SCE elicits certain anti-fibrotic effects. Compared to free SCU, SCE more effectively reduces histological fibrosis, lowers fibrotic scar collagen, deactivates HSCs, and improves liver function. The superior therapeutic outcome is likely attributable to enhanced bioavailability and targeted delivery of SCU achieved via the nanoemulsion system. These findings highlight the promise of the SCE formulation as a potent anti-fibrotic intervention.

2.7. SCE Safety and Biocompatibility

Beyond efficacy, we assessed the safety of the SCE formulation, which is critical for its potential clinical application.

In vitro cytotoxicity tests using the CCK-8 assay indicated that SCE is non-toxic to both target cells (LX-2 cells) and off-target cells (LO2 hepatocytes) at relevant concentrations (Figure 9A,B). LO2 cells incubated with either free SCU or SCE (at 5–30 μM SCU equivalents) maintained high viability (consistently above 85%). At the highest concentration (30 μM), SCE-treated LO2 cells showed slightly higher viability than those treated with free SCU. LX-2 cells exhibited a similar trend: free SCU did not reduce viability significantly, and SCE-treated cells maintained or slightly exceeded 100% viability relative to control. These data confirm that the nanoformulation itself (composed of phospholipid, MCT oil, Cremophor, and chitosan oligosaccharide) is biocompatible with liver cells and that SCU at therapeutic doses is safe for normal hepatocytes.

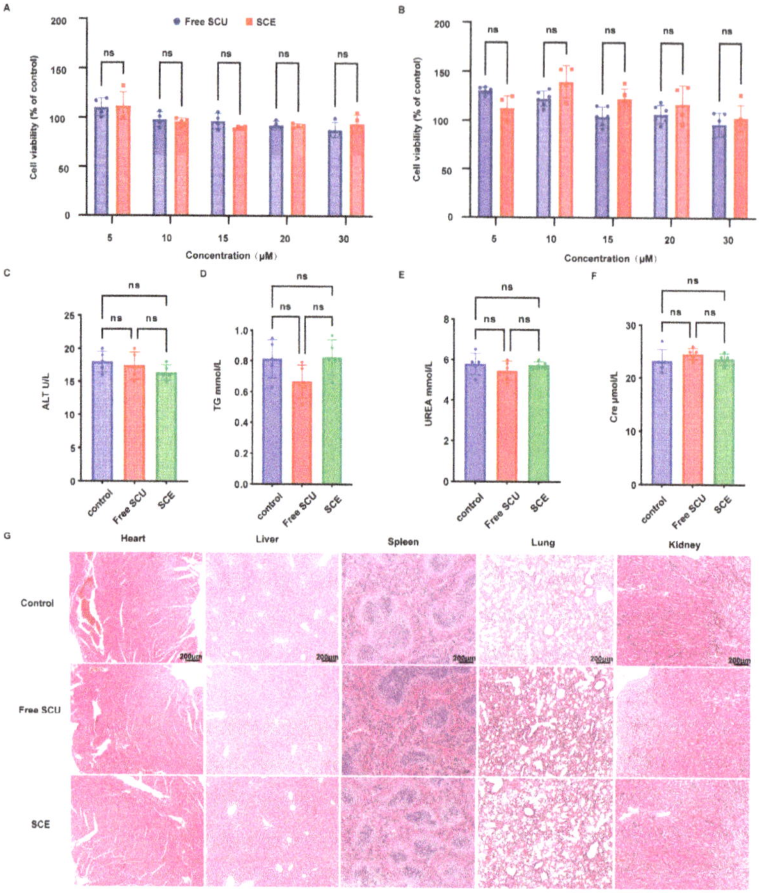

Figure 9. Biocompatibility assessment of the SCU nanoemulsion (SCE) in vitro and in vivo. (**A**,**B**) Cell viability of LO2 (normal liver) cells and LX-2 (HSC) cells after 24 h exposure to increasing concentrations of free SCU vs. SCE (SCU 5–30 μM), measured by CCK-8 assay. Data are presented as mean ± SD (n = 4). (**C**–**F**) Serum chemistry of mice following 14 days of treatment with saline (control), free SCU, or SCE (n = 5). (**G**) Histological examination (H&E staining) of major organs (heart, liver, spleen, lung, and kidney) from control, free SCU, and SCE-treated mice. ns, not significant ($p \geq 0.05$).

For in vivo safety assessment, we administered daily doses of free SCU or SCE (10 mg/kg SCU) to healthy ICR mice (non-BDL) for 14 days, mimicking the treatment regimen. Serum analyses (Figure 9C–F) showed no significant differences among saline control, free SCU, and SCE groups in liver and kidney function markers. ALT and AST remained in the normal low range for all groups, indicating that neither SCU nor SCE caused hepatocellular injury in normal mice. Similarly, triglyceride (TG) levels were not elevated; in fact, they were similar across groups, suggesting the lipid-based nanoemulsion did not disrupt lipid metabolism. Renal function markers, including blood urea nitrogen (UREA) and creatinine (Cre) levels, were also comparable among the groups, implying that renal function was unaffected and there was no nephrotoxicity. These results suggest that SCE does not produce adverse effects on critical organ functions at the given dose.

Histopathological examination supported the safety profile of SCE (Figure 9G). H&E staining of major organs—including the heart, liver, spleen, lungs, and kidneys—revealed no pathological changes in SCE-treated mice. Specifically, liver sections displayed normal lobular architecture with no signs of inflammation or degeneration; kidney glomeruli and tubules were intact; cardiac myocytes, pulmonary alveoli, and splenic white pulp/red pulp were all unremarkable and similar to controls. We observed no signs of organ toxicity, such as cellular infiltration, tissue damage, or lipid accumulation, after SCE administration. Notably, although some nanoemulsion components (e.g., Cremophor EL) can occasionally provoke systemic reactions, we detected no such adverse effects at the administered dose.

In summary, both SCU and its SCE demonstrated excellent safety and tolerability in vitro and in vivo. The nanoemulsion did not introduce any detectable toxicity. These findings indicate a favorable therapeutic index for SCE, meaning we can achieve efficacious concentrations in the target organ (liver) without causing harm to normal cells or other organs. This is a key requirement for advancing such a therapy towards clinical consideration.

3. Discussion

Liver fibrosis is characterized by excessive deposition of extracellular matrix components, particularly collagen, leading to distortion of liver architecture and function. If unresolved, fibrosis progresses to cirrhosis and eventually hepatocellular carcinoma [38]. Although researchers have extensively elucidated the mechanisms of fibrogenesis, clinically approved anti-fibrotic therapies remain scarce. Current treatment strategies primarily target underlying etiologies (e.g., antiviral therapy for viral hepatitis) and provide supportive care, but they show limited efficacy in reversing established fibrosis.

SCU has emerged as a potential multi-target agent due to its broad pharmacological activities. Prior studies have documented its anti-fibrotic properties in non-hepatic tissues. In this context, multi-target approaches that simultaneously address different aspects of fibrogenesis are highly desirable. Our study focused on scutellarin (SCU) because of its well-documented pharmacological profile and its critical limitation of poor solubility and bioavailability, which made it an ideal candidate to demonstrate the utility of our nanoemulsion platform. Other flavonoids such as silybin and quercetin are also promising anti-fibrotic candidates, and the nanoemulsion strategy described here could, in principle, be extended to these compounds in future work.

Scutellarin (SCU) has emerged as a promising candidate due to its multitarget pharmacological profile. Previous studies have shown that SCU reduces fibrosis in non-hepatic organs [16,17] and protects the liver [9,10]. Other researchers demonstrated that SCU reshapes gut microbiota in liver disease models [5], matters because gut-derived factors drive liver inflammation. In this study, we showed that SCU treatment altered both intestinal and hepatic microbiota under fibrotic conditions to some extent. SCU reduced the abundance of several potentially pathogenic taxa (such as *Actinobacteriota* and *Desulfobacterota* in the gut), which BDL had elevated. Elevated *Actinobacteriota* abundance worsens liver pathology in steatosis and cancer models [19,20], whereas elevated *Desulfobacterota* abundance is linked to hepatic inflammation [21]. Although the nanoemulsion primarily facilitated efficient hepatic targeting, its oral route of administration also allowed for transient interaction with the gut microbiota. This was associated with partial normalization of dysbiotic taxa, which—while not the principal mechanism of action—may have contributed additively to the observed therapeutic benefits. By attenuating these dysbiotic shifts, SCU likely blocked the translocation of pro-fibrogenic microbial products such as endotoxins from the gut to the liver. SCU also restored normal levels of *Erysipelotrichales* and *Oscillibacter* in the hepatic microbiota, which indicates that SCU directly or indirectly regulates bacteria that colonize or translocate to the liver during fibrogenesis. The liver microbiome is a relatively new research frontier. We observed that SCU shifted microbial abundance, but we have not yet determined how these changes drive anti-fibrotic outcomes. The liver microbiota results were obtained from low-biomass samples and should therefore be regarded as highly exploratory and inherently prone to contamination—a well-recognized limitation in this field. Although we implemented stringent precautions, including processing all samples under a biosafety cabinet with sterile, DNA-free reagents, the absence of dedicated negative controls and formal bioinformatic decontamination necessitates cautious interpretation. Accordingly, we do not present these findings as definitive evidence of a resident liver microbiota, but rather as preliminary, hypothesis-generating observations that are consistent with the emerging concept of a gut–liver axis in fibrosis. We classify our hepatic microbiota findings as exploratory. Future studies using germ-free or microbiota-depleted models containing adequate mice must establish whether microbiome modulation directly improves fibrosis. Despite this limitation, our results strengthen the concept of a gut–liver axis in fibrosis and show that SCU restores microbial balance to improve this axis. Furthermore, we acknowledge that microbiota abundance data may not follow a normal distribution, which limits the robustness of parametric comparisons.

SCU shows poor water solubility and low oral bioavailability, which likely hinders its therapeutic effectiveness in vivo [12,18]. In this study, we overcame this barrier by developing the scutellarin–phospholipid complex (SPC). Characterization techniques, including FTIR, DSC, and XRD, confirmed that SCU was present in an amorphous state within the lipid matrix, likely at least partially molecularly dispersed. Amorphization is known to enhance the dissolution of hydrophobic drugs [39]. By incorporating SPC into a nanoemulsion (SCE), we formulated SCU at therapeutically relevant concentrations with greater solubility and absorption. Nanoemulsions commonly increase the oral bioavailability of lipophilic compounds by promoting lymphatic transport and preventing precipitation or metabolism [40]. Our findings support this mechanism: the SCE nanoemulsion likely promoted intestinal absorption via lymphatic transport pathways, thereby enhancing hepatic delivery of SCU, as corroborated by fluorescence imaging.

SCE significantly enhanced the pharmacokinetic and tissue distribution profiles of SCU. In vivo imaging results showed that SCE achieved higher and more sustained liver concentrations than free SCU. This liver-targeting effect benefits anti-fibrotic therapy by concentrating the drug at the site of action and reducing systemic exposure, thereby lowering the risk of off-target effects. SCE also prolonged hepatic retention, maintaining significant levels even 24 h after treatment. This sustained presence may extend drug action and permit less frequent dosing in clinical applications.

At the cellular level, the nanoemulsion also proved advantageous. Our results showed that SCE is readily taken up by liver cells (stellate cells, hepatocytes, and possibly liver macrophages) once it reaches the liver. The mechanism involved caveolae-mediated endocytosis, a pathway that nanoparticles often exploit to enter cells efficiently. Through this process, the nanoemulsion allowed SCU to cross the cellular barriers of fibrotic tissue more effectively. In fibrotic livers, the dense extracellular matrix typically impedes drug diffusion. However, nanoemulsions of suitable size can exploit disrupted sinusoidal endothelium and increased vascular permeability to access activated HSCs, which overexpress endocytic receptors. This mechanism likely explains why SCE suppressed HSC activation more effectively than free SCU in vivo—greater intracellular delivery enabled the drug to exert its pharmacological effect.

Therapeutically, SCE treatment produced certain outcomes in the BDL-induced fibrosis model. SCE improved the progression of fibrosis, as indicated by the lower collagen content and improved histology compared to untreated fibrotic mice. In contrast, free SCU exhibited only minimal effects, underscoring the critical role of the nanoemulsion delivery system in unlocking the therapeutic potential of SCU. The limited efficacy of free SCU at a dose of

10 mg/kg is consistent with previous reports that its poor bioavailability hinders in vivo activity [12,41,42]. Using SCE, we effectively increased the bioavailability and hepatic concentration of SCU, thereby achieving the desired anti-fibrotic action.

Mechanistically, SCU exerts anti-fibrotic effects through several interconnected pathways. Known for its antioxidant and anti-inflammatory properties [10], SCU disrupts TGF-β/SMAD signaling, the central fibrogenic pathway, and downregulates key fibrotic markers such as α-SMA and collagen I. SCU also modulates inflammatory cascades, likely through NF-κB signaling, as evidenced by the pronounced reduction in hepatic IL-6 expression in SCE-treated mice. The downregulation is particularly significant, since IL-6 not only signals inflammation but also drives fibrosis and carcinogenesis in chronic liver disease. By suppressing IL-6, SCU helps dampen the inflammatory milieu that fuels fibrogenesis. We also found that SCU reduced MMP2 expression in HSCs in vitro. This modulation of MMP2 suggests that SCU influences matrix remodeling dynamics, promoting a more balanced environment where collagen deposition and degradation can proceed toward the resolution of fibrosis.

Our study also highlights the favorable safety of the SCE system. We were careful to demonstrate that the formulation components (lipid, surfactant, and COS polymer) did not introduce toxicity. COS is generally regarded as biocompatible and has been reported to confer additional biological benefits, such as promoting intestinal health and enhancing mucosal permeability. The surfactant (Cremophor EL), while sometimes causing hypersensitivity at high doses in intravenous formulations, is in a relatively low dose orally and was well-tolerated. Furthermore, the negative surface charge and nanoscale size of SCE contributed to its stability and low immune recognition, preventing unwanted immune responses.

From a translational perspective, SCE shows promise as a therapy for liver fibrosis. Nevertheless, several considerations and future directions must be addressed. First, although our results in the BDL model were compelling, liver fibrosis is a heterogeneous condition with diverse etiologies. Future work should evaluate the efficacy of SCE in additional models, such as those induced by carbon tetrachloride or NASH, to confirm its broader applicability. Second, the observed microbiome-modulating effects of SCU raise the possibility of combining SCU with specific probiotics or prebiotics to enhance therapeutic outcomes via synergistic modulation of the gut–liver axis. Third, while nanoemulsions are relatively straightforward to scale up, we must confirm long-term stability beyond seven days. Transforming the nanoemulsion into a solid dosage form, such as a freeze-dried powder for reconstitution, could improve practicality for clinical use. Fourthly, another limitation of this study is the use of a single dose of SCE. While this dose was effective and informed by previous studies [41,42], future work will include a comprehensive dose–response evaluation to determine the optimal therapeutic window and maximize the potential of this formulation. Finally, another limitation of this study is the absence of a blank nanoemulsion control in the in vitro experiments. Although our primary objective was to evaluate the therapeutic potential of SCU when formulated into a nanoemulsion, we acknowledge that vehicle-only controls would have further strengthened the interpretation of these findings. Without this control, we cannot fully exclude the possibility that some of the observed effects may be partially attributable to the components of the vehicle rather than SCU itself. Future studies will incorporate such vehicle controls to rigorously confirm the specificity of the nanoemulsion-mediated effects.

In addition, we acknowledge that effect sizes and confidence intervals were not provided, which limits the interpretability of some results.

The strategy of targeting both fibrotic processes and the microbiome is still relatively novel. Our work provides proof-of-concept that a single agent, when properly formulated, may simultaneously modulate HSC activity and the gut–liver axis. SCU exerts a dual mechanism of action by directly suppressing fibrotic processes and simultaneously ancillary modulating the microbiota, thereby exemplifying a polypharmacological strategy [7]. In complex diseases such as liver fibrosis, where inflammation, cell activation, and gut–liver signaling interact through tightly linked pathways, multi-pronged agents like SCU achieve greater therapeutic effectiveness.

In conclusion, this work demonstrates that formulating SCU into a nanoemulsion markedly enhances its therapeutic efficacy against liver fibrosis. SCE improves pharmacokinetics and liver targeting, which increases anti-fibrotic effects by reducing collagen deposition and HSC activation. Additionally, SCE preserves the ability of SCU to modulate the gut–liver axis through microbiota changes. These results highlight the critical role of drug delivery systems in unlocking the clinical potential of poorly soluble natural products like SCU. The nanoemulsion approach described herein may apply to other phytochemicals or therapeutic agents facing similar solubility and bioavailability challenges. Overall, our findings contribute to the development of an effective, multi-target therapy for liver fibrosis and highlight the innovative angle of targeting the microbiome as part of the therapeutic mechanism.

4. Materials and Methods

4.1. Materials

Scutellarin (SCU, >98% purity, CAS No. 27740-01-8) was purchased from J&K Scientific (Wuhan, China). Soybean phospholipid (lecithin) served as the excipient for phospholipid complex preparation. Caprylic/capric triglycerides (medium-chain triglycerides) and Cremophor EL (polyoxyethylated castor oil) were used as the oil phase and surfactant, respectively, in the nanoemulsion. Chitosan oligosaccharide (COS) was used as a stabilizer. Recombinant human TGF-β1 (transforming growth factor beta 1) was purchased from R&D Systems (Minneapolis, MN, USA) for HSC activation in vitro. The human HSC line LX-2 was purchased from Shanghai Mingjin Biotechnology Co., Ltd. (Shanghai, China) (RRID:CVCL_5792) on 16 May 2023. The human normal liver cell line LO2 was kindly provided by Professor He (Institute of Medicinal Biotechnology, Chinese Academy of Medical Sciences & Peking Union Medical College). Antibodies against α-SMA (α-smooth muscle actin, Cat# 14395-1-AP), MMP2 (matrix metalloproteinase-2, Cat# 66366-1-Ig), and collagen type I (COL1A1, Cat# 14695-1-AP) were purchased from Proteintech (Wuhan, China). GAPDH antibody (Cat# ABL-1021) was purchased from Abbkine (Wuhan, China). All other reagents and chemicals were of analytical reagent grade and were used as received without further purification.

4.2. Preparation of Scutellarin–Phospholipid Complex (SPC)

A scutellarin–phospholipid complex (SPC) was prepared to enhance the lipophilicity of SCU, following a previously reported method with slight modifications [39]. Briefly, SCU and soybean phospholipid (mass ratio 1:5) were co-dissolved in absolute ethanol to obtain a solution containing 0.5 mg/mL of SCU. The solvent was subsequently removed by rotary evaporation under reduced pressure at approximately 40 °C. During evaporation, the mixture was sonicated to facilitate molecular interaction between SCU and the phospholipid. The resulting solid residue was dried to constant weight to yield the SPC. For comparison, a physical mixture of SCU and phospholipid was prepared by manually blending the two components at the same 1:5 mass ratio using a mortar and pestle, without the use of any solvent.

4.3. Preparation of SCU-Loaded Nanoemulsion (SCE)

The SCU-loaded nanoemulsion (SCE) was prepared using the previously prepared SPC. First, a primary coarse emulsion was prepared using the phase inversion method. Caprylic/capric triglyceride (1 mL) was mixed with an equal mass of Cremophor EL to form the oil-surfactant phase. Subsequently, 48 mg of SPC (containing SCU) was added to the oil-surfactant phase and dissolved with gentle heating and sonication until a clear solution was obtained. Separately, chitosan oligosaccharide (15 mg) was dissolved in 3 mL of deionized water to prepare the aqueous phase. The oil phase was added dropwise into the aqueous phase under magnetic stirring at approximately 1500 rpm at room temperature, resulting in a crude oil-in-water emulsion. This primary emulsion was then subjected to high-energy ultrasonication to reduce droplet size, using an ultrasonic cell disruptor (Xinzhi, Ningbo, China) (200 Hz) in an ice-water bath with pulse cycles of 10 s on and 5 s off for a total of 15 min. The resulting nanoemulsion was equilibrated to room temperature and stored at 4 °C until use.

4.4. Characterization of SPC and SCE

To confirm SPC formation, differential scanning calorimetry (DSC), Fourier-transform infrared spectroscopy (FTIR), and X-ray powder diffraction (XRD) analyses were conducted on SCU, phospholipid, their physical mixture, and the SPC product. DSC thermograms were acquired using a DSC analyzer (Mettler-Toledo, Switzerland) by heating samples from 30 °C to 400 °C at a programmed rate to detect alterations in melting or crystallization behavior indicative of complex formation. FTIR spectra were recorded using a Nicolet 5700 FTIR spectrometer (Thermo, Waltham, MA, USA) over the range 4000–400 cm^{-1} to identify chemical interactions, such as hydrogen bonding, between SCU and the phospholipid. XRD patterns were recorded using a Bruker D8 Advance diffractometer (Billerica, MA, USA) (Cu Kα radiation, λ = 1.5406 Å, 40 kV, 40 mA) to assess the crystallinity of SCU in each sample; the disappearance of SCU's characteristic crystalline peaks in the SPC would suggest an amorphous or molecularly dispersed state.

SCE was characterized in terms of particle size, size distribution, zeta potential, morphology, and drug content. The hydrodynamic particle size (mean diameter) and polydispersity index (PDI) were measured by dynamic light scattering (DLS) using a Zetasizer Nano ZS (Malvern Instruments, Malvern, UK) at 25 °C. Zeta potential, indicating the surface charge of nanoemulsion droplets, was determined by electrophoretic light scattering with the same instrument. The morphology of SCE was observed using transmission electron microscopy (TEM). A drop of diluted SCE was placed on a carbon-coated copper grid, allowed to sit for 1 min, and excess fluid was removed by blotting. The grid was then air-dried and examined under a TEM (JEM-2100, JEOL, Akishima, Japan) operated at an accelerating voltage of 200 kV. The TEM images provided visual confirmation of particle size and morphology. The SCU concentration in the nanoemulsion was determined using a validated high-performance liquid chromatography (HPLC) method (Shimadzu, Kyoto, Japan) with UV detection at 335 nm. Briefly, SCE samples were diluted in methanol and analyzed on a C18 column. The mobile phase consisted of methanol and 0.5% acetic acid (4:6, v/v) and was delivered at a flow rate of 1 mL/min. The column temperature was maintained at 30 °C. Quantification was achieved using a standard calibration curve.

4.5. Storage Stability Study

The short-term physical and chemical stability of SCE was assessed over 7 days at 4 °C. Aliquots were stored in sealed vials and sampled on days 0 (initial), 1, 3, 5, and 7. At each time point, mean particle size, PDI, and zeta potential were measured by DLS as described in Section 4.4. Additionally, the SCU concentration within the formulation was analyzed by HPLC to monitor any potential drug degradation or precipitation. All measurements were performed in triplicate. Stability was considered acceptable if no significant particle growth or aggregation occurred and if the SCU content remained above 90% of the initial drug content over the storage period.

4.6. Cell Culture and Cytotoxicity Assay

LX-2 and LO2 cells were cultured in Dulbecco's Modified Eagle Medium (DMEM, Gibco, Waltham, MA, USA) and Roswell Park Memorial Institute (RPMI) 1640 (Gibco, USA), respectively. Both media were supplemented with 10% fetal bovine serum (FBS) and 1% penicillin–streptomycin. Cells were maintained at 37 °C in a humidified atmosphere with 5% CO_2. The culture medium was replaced every 2–3 days, and cells were subcultured using trypsin-EDTA upon reaching 80–90% confluence.

The cytotoxicity of SCE was evaluated in vitro using the Cell Counting Kit-8 (CCK-8, Meilunbio, Dalian, China) assay in both LX-2 and LO2 cell lines. Cells were seeded into 96-well plates at a density of 5×10^3 cells per well and allowed to adhere for 12 h. Subsequently, the medium was replaced with 100 µL of fresh medium containing either free SCU or SCE at various concentrations (ranging from 5 µM to 30 µM SCU equivalent). Control wells received only drug-free medium. After 24 h of treatment, 10 µL of CCK-8 reagent was added to each well, followed by an additional 4 h incubation period. Absorbance was measured at 450 nm using a microplate reader (Synergy H1, BioTek, Winooski, VT, USA). Cell viability was expressed as a percentage of the untreated control. Each concentration was tested in quadruplicate, and data are presented as mean ± SD. For all cell-based assays (cytotoxicity, cellular uptake, gene/protein expression, and migration), at least three independent experiments were performed on different days, and each experiment contained multiple technical replicates.

4.7. Cellular Uptake Study in LX-2 Cells

The cellular uptake of SCU delivered via SCE versus free drug was examined in LX-2 cells using a fluorescent probe. Nile Red (NR), a hydrophobic fluorescent dye, was used as a surrogate to visualize and quantify uptake. NR-loaded SCE was prepared by adding a small amount of NR to the SPC (at 10% *w*/*w* of SCU) during the nanoemulsion preparation process (Section 4.3). A solution of free NR in PBS (containing a small amount of DMSO to aid solubilization) was used as a control corresponding to free SCU.

For qualitative uptake visualization, LX-2 cells were seeded in glass-bottomed confocal dishes at 2×10^5 cells per dish and cultured for 24 h. Cells were then treated with either free NR (in PBS) or NR-loaded SCE (NR/SCE) at an equivalent NR concentration. After incubation for predetermined periods (5 min, 15 min, 30 min, 1 h, 2 h, and 4 h) at 37 °C, the cells were washed three times with PBS to remove extracellular NR, fixed with 4% paraformaldehyde for 15 min, and stained with DAPI (4′,6-diamidino-2-phenylindole) to label nuclei. Fluorescent images were acquired using a laser scanning confocal microscope (Zeiss LSM 710, Oberkochen, Germany). NR (red) and DAPI (blue) signals were visualized to evaluate intracellular localization and uptake intensity.

For quantitative analysis, cellular uptake was evaluated by flow cytometry. LX-2 cells were seeded in 6-well plates (3×10^5 cells/well) and treated with free NR or NR-loaded SCE as described above. At selected time points (up to 4 h), cells were washed with ice-cold PBS, detached with trypsin, and resuspended in PBS. Intracellular fluorescence was measured using a flow cytometer (BD FACSCalibur, Franklin Lakes, NJ, USA) in the FL2 channel. At least 10,000 events were recorded per sample. The mean fluorescence intensity (MFI) was calculated to compare the cellular uptake efficiency of NR/SCE versus free NR.

To elucidate the cellular internalization pathways of SCE, LX-2 cells were pre-treated for 15 min with one of several pharmacological endocytosis inhibitors before exposure to NR-loaded SCE (NR/SCE). The inhibitors included: chlorpromazine (10 μg/mL, inhibitor of clathrin-mediated endocytosis), nystatin (25 μg/mL, inhibits caveolae-mediated endocytosis by cholesterol sequestration), methyl-β-cyclodextrin (MβCD, 5 mM, depletes membrane cholesterol and disrupts caveolae), genistein (100 μM, tyrosine kinase inhibitor that also disrupts caveolae-mediated endocytosis), imipramine (10 μg/mL, reported to inhibit caveolae pathway), and 5-(N-ethyl-N-isopropyl) amiloride (EIPA, 50 μM, inhibitor of macropinocytosis). Following inhibitor pretreatment, cells were washed with PBS and incubated with NR/SCE for 2 h at a concentration previously determined to produce intense fluorescence. Cells were then processed for flow cytometry as described in the cellular uptake section. A significant reduction in MFI in the presence of a specific inhibitor, relative to the untreated control, was interpreted as indicative of that pathway's involvement in SCE internalization.

4.8. In Vitro Anti-Fibrotic Activity in LX-2 Cells

LX-2 cells were employed as an in vitro model to assess HSC activation and evaluate the anti-fibrotic potential of SCU formulations. Cells were seeded in 6-well plates at a density of approximately 1×10^5 cells per well and cultured to 80–90% confluence. To induce a fibrogenic phenotype, the cells were serum-starved in DMEM containing 2% FBS for 24 h, followed by stimulation with TGF-β1 (2 ng/mL) for an additional 24 h. This treatment activates LX-2 cells, resulting in upregulation of fibrotic markers. After activation, the medium was replaced, and SCE was added. The cells were then incubated for 12 h. Subsequently, the cells were harvested for analysis of fibrogenic gene and protein expression by quantitative real-time PCR (qPCR) and Western blot.

Total RNA was isolated using the RaPure Total RNA Kit (Magen, Guangzhou, China) according to the manufacturer's protocol. RNA concentration and purity were verified by spectrophotometry. cDNA was synthesized and amplified using a one-step RT-qPCR SYBR Green kit (Vazyme, Nanjing, China) on a 7500 Fast Real-Time PCR System (Applied Biosystems, Waltham, MA, USA). The primer sequences for target genes (collagen I [COL1A1], α-SMA [ACTA2], and β-actin [ACTB] as a housekeeping gene) were as follows: COL1A1: forward 5′-TGACCTTCCTGCGCCTAATG-3′, reverse 5′-GCTACGCTGTTCTTGCAGTG-3′; ACTA2 (α-SMA): forward 5′-CTCTGTCTGGATCGGTGGC-3′, reverse 5′-TTCGTCGTATTCCTGTTTGCT-3′; ACTB: forward 5′-CCTGGACTTCGAGCAAGAGATGG-3′, reverse 5′-GTGGTTTCGCTCGGCACATT-3′. The thermal cycling conditions were set according to kit protocols. The relative gene expression was calculated using the $2^{(-\Delta\Delta Ct)}$ method, with normalization to ACTB and comparison to the untreated control group.

For Western blot analysis, we lysed LX-2 cells using radioimmunoprecipitation assay (RIPA) buffer supplemented with protease and phosphatase inhibitors to extract total protein. We measured protein concentrations and loaded equal amounts (20 μg per sample) onto SDS-PAGE gels, then transferred the separated proteins onto polyvinylidene difluoride (PVDF) membranes. We blocked the membranes with 5% non-fat milk for 1 h, followed by overnight incubation at 4 °C with primary antibodies against key fibrogenic proteins: collagen I (1:2000 dilution) and MMP2 (1:2000). GAPDH (1:5000) was used as an internal loading control. After washing, we incubated the membranes with horseradish peroxidase-conjugated secondary antibodies (1:10,000) for 1 h at room temperature. We visualized protein bands using enhanced chemiluminescence (ECL) substrate and captured images with a ChemiDoc imaging system (Bio-Rad, Hercules, CA, USA). We quantified band intensities using ImageJ software (1.54) and normalized target protein levels to GAPDH.

4.9. Cell Migration Assay

We evaluated the effect of SCU on HSC migration, a characteristic of activated HSCs that contributes to fibrotic tissue remodeling, using a wound healing (scratch) assay. LX-2 cells were seeded in 6-well plates and cultured until they formed a nearly confluent monolayer. We created a linear scratch approximately 1 mm wide through the cell monolayer using a sterile pipette tip. After gently washing the wells with PBS to remove detached cells and debris, we treated the remaining cells in serum-reduced medium containing TGF-β1 and SCU. We captured images of the wound area at 0 h (immediately after scratching), 12 h, and 24 h post-treatment using an inverted phase-contrast microscope. We quantified cell migration by measuring the remaining wound width or calculating the wound area at each time point relative to the initial wound area at 0 h.

4.10. In Vivo Tissue Distribution and Liver Targeting

We assessed the liver-targeting efficiency of SCE using a near-infrared fluorescent probe, DIR (1,1′-dioctadecyl-3,3,3′,3′-tetramethylindotricarbocyanine iodide), which is lipophilic and suitable for tracking nanoemulsion distribution via fluorescence imaging. We prepared DIR-loaded SCE similarly to NR/SCE by dissolving DIR in the oil phase before emulsification. As a control, we prepared free DIR by dissolving the dye in a Cremophor/ethanol mixture and subsequently diluting it with saline to mimic the free drug formulation.

Male ICR mice (6–8 weeks old, approximately 25 g) were obtained from Beijing Vital River Labs (Beijing, China) and acclimatized for one week under standard laboratory conditions. All animal experiments were approved by the Institutional Animal Care and Use Committee (IACUC) of the Institute of Medicinal Biotechnology, CAMS & PUMC (Approval No. IMB-20231109D102), and conducted per national ethical guidelines. The mice were fasted for 12 h before the experiment and randomly divided into two treatment groups (n = 3 per group per time point): free DIR and DIR-loaded SCE groups. Each mouse received a single oral gavage of DIR (0.5 mg/kg). At 1, 3, 6, 12, and 24 h post-administration, we anesthetized and euthanized three mice from each group and excised their major organs (heart, liver, spleen, lung, and kidneys). We immediately performed ex vivo fluorescence imaging using the IVIS system (PerkinElmer, Waltham, MA, USA) with DIR-appropriate filter settings. We drew regions of interest (ROI) over each organ to quantify fluorescence intensity (radiant efficiency), with particular focus on the liver to assess targeting efficiency. The mean fluorescence intensity of the liver at each time point was calculated and compared between the two groups to evaluate the extent and duration of hepatic accumulation under the appropriate excitation/emission filter settings for DIR.

4.11. Bile Duct Ligation-Induced Liver Fibrosis Model and Treatment Protocol

We employed a bile duct ligation (BDL) mouse model to induce liver fibrosis and evaluate the therapeutic effects of SCU formulations. Male ICR mice (6–8 weeks old) were randomly divided into four groups (n = 5–6 per group): sham (sham operation + vehicle treatment), BDL model (BDL + vehicle), free SCU (BDL + free SCU treatment), and SCE (BDL + SCU nanoemulsion treatment). The BDL procedure was performed under anesthesia (isoflurane, gas anesthesia) by double ligation and transection of the common bile duct using sterile technique. In the sham-operated group, we exposed the bile duct but left it intact. Following surgery, we administered buprenorphine for postoperative analgesia and monitored all animals closely until full recovery. Each experimental group included 5 animals (biological replicates). For each endpoint, the exact number of samples analyzed is indicated in the figure legends. All data are presented as mean ± SD, and statistical analyses were performed using the full dataset (n = 5), unless otherwise specified.

We began treatment on postoperative day 2. Free SCU was freshly prepared by dissolving scutellarin in saline containing a minimal amount of NaOH to enhance solubility. This solution was then administered at a dose of 10 mg/kg. We formulated SCE to deliver an equivalent SCU dose. The Sham and BDL control groups received an equivalent volume of normal saline. We administered all treatments once daily by oral gavage for 14 consecutive days, recorded body weights regularly, and noted any signs of distress.

At the end of the treatment period, we anesthetized the mice and collected blood samples via the orbital sinus. Serum was separated by centrifugation at 3500 rpm for 10 min at 4 °C and stored at −80 °C for subsequent biochemical analyses. We then euthanized the mice and harvested liver tissues. Portions of the liver were fixed in 10% neutral-buffered formalin for histological and immunofluorescence analysis, whereas other portions were snap-frozen in liquid nitrogen and stored at −80 °C for biochemical assays, including Western blotting and hydroxyproline quantification. Additionally, we aseptically collected 50 mg of fresh liver tissue and fecal pellets from the colon and immediately frozen for microbiome analysis via 16S rRNA gene sequencing.

We performed quantitative real-time PCR (qPCR) and Western blot analyses to assess the expression of fibrogenic genes and proteins in liver tissues, following the procedures described in Section 4.8. Notably, the primer sequences used for liver tissue qPCR differed from those used for cellular analysis. The primer sequences were as follows: COL1A1: forward 5′-CATGTTCAGCTTTGTGGACCT-3′, reverse 5′-GCAGCTGACTTCAGGGATGT-3; ACTA2: 5′-TTCCTTCGTGACTACTGCCG-3′, reverse 5′-TATAGGTGGTTTCGTGGATGCC-3′; ACTB: forward 5′-CGTTCAATACCCCAGCCATG-3′, reverse 5′-GACCCCGTCACCAGAGTCC-3′.

4.12. Serum Biochemistry Analysis

To evaluate liver injury and systemic responses to treatment, we quantified serum levels of key liver function markers. We measured alanine aminotransferase (ALT), aspartate aminotransferase (AST), and alkaline phosphatase (ALP) activities using an automated biochemical analyzer or colorimetric assay kits (Nanjing Jiancheng Bioengineering, Nanjing, China) following the manufacturer's protocols. We also assessed serum total bile acids (TBA) as an indicator of BDL-induced cholestasis. We performed all assays in duplicate, and results were reported as mean ± SD for each group.

4.13. Histological Analysis

For histopathological examination, liver and other major organs—including the heart, spleen, lungs, and kidneys—were fixed in formalin, embedded in paraffin, and sectioned at a thickness of 4 μm. We stained liver sections with hematoxylin and eosin (H&E) to evaluate general liver architecture and injury. To visualize collagen deposition and fibrosis, we performed Masson's trichrome and Sirius Red staining. Masson's stain renders collagen fibers blue, whereas Sirius Red binds specifically to collagen, appearing red under light microscopy and exhibiting birefringence under polarized light.

To evaluate potential off-target toxicity, we stained heart, lung, spleen, and kidney sections from mice treated with saline, free SCU, or SCE (without BDL surgery) with H&E. A pathologist blinded to the treatment groups examined all tissues for signs of inflammation, necrosis, or other histological abnormalities.

4.14. Hydroxyproline Assay

We quantified hepatic hydroxyproline content as a surrogate marker for collagen accumulation and liver fibrosis. For each mouse, we assayed ~50 mg of liver tissue using a commercial hydroxyproline assay kit (Nanjing Jiancheng Bioengineering Institute, China), following the manufacturer's instructions. Briefly, liver samples were hydrolyzed in concentrated hydrochloric acid at 110 °C for several hours to release free hydroxyproline from collagen. After hydrolysis, the samples were neutralized and treated with, followed by reaction with Ehrlich's reagent (p-dimethylaminobenzaldehyde) to generate a chromogenic complex. The absorbance of the resulting solution was then measured at 550 nm and compared to a standard hydroxyproline calibration curve. Hydroxyproline content was expressed as micrograms per gram of liver tissue. All measurements were performed in duplicate.

4.15. 16S rRNA Gene Sequencing for Microbiota Analysis

We assessed the impact of SCU on gut and liver microbiota composition by performing 16S rRNA gene sequencing on fecal and liver samples from the Sham, BDL, and SCU-treated groups [43,44]. Genomic DNA was extracted from 100 mg of fecal material using the QIAamp Fast DNA Stool Mini Kit (Qiagen, Hilden, Germany) and from approximately 25 mg of liver tissue using the DNeasy Blood & Tissue Kit (Qiagen) per the respective manufacturer's protocols. We amplified the hypervariable V3–V4 regions of the bacterial 16S rRNA gene by PCR using universal primers (338F/806R) with Illumina adapter overhangs. After purification with the AxyPrep DNA Gel Extraction Kit (Axygen, Corning, NY, USA), we quantified the PCR products. Equimolar amounts of each sample's PCR product were pooled to construct sequencing libraries using the TruSeq DNA Sample Prep Kit (Illumina, San Diego, CA, USA). High-throughput sequencing was performed on an Illumina HiSeq 2500 platform, generating paired-end reads of 2 × 250 bp.

We processed raw sequencing reads using the QIIME pipeline. After quality filtering and merging, we removed chimeric sequences to obtain high-quality reads. We clustered the reads into operational taxonomic units (OTUs) at 97% sequence similarity and assigned taxonomic identities to representative sequences using the Greengenes or Silva reference database with the RDP classifier algorithm. To standardize sequencing depth across samples, we normalized the OTU abundance table before downstream analyses. We assessed within-sample microbial diversity using alpha diversity metrics, including Chao1 richness and Shannon diversity indices. To evaluate between-group variation, we analyzed beta diversity by principal coordinates analysis (PCoA) and partial least squares discriminant analysis (PLS-DA) based on Bray–Curtis distances. Finally, we identified differentially abundant taxa at the phylum, family, and genus levels and correlated them with disease status or treatment.

All liver samples were collected under aseptic conditions. DNA extraction and PCR reagents were handled in a sterile environment to minimize contamination. Although no dedicated blank extraction controls were included, all experimental groups were processed in parallel under identical conditions. Data interpretation was based on relative differences between treatment and control groups rather than absolute abundance. Notably, due to the relatively low bacterial biomass in liver tissue, the liver microbiome analysis required careful contamination control and validation, and the results from liver 16S sequencing should be considered exploratory.

4.16. Immunofluorescence Staining of Liver Sections

Immunofluorescence was performed to visualize key fibrosis markers (collagen I and α-SMA) in liver tissue sections. Paraffin-embedded liver sections (4 μm) were deparaffinized, rehydrated, and subjected to antigen retrieval (heating in citrate buffer, pH 6.0). After blocking with 5% bovine serum albumin and 0.3% Triton X-100 in PBS for 1 h, sections were incubated overnight at 4 °C with primary antibodies against collagen I (rabbit polyclonal, 1:200 dilution) and α-SMA (mouse monoclonal, 1:200). The next day, sections were washed and incubated for 1 h at room temperature in the dark with species-specific secondary antibodies (Alexa Fluor 594–conjugated goat anti-rabbit for collagen I, Alexa Fluor 488–conjugated goat anti-mouse for α-SMA). Nuclei were counterstained with DAPI, and Slides were mounted with antifade medium before fluorescence microscopy.

4.17. Statistical Analysis

For normally distributed data, results are expressed as mean ± SD and analyzed using one-way ANOVA with Tukey's post hoc test. For data not following a normal distribution (e.g., microbiota abundances), values are presented as median with interquartile range (IQR), and non-parametric tests were applied as appropriate. $p < 0.05$ (corrected with q value if necessary) was considered statistically significant.

5. Conclusions

In summary, we developed SCE to overcome the bioavailability limitations of SCU, achieve targeted hepatic delivery, and produce pronounced anti-fibrotic effects. This nanocarrier system enabled SCU to directly inhibit HSC activation and fibrogenic pathways. In vitro, SCE markedly improved SCU uptake by HSCs and suppressed the expression of fibrotic markers, including collagen I and MMP2. In vivo, SCE treatment in BDL-induced fibrotic mice improved histological and biochemical outcomes, reducing collagen deposition, lowering hydroxyproline content, and restoring liver function. The oral administration of SCU also allowed transient interaction with gut microbiota and liver microbiota, partially normalizing dysbiotic taxa. While this effect was not the main mechanism, it likely contributed additively to therapeutic outcomes. Thus, SCE exerts dual-site activity by modulating both hepatic fibrogenesis and gut microbiota imbalance. Collectively, our findings indicate that SCE is a safe and effective nanomedicine with dual activity: targeting both hepatic fibrogenesis and microbiota imbalance. This strategy exemplifies the potential of oral nanoformulations for multi-target treatment of chronic liver diseases, particularly liver fibrosis.

Supplementary Materials: The following supporting information can be downloaded at: https://www.mdpi.com/article/10.3390/ijms26199746/s1.

Author Contributions: Conceptualization, L.W., G.L. and H.Y.; methodology, H.Y. and X.N.; software, H.Y.; validation, B.N., P.L. and N.X.; formal analysis, S.Y. and B.N.; investigation, H.Y. and X.N.; resources, L.W. and G.L.; data curation, H.Y.; writing—original draft preparation, H.Y.; writing—review and editing, L.W. and G.L.; visualization, Q.Y.; supervision, L.W. and G.L.; project administration, L.W.; funding acquisition, L.W. All authors have read and agreed to the published version of the manuscript.

Funding: This work was supported by the CAMS Innovation Fund for Medical Sciences (2023-I2M-2-006; 2024-I2M-ZH-012; 2025-I2M-KJ-018).

Institutional Review Board Statement: All experimental procedures were approved by the Ethics Committee of the Institute of Medicinal Biotechnology, Chinese Academy of Medical Sciences & Peking Union Medical College (Beijing, China; No. IMB-20231109D102, approved on 9 November 2023).

Informed Consent Statement: Not applicable.

Data Availability Statement: The datasets used and/or analyzed during the current study are available from the corresponding author upon reasonable request. The 16S rRNA data generated in this study will be deposited in the same repository and made publicly available upon acceptance of the manuscript for publication. Further details can be accessed at PRJNA1265637 and PRJNA1265571.

Conflicts of Interest: The authors declare no conflicts of interest.

Abbreviations

SCU	Scutellarin
SCE	SCU-loaded nanoemulsion
SPC	SCU-phospholipid complex
HSCs	Hepatic stellate cells
TGF-β1	Transforming growth factor-β1
BDL	bile duct ligation
NASH	Nonalcoholic steatohepatitis
COS	Chitosan oligosaccharide
αSMA	α-smooth muscle actin
MMP2	Matrix metalloproteinase-2
Col1	Collagen type I
DSC	Differential scanning calorimetry
FTIR	Fourier-transform infrared spectroscopy
XRD	X-ray powder diffraction
PDI	Polydispersity index
DLS	Dynamic light scattering
TEM	Transmission electron microscopy
HPLC	High-performance liquid chromatography
DMEM	Dulbecco's Modified Eagle Medium
RPMI	Roswell Park Memorial Institute
FBS	Fetal bovine serum
CCK-8	Cell Counting Kit-8
NR	Nile Red
NR/SCE	NR-loaded SCE
DAPI	4′,6-diamidino-2-phenylindole
MFI	Mean fluorescence intensity
CPZ	Chlorpromazine
NYS	Nystatin
MβCD	Methyl-β-cyclodextrin
Gen	Genistein
IMI	Imipramine
EIPA	5-(N-ethyl-N-isopropyl) amiloride
qPCR	Quantitative real-time PCR
RIPA	Radioimmunoprecipitation assay
SDS-PAGE	Sulfate–polyacrylamide gel electrophoresis
PVDF	Polyvinylidene difluoride
ECL	Enhanced chemiluminescence
DIR	1,1′-dioctadecyl-3,3,3′,3′-tetramethylindotricarbocyanine iodide
IACUC	Institutional Animal Care and Use Committee
ROI	Regions of interest
ALT	Alanine aminotransferase
AST	Aspartate aminotransferase
ALP	Alkaline phosphatase
TBA	Total bile acids
H&E	Hematoxylin and eosin

OTUs	Operational taxonomic units
PCoA	Principal coordinates analysis
PLS-DA	Partial least squares discriminant analysis
ANOVA	One-way analysis of variance
TG	Triglyceride
UREA	Blood urea nitrogen
Cre	Creatinine

References

1. Friedman, S.L.; Neuschwander-Tetri, B.A.; Rinella, M.; Sanyal, A.J. Mechanisms of NAFLD development and therapeutic strategies. *Nat. Med.* **2018**, *24*, 908–922. [CrossRef]
2. Bataller, R.; Brenner, D.A. Liver fibrosis. *J. Clin. Investig.* **2005**, *115*, 209–218. [CrossRef]
3. Harrison, S.A.; Bedossa, P.; Guy, C.D.; Schattenberg, J.M.; Loomba, R.; Taub, R.; Labriola, D.; Moussa, S.E.; Neff, G.W.; Rinella, M.E.; et al. A Phase 3, Randomized, Controlled Trial of Resmetirom in NASH with Liver Fibrosis. *N. Engl. J. Med.* **2024**, *390*, 497–509. [CrossRef] [PubMed]
4. Tilg, H.; Adolph, T.E.; Trauner, M. Gut-liver axis: Pathophysiological concepts and clinical implications. *Cell Metab.* **2022**, *34*, 1700–1718. [CrossRef] [PubMed]
5. Miao, Z.; Lai, Y.; Zhao, Y.; Chen, L.; Zhou, J.; Li, C.; Wang, Y. Protective Property of Scutellarin Against Liver Injury Induced by Carbon Tetrachloride in Mice. *Front. Pharmacol.* **2021**, *12*, 710692. [CrossRef] [PubMed]
6. Zhou, Z.; He, W.; Tian, H.; Zhan, P.; Liu, J. Thyme (*Thymus vulgaris* L.) polyphenols ameliorate DSS-induced ulcerative colitis of mice by mitigating intestinal barrier damage, regulating gut microbiota, and suppressing TLR4/NF-κB-NLRP3 inflammasome pathways. *Food Funct.* **2023**, *14*, 1113–1132. [CrossRef]
7. Kabir, A.; Muth, A. Polypharmacology: The science of multi-targeting molecules. *Pharmacol. Res.* **2022**, *176*, 106055. [CrossRef]
8. Niu, X.; Meng, Y.; Cui, J.; Li, R.; Ding, X.; Niu, B.; Chang, G.; Xu, N.; Li, G.; Wang, Y.; et al. Hepatic Stellate Cell- and Liver Microbiome-Specific Delivery System for Dihydrotanshinone I to Ameliorate Liver Fibrosis. *ACS Nano* **2023**, *17*, 23608–23625. [CrossRef]
9. Shi, X.; Hu, Y.; Jiang, Y.; Wu, J.; Zhang, C.; Zhang, J.; Wu, S.; Wu, Y.; Dong, W.; Li, J. Scutellarein protects against cardiac hypertrophy via suppressing TRAF2/NF-κB signaling pathway. *Mol. Biol. Rep.* **2022**, *49*, 2085–2095. [CrossRef]
10. Wu, H.; Jia, L. Scutellarin attenuates hypoxia/reoxygenation injury in hepatocytes by inhibiting apoptosis and oxidative stress through regulating Keap1/Nrf2/ARE signaling. *Biosci. Rep.* **2019**, *39*, BSR20192501. [CrossRef]
11. Ibrahim, M.A.A.; Elwan, W.M.; Elgendy, H.A. Role of Scutellarin in Ameliorating Lung Injury in a Rat Model of Bilateral Hind Limb Ischemia-Reperfusion. *Anat. Rec.* **2019**, *302*, 2070–2081. [CrossRef] [PubMed]
12. Wang, L.; Ma, Q. Clinical benefits and pharmacology of scutellarin: A comprehensive review. *Pharmacol. Ther.* **2018**, *190*, 105–127. [CrossRef] [PubMed]
13. Malekmohammad, K.; Rafieian-Kopaei, M. Mechanistic Aspects of Medicinal Plants and Secondary Metabolites against Severe Acute Respiratory Syndrome Coronavirus 2 (SARS-CoV-2). *Curr. Pharm. Des.* **2021**, *27*, 3996–4007. [CrossRef] [PubMed]
14. Zhu, J.; Chen, L.; Qi, Y.; Feng, J.; Zhu, L.; Bai, Y.; Wu, H. Protective effects of Erigeron breviscapus Hand.- Mazz. (EBHM) extract in retinal neurodegeneration models. *Mol. Vis.* **2018**, *24*, 315–325.
15. Chen, S.; Li, R.; Chen, Y.; Chou, C.K.; Zhang, Z.; Yang, Y.; Liao, P.; Wang, Q.; Chen, X. Scutellarin enhances anti-tumor immune responses by reducing TNFR2-expressing CD4(+)Foxp3(+) regulatory T cells. *Biomed. Pharmacother.* **2022**, *151*, 113187. [CrossRef]
16. Pan, Z.; Zhao, W.; Zhang, X.; Wang, B.; Wang, J.; Sun, X.; Liu, X.; Feng, S.; Yang, B.; Lu, Y. Scutellarin alleviates interstitial fibrosis and cardiac dysfunction of infarct rats by inhibiting TGFβ1 expression and activation of p38-MAPK and ERK1/2. *Br. J. Pharmacol.* **2011**, *162*, 688–700. [CrossRef]
17. Peng, L.; Wen, L.; Shi, Q.F.; Gao, F.; Huang, B.; Meng, J.; Hu, C.P.; Wang, C.M. Scutellarin ameliorates pulmonary fibrosis through inhibiting NF-κB/NLRP3-mediated epithelial-mesenchymal transition and inflammation. *Cell Death Dis.* **2020**, *11*, 978. [CrossRef]
18. Shi, Y.; Liu, Y.; Wang, S.; Huang, J.; Luo, Z.; Jiang, M.; Lu, Y.; Lin, Q.; Liu, H.; Cheng, N.; et al. Endoplasmic reticulum-targeted inhibition of CYP2E1 with vitamin E nanoemulsions alleviates hepatocyte oxidative stress and reverses alcoholic liver disease. *Biomaterials* **2022**, *288*, 121720. [CrossRef]
19. Yuan, T.; Cheng, X.; Shen, L.; Liu, Z.; Ye, X.; Yan, Z.; Wei, W.; Wang, X. Novel Human Milk Fat Substitutes Based on Medium- and Long-Chain Triacylglycerol Regulate Thermogenesis, Lipid Metabolism, and Gut Microbiota Diversity in C57BL/6J Mice. *J. Agric. Food Chem.* **2024**, *72*, 6213–6225. [CrossRef]
20. Huang, J.H.; Wang, J.; Chai, X.Q.; Li, Z.C.; Jiang, Y.H.; Li, J.; Liu, X.; Fan, J.; Cai, J.B.; Liu, F. The Intratumoral Bacterial Metataxonomic Signature of Hepatocellular Carcinoma. *Microbiol. Spectr.* **2022**, *10*, e0098322. [CrossRef]

21. Li, Q.; Guo, P.; Wang, S.; Su, L.; Liang, T.; Yu, W.; Guo, J.; Yang, Q.; Tang, Z.; Liao, J. Gut microbiota disorders aggravate terbuthylazine-induced mitochondrial quality control disturbance and PANoptosis in chicken hepatocyte through gut-liver axis. *Sci. Total. Environ.* **2024**, *913*, 169642. [CrossRef]
22. Bárcena, C.; Valdés-Mas, R.; Mayoral, P.; Garabaya, C.; Durand, S.; Rodríguez, F.; Fernández-García, M.T.; Salazar, N.; Nogacka, A.M.; Garatachea, N.; et al. Healthspan and lifespan extension by fecal microbiota transplantation into progeroid mice. *Nat. Med.* **2019**, *25*, 1234–1242. [CrossRef] [PubMed]
23. Qiu, J.J.; Liu, Z.; Zhao, P.; Wang, X.J.; Li, Y.C.; Sui, H.; Owusu, L.; Guo, H.S.; Cai, Z.X. Gut microbial diversity analysis using Illumina sequencing for functional dyspepsia with liver depression-spleen deficiency syndrome and the interventional Xiaoyaosan in a rat model. *World J. Gastroenterol.* **2017**, *23*, 810–816. [CrossRef] [PubMed]
24. Liu, X.; Ma, Q.; Feng, Y.; Wang, F.; Wang, W.; Wang, J.; Sun, J. Potato resistant starch improves type 2 diabetes by regulating inflammation, glucose and lipid metabolism and intestinal microbial environment. *Int. J. Biol. Macromol.* **2024**, *281*, 136389. [CrossRef] [PubMed]
25. Lu, Z.; Zheng, Y.; Zheng, J.; Liang, Q.; Zhen, Q.; Cui, M.; Yang, H.; Wu, H.; Tian, C.; Zhu, K.; et al. Theabrownin from Fu Brick tea ameliorates high-fat induced insulin resistance, hepatic steatosis, and inflammation in mice by altering the composition and metabolites of gut microbiota. *Food Funct.* **2024**, *15*, 4421–4435. [CrossRef]
26. Pan, X.; Kaminga, A.C.; Liu, A.; Wen, S.W.; Luo, M.; Luo, J. Gut Microbiota, Glucose, Lipid, and Water-Electrolyte Metabolism in Children With Nonalcoholic Fatty Liver Disease. *Front. Cell. Infect. Microbiol.* **2021**, *11*, 683743. [CrossRef]
27. Kang, D.J.; Betrapally, N.S.; Ghosh, S.A.; Sartor, R.B.; Hylemon, P.B.; Gillevet, P.M.; Sanyal, A.J.; Heuman, D.M.; Carl, D.; Zhou, H.; et al. Gut microbiota drive the development of neuroinflammatory response in cirrhosis in mice. *Hepatology* **2016**, *64*, 1232–1248. [CrossRef]
28. Zhang, X.; Coker, O.O.; Chu, E.S.; Fu, K.; Lau, H.C.H.; Wang, Y.X.; Chan, A.W.H.; Wei, H.; Yang, X.; Sung, J.J.Y.; et al. Dietary cholesterol drives fatty liver-associated liver cancer by modulating gut microbiota and metabolites. *Gut* **2021**, *70*, 761–774. [CrossRef]
29. Hong, Y.; Sheng, L.; Zhong, J.; Tao, X.; Zhu, W.; Ma, J.; Yan, J.; Zhao, A.; Zheng, X.; Wu, G.; et al. Desulfovibrio vulgaris, a potent acetic acid-producing bacterium, attenuates nonalcoholic fatty liver disease in mice. *Gut Microbes* **2021**, *13*, 1930874. [CrossRef]
30. Yan, F.; Zhang, Q.; Shi, K.; Zhang, Y.; Zhu, B.; Bi, Y.; Wang, X. Gut microbiota dysbiosis with hepatitis B virus liver disease and association with immune response. *Front. Cell. Infect. Microbiol.* **2023**, *13*, 1152987. [CrossRef]
31. Ponziani, F.R.; Bhoori, S.; Castelli, C.; Putignani, L.; Rivoltini, L.; Del Chierico, F.; Sanguinetti, M.; Morelli, D.; Paroni Sterbini, F.; Petito, V.; et al. Hepatocellular Carcinoma Is Associated With Gut Microbiota Profile and Inflammation in Nonalcoholic Fatty Liver Disease. *Hepatology* **2019**, *69*, 107–120. [CrossRef] [PubMed]
32. Boursier, J.; Mueller, O.; Barret, M.; Machado, M.; Fizanne, L.; Araujo-Perez, F.; Guy, C.D.; Seed, P.C.; Rawls, J.F.; David, L.A.; et al. The severity of nonalcoholic fatty liver disease is associated with gut dysbiosis and shift in the metabolic function of the gut microbiota. *Hepatology* **2016**, *63*, 764–775. [CrossRef] [PubMed]
33. Huang, W.; Chen, H.; He, Q.; Xie, W.; Peng, Z.; Ma, Q.; Huang, Q.; Chen, Z.; Liu, Y. Nobiletin protects against ferroptosis to alleviate sepsis-associated acute liver injury by modulating the gut microbiota. *Food Funct.* **2023**, *14*, 7692–7704. [CrossRef] [PubMed]
34. Zhang, Q.; Fan, X.Y.; Cao, Y.J.; Zheng, T.T.; Cheng, W.J.; Chen, L.J.; Lv, X.C.; Ni, L.; Rao, P.F.; Liang, P. The beneficial effects of Lactobacillus brevis FZU0713-fermented Laminaria japonica on lipid metabolism and intestinal microbiota in hyperlipidemic rats fed with a high-fat diet. *Food Funct.* **2021**, *12*, 7145–7160. [CrossRef]
35. Carter, J.K.; Bhattacharya, D.; Borgerding, J.N.; Fiel, M.I.; Faith, J.J.; Friedman, S.L. Modeling dysbiosis of human NASH in mice: Loss of gut microbiome diversity and overgrowth of Erysipelotrichales. *PLoS ONE* **2021**, *16*, e0244763. [CrossRef]
36. Zhou, D.; Zhang, J.; Xiao, C.; Mo, C.; Ding, B.S. Trimethylamine-N-oxide (TMAO) mediates the crosstalk between the gut microbiota and hepatic vascular niche to alleviate liver fibrosis in nonalcoholic steatohepatitis. *Front. Immunol.* **2022**, *13*, 964477. [CrossRef]
37. Scorletti, E.; Afolabi, P.R.; Miles, E.A.; Smith, D.E.; Almehmadi, A.; Alshathry, A.; Childs, C.E.; Del Fabbro, S.; Bilson, J.; Moyses, H.E.; et al. Synbiotics Alter Fecal Microbiomes, But Not Liver Fat or Fibrosis, in a Randomized Trial of Patients With Nonalcoholic Fatty Liver Disease. *Gastroenterology* **2020**, *158*, 1597–1610.e1597. [CrossRef]
38. Friedman, S.L. Mechanisms of hepatic fibrogenesis. *Gastroenterology* **2008**, *134*, 1655–1669. [CrossRef]
39. Zhou, H.; Wan, J.; Wu, L.; Yi, T.; Liu, W.; Xu, H.; Yang, X. A new strategy for enhancing the oral bioavailability of drugs with poor water-solubility and low liposolubility based on phospholipid complex and supersaturated SEDDS. *PLoS ONE* **2013**, *8*, e84530; Correction in *PLoS ONE* **2014**, *9*, e91605. [CrossRef]
40. Böttger, R.; Pauli, G.; Chao, P.H.; Al Fayez, N.; Hohenwarter, L.; Li, S.D. Lipid-based nanoparticle technologies for liver targeting. *Adv. Drug. Deliv. Rev.* **2020**, *154–155*, 79–101. [CrossRef]

41. Huo, Y.; Mijiti, A.; Cai, R.; Gao, Z.; Aini, M.; Mijiti, A.; Wang, Z.; Qie, R. Scutellarin alleviates type 2 diabetes (HFD/low dose STZ)-induced cardiac injury through modulation of oxidative stress, inflammation, apoptosis and fibrosis in mice. *Hum. Exp. Toxicol.* **2021**, *40*, S460–S474. [CrossRef]
42. Huang, B.; Han, R.; Tan, H.; Zhu, W.; Li, Y.; Jiang, F.; Xie, C.; Ren, Z.; Shi, R. Scutellarin ameliorates diabetic nephropathy via TGF-β1 signaling pathway. *Nat. Prod. Bioprospect* **2024**, *14*, 25. [CrossRef]
43. Leinwand, J.C.; Paul, B.; Chen, R.; Xu, F.; Sierra, M.A.; Paluru, M.M.; Nanduri, S.; Alcantara, C.G.; Shadaloey, S.A.; Yang, F.; et al. Intrahepatic microbes govern liver immunity by programming NKT cells. *J. Clin. Investig.* **2022**, *132*, S463. [CrossRef]
44. Suppli, M.P.; Bagger, J.I.; Lelouvier, B.; Broha, A.; Demant, M.; Kønig, M.J.; Strandberg, C.; Lund, A.; Vilsbøll, T.; Knop, F.K. Hepatic microbiome in healthy lean and obese humans. *JHEP Rep.* **2021**, *3*, 100299. [CrossRef]

International Journal of
Molecular Sciences

MDPI

Article

D-Tryptophan Promotes Skin Wound Healing via Extracellular Matrix Remodeling in Normal and Diabetic Models

Dawit Adisu Tadese [1,2,3], James Mwangi [1,2,3], Brenda B. Michira [1,2,3], Yi Wang [4], Kaixun Cao [5], Min Yang [1,2,3], Mehwish Khalid [1,2,3], Ziyi Wang [1], Qiumin Lu [1] and Ren Lai [1,2,3,4,5,*]

1 Key Laboratory of Genetic Evolution & Animal Models, Engineering Laboratory of Peptides of Chinese Academy of Sciences, Key Laboratory of Bioactive Peptides of Yunnan Province, KIZ-CUHK Joint Laboratory of Bioresources and Molecular Research in Common Diseases, National Resource Center for Non-Human Primates, and Sino-African Joint Research Center, New Cornerstone Science Laboratory, Kunming Institute of Zoology, The Chinese Academy of Sciences, Kunming 650201, China; davadisu@gmail.com (D.A.T.); jams@mail.kiz.ac.cn (J.M.); brenda@mail.kiz.ac.cn (B.B.M.); yangmin@mail.kiz.ac.cn (M.Y.); mehwishkhalidd120@gmail.com (M.K.); wangziyi@mail.kiz.ac.cn (Z.W.); lvqm@mail.kiz.ac.cn (Q.L.)
2 University of Chinese Academy of Sciences, Beijing 100049, China
3 Kunming College of Life Science, University of Chinese Academy of Sciences, Beijing 100049, China
4 Center for Evolution and Conservation Biology, Southern Marine Science and Engineering Guangdong Laboratory (Guangzhou), Guangzhou 511458, China; 13477288321@163.com
5 College of Life Sciences, Nanjing Agricultural University, Nanjing 210095, China; kaixuncao@gmail.com
* Correspondence: rlai@mail.kiz.ac.cn; Fax: +86-871-6513-0513

Abstract

Diabetic wounds are a devastating complication that cause chronic pain, recurrent infections, and limb amputations due to impaired healing. Despite advances in wound care, existing therapies often fail to address the underlying molecular dysregulation, highlighting the need for innovative and safe therapeutic approaches. Among these, D-amino acids such as D-tryptophan (D-Trp) have emerged as key regulators of cellular processes; however, their therapeutic potential in diabetic wounds remains largely unexplored. Here, we investigate the therapeutic potential of D-Trp in streptozotocin (STZ)-induced diabetic mice, comparing it with phosphate-buffered saline (PBS) controls and vascular endothelial growth factor (VEGF) as a positive control. Wound healing, inflammation, and histopathology were assessed. Protein and gene expression were analyzed via Western blot and RT-qPCR, respectively. Biolayer interferometry (BLI) measured the binding of D-Trp to hypoxia-inducible factor-1α (HIF-1α). D-Trp accelerated wound healing by modulating extracellular matrix (ECM) remodeling, signaling, and apoptosis. It upregulated matrix metalloproteinases (MMP1, MMP3, MMP-9), Janus kinase 2 (JAK2), and mitogen-activated protein kinase (MAPK) proteins while reducing pro-inflammatory cytokines (tumor necrosis factor-α [*TNF-α*], interleukin-1β [*IL-1β*], *IL-6*). D-Trp also suppressed caspase-3 and enhanced angiogenesis through HIF-1α activation. These findings suggest that D-Trp promotes healing by boosting ECM turnover, reducing inflammation, and activating MAPK/JAK pathways. Thus, D-Trp is a promising therapeutic for diabetic wounds.

Keywords: D-tryptophan; diabetic wound healing; hypoxia-inducible factor 1-alpha; extracellular matrix; inflammation

1. Introduction

Wound healing is an intricate and dynamic process crucial for restoring tissue integrity and function following injury or skin damage [1]. This complex biological process involves a meticulously coordinated sequence of steps that aim to repair injured tissue, restore the skin barrier, and achieve complete tissue regeneration [2]. Rapid and efficient wound healing of skin is critical for homeostasis, preventing infections, and restoring the structural and functional integrity of the skin, the largest organ of body [1]. There are several discrete but interrelated phases of wound healing, including hemostasis, inflammation, proliferation, and remodeling [3]. Each phase is defined by distinct cellular and molecular occurrences controlled by several signaling pathways, growth factors, cytokines, and immune cells [4]. Impaired wound healing, especially in individuals with chronic diseases such as type 2 diabetes, poses a significant clinical challenge [5,6], resulting in delayed closure [7], increased risk of infection [8], and prolonged consequences [9]. Comprehending the complex principles of wound healing is crucial for formulating efficient therapeutic techniques to improve the healing process and increase tissue restoration results.

Diabetic wounds, a common and serious complication of diabetes mellitus (DM), can cause serious problems such as amputation and infection-related death, primarily due to poor circulation and sensory loss on the feet [10]. Refractory diabetic foot unclears (DFUs) are particularly concerning, as they frequently cause non-traumatic amputations and are a major health risk for diabetics [11,12]. Delayed wound healing is often due to chemokine deficiency, abnormal inflammation, and insufficient angiogenesis and epithelial regeneration [13,14]. Current treatments, such as wound debridement, anti-inflammatory medications, and topical antibiotics or growth factors, offer limited effectiveness [2]. Examining diabetic wounds from various perspectives could help identify new targets and strategies for treating refractory diabetic wounds [15–21]. Likewise, advances in wound healing research, including stem cell therapies, tissue engineering methods, and personalized medicine strategies, have significant potential to transform the wound care domain [22]. Researchers seek to develop innovative therapies that enhance healing speed, minimize scarring, and improve patient quality of life by elucidating the molecular mechanisms underlying wound healing, identifying novel therapeutic targets, and leveraging the regenerative capabilities of stem cells [23–25].

D-amino acids, the enantiomers of L-amino acids, were historically considered to be non-functional. However, recent studies have demonstrated that certain D-amino acids present in mammalian tissues play a significant role in various physiological processes [26]. Among these, D-Trp has gained attention for its diverse metabolic functions and potential therapeutic applications. A separate study reports that D-Trp suppresses colitis progression by reducing specific intestinal microorganisms. Moreover, D-Trp augments intracellular levels of indole acrylic acid (IA), a critical molecule that modulates the susceptibility of enteric microorganisms to D-trp. Administration of IA has been shown to improve the survival rates of mice infected with *C. rodentium* [27]. Moreover, research has demonstrated that D-Trp can reduce the initial adhesion of pathogenic cells and alter the extracellular environment, leading to a notable decrease in pathogen viability [28]. Additionally, D-Trp inhibits the formation of biofilms through similar mechanisms and acts as an auto-inhibitory compound that prevents the germination of spores and the growth of harmful microorganisms [27–30]. Extensive studies have been conducted on the interaction between signaling pathways and important proteins in wound healing [31]; however, the potential therapeutic role of D-Trp, as an amino acid, remains largely unexplored. Thus, we aimed to evaluate the therapeutic efficacy of D-Trp in improving extracellular matrix (ECM) remodeling and facilitating wound closure in both normal and diabetic models. The findings underscore D-Trp as a potentially effective metabolite-based therapeutic agent for the promotion of wound healing. This research lays the foundation for the advancement of innovative metabolite-driven treatments that offer not only enhanced efficacy, but also a favorable safety profile, representing a significant advancement in wound care.

2. Results

2.1. D-Trp Promotes the Migration of HaCaT Cells

To investigate the role of D-Trp in wound healing, we performed a wound healing assay using HaCaT cells, a well-established human keratinocyte cell line that is frequently used in wound healing research [32]. Figure 1A illustrates the cell migration process, detailing the key steps involved in the movement of cells from one location to another. This assay replicated a wound environment by introducing a linear scratch into a confluent monolayer of cells and subsequently monitoring the capacity of the cells to migrate and bridge the gap. The findings revealed that D-Trp treatment markedly augmented the migratory capacity of HaCaT cells compared to the control group (Figure 1B,C). Specifically, a marked increase in the rate at which D-Trp-treated cells migrated across the wound area was observed, effectively closing the gap more quickly than control cells, while amino acid treatment significantly accelerated wound closure, suggesting that D-Trp played a crucial role in promoting cell migration, a critical process for wound closure.

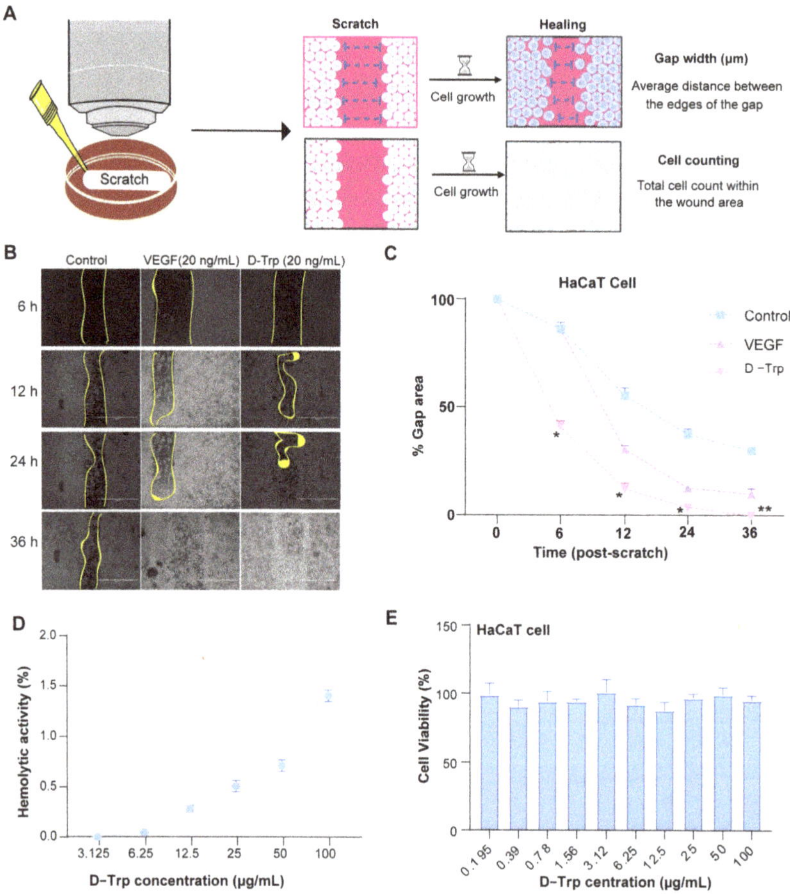

Figure 1. Effects of D-tryptophan on HaCaT cell migration. (**A**) Schematic figure of wound healing assay measurements. (**B**) The impact of D-Trp on HaCaT cell migration is illustrated at various time points, with a scale bar of 650 μm. (**C**) Quantification of the number of HaCaT cells that migrated after treatment (**D**) % hemolytic activity in human blood (**E**) % cell viability. Data are presented as mean ± standard error (SE) with a sample size of $n = 3$. Statistical significance is denoted as follows: * $p < 0.05$, ** $p < 0.01$.

These findings strongly validated that D-Trp actively promoted cell migration, a key aspect of wound healing, further highlighting its potential therapeutic value in accelerating tissue repair. As illustrated in Figure 1D, the hemolysis rate of D-Trp at a concentration of 100 μg/mL was found to be less than 2%, indicating that D-Trp exhibited minimal hemolytic activity and is unlikely to induce rupture or lysis of human red blood cells under these conditions. Furthermore, Figure 1E confirmed that the cytotoxicity of D-Trp at the same concentration (100 μg/mL) was less than 1%, suggesting its negligible toxic effect on HaCaT cells at this concentration.

2.2. D-Trp Accelerates Wound Healing in a Non-Diabetic Model

Wound healing is a dynamic and multi-phase process that involves inflammation, tissue regeneration, and remodeling to restore skin integrity [33]. Effective wound healing depends on adequate epithelialization, granulation tissue formation, and a well-regulated inflammatory response [34]. In this study, we used a C57BL/6J mouse model with severe skin wounds. Figure 2A provides a schematic representation of the key processes involved in wound healing. The findings indicated that the D-Trp treatment group showed a notably faster wound healing rate compared to the control group (Figure 2B). Figure 2C illustrates the percentage of wounds that achieved complete closure over time, with the D-Trp group demonstrating a significantly higher proportion of fully healed wounds compared to controls. All experimental groups showed typical wound healing responses, including wound contraction and the formation of neo-tissue-supported wound beds. However, on days 3 and 7, the D-Trp group demonstrated a faster reduction in the wound area and an increase in tissue thickness compared to the control, as assessed by measuring the area of the injury (Figure 2D). In particular, on day 3, the D-Trp group exhibited substantial epithelial turnover and keratinization tissue, along with the appearance of new white hair, suggesting elevated epithelialization and hair follicle regeneration. On day 7, the D-Trp group was the first to develop a newly formed epidermis and surrounding hair growth, indicating more complete and accelerated wound closure. Histological analysis of skin sections using H&E staining revealed significant improvements in wound healing for the group treated with D-Trp relative to the control (Figure 2E).

Specifically, D-Trp treatment led to an increase in granulation tissue formation and epithelialization, along with a marked reduction in inflammatory cell infiltration at the wound site, as evidenced by H&E staining. These findings are consistent with previous studies that have demonstrated the role of tryptophan metabolites in improving wound healing, promoting tissue regeneration, and modulating the inflammatory response [35–37].

2.3. D-Trp Modulates Key Protein Expression Involved in Non-Wound Healing

The administration of D-Trp exerted significant modulatory effects on the expression of proteins implicated in wound healing (Figure 3). As shown in the representative Western blot images (Figure 3A), D-Trp treatment notably enhanced the levels of MMP1 and MMP3 (Figure 3B,C), indicative of improved extracellular matrix (ECM) remodeling through increased proteolytic activity [38]. Furthermore, D-Trp significantly elevated the expression of total MAPK and phosphorylated MAPK (p-MAPK) compared to control and VEGF-treated groups (Figure 3D,E). While p-MAPK is often normalized to total MAPK to evaluate pathway activation, in this analysis, both MAPK and p-MAPK were independently quantified relative to GAPDH. The observed parallel increase in both forms supports MAPK pathway involvement in promoting keratinocyte proliferation and tissue regeneration.

Additional Western blot analyses revealed marked upregulation of JAK2 (Figure 3G), highlighting potential activation of the JAK/STAT signaling pathway, which is central to inflammatory modulation and tissue repair [39]. Similarly, the expression of SMAD2

(Figure 3H) was increased in D-Trp-treated wounds, suggesting participation of the TGF-β/SMAD pathway in enhancing ECM deposition and wound contraction. Notably, caspase-3 levels were significantly reduced (Figure 3I), indicating a potent anti-apoptotic effect that may contribute to enhanced cellular survival in the wound bed [40]. Collectively, these results suggest that D-Trp promotes wound healing by modulating multiple molecular pathways involved in matrix remodeling, proliferation, and cell survival.

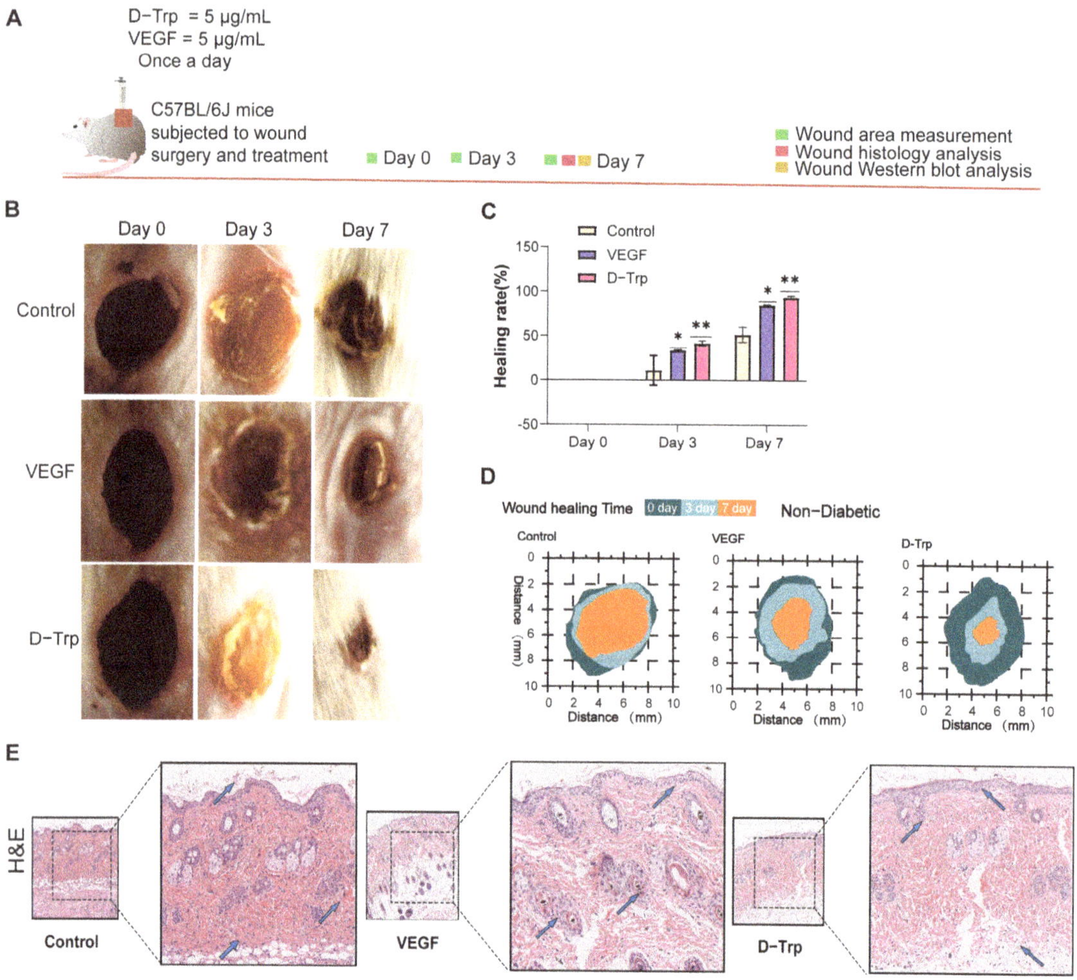

Figure 2. An evaluation of non-diabetic wound healing in vivo. (**A**) A schematic illustration of the treatment process alongside the progression of wound healing. (**B**) Quantification of wound healing processes on days 0, 3 and 7. (**C**) Representative images of chronic diabetic wounds that were treated with PBS, VEGF, and D-Trp. A schematic diagram in panel (**D**) further illustrates the percentage of wound closure for each group across the three time points. (**E**) Representative images of hematoxylin and eosin staining (H&E), which emphasize blood vessels, with blue arrows denoting newly formed vessels and hair follicles (scale bars: 100 μm and 20 μm). Data are expressed as mean ± standard error (SE) with a sample size of $n = 5$, and statistical significance is indicated as follows: * $p < 0.05$; ** $p < 0.01$.

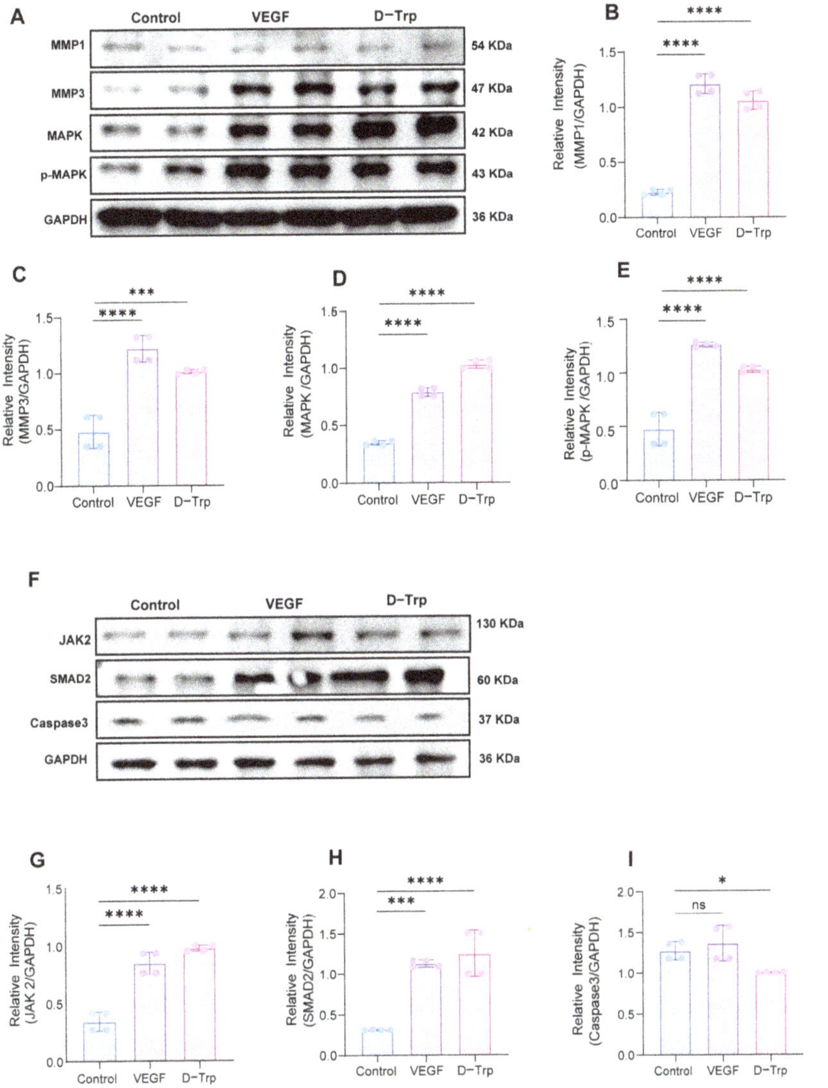

Figure 3. D-Trp regulates the expression of proteins associated with impaired wound healing. (**A**) Representative Western blot images of key proteins in skin wound tissues. (**B–E**) Quantitative analysis of MMP1, MMP3, MAPK, and phosphorylated MAPK (p-MAPK) expression levels, respectively. (**F**) Additional Western blot images of wound tissue samples. (**G–I**) Densitometric quantification of JAK2, SMAD2, and Caspase-3 protein levels in skin and tissues, respectively, from mice treated with phosphate-buffered saline (PBS), vascular endothelial growth factor (VEGF), or D-tryptophan (D-Trp). Data are presented as mean ± standard error (SE) (n = 4). Statistical significance is indicated as follows: ns ($p > 0.05$), * $p < 0.05$, *** $p < 0.001$, **** $p < 0.0001$.

2.4. D-Trp Accelerates Wound Healing Diabetic Model

In an experimental framework of 14 days employing streptozotocin (STZ)-induced diabetic mice (Figure 4A), it was discovered that D-Trp significantly enhanced wound healing in this diabetic model at a concentration of 5 μg/mL, surpassing the efficacy observed in the control groups (Figure 4B). While all groups exhibited wound contraction and new tissue formation, the D-Trp group demonstrated a notably faster reduction in wound area and developed thicker, more robust tissue (Figure 4B), indicative of a more effective and comprehensive healing process. Furthermore, D-Trp-treated mice exhibited early signs of advanced epithelialization and tissue regeneration, marked by the collapse of newly formed tissue and the emergence of new hair growth (Figure 4C).

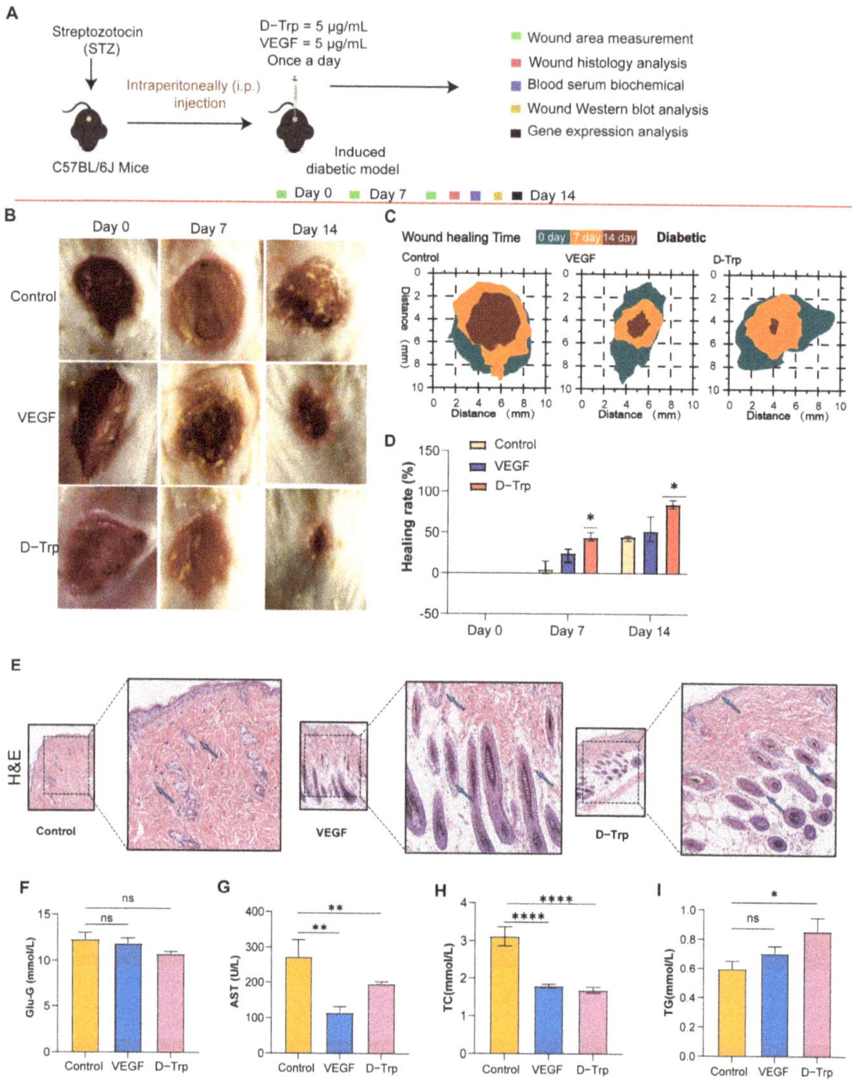

Figure 4. An evaluation of diabetic wound healing in vivo. (**A**) A schematic illustration of the treatment process alongside the progression of wound healing. (**B**) Wound healing processes at days 0, 7, and 14. (**C**) Representative images of chronic diabetic wounds that were treated with PBS, VEGF, and D-trp. (**D**) Schematic diagram illustrating the percentage of wound closure for each group at three different time points. (**E**) Representative images of hematoxylin and eosin (H&E) staining, highlighting blood vessels with blue arrows indicating newly formed vessels and hair follicles (scale bars: 100 μm and 20 μm). Analysis of biochemical plasma parameters following 14 days of intervention, (**F**) glucose levels, (**G**) aspartate aminotransferase (AST), (**H**) triglycerides (TG), and (**I**) total cholesterol (TC). Data are expressed as mean ± standard error (SE) with a sample size of $n = 5$, and statistical significance is indicated as follows: ns, $p > 0.05$; * $p < 0.05$; ** $p < 0.01$; **** $p < 0.0001$.

After 14 days, the D-Trp group showed the most significant healing response, with a fully restored epidermis and complete hair regrowth. These findings were consistent with existing research suggesting that tryptophan metabolites, including D-Trp, may play a role in promoting tissue repair and accelerating wound healing, particularly in diabetic conditions. Similar studies have demonstrated that amino acids like tryptophan enhance collagen deposition, fibroblast activity, and keratinocyte proliferation, all of which are crucial for effective wound closure and regeneration [41]. Moreover, the accelerated epithelialization and tissue regeneration observed in our study aligned with the findings of previous work on amino acid-based therapies that underscore their potential to improve wound healing in diabetic models [42,43]. Collectively, our results underscore the therapeutic potential of D-Trp as a promising candidate for enhancing wound healing outcomes in diabetic conditions.

2.5. D-Trp Modulates Key Protein Expression Involved in Diabetic Wound Healing

D-Trp treatment markedly improved wound healing in diabetic mice through coordinated activation of critical molecular pathways. Western blot analysis revealed a 2.39-fold upregulation of TGF-β (Figure 5B), indicating robust activation of this master regulator of fibrogenesis that stimulates fibroblast differentiation and collagen production. Concurrently, we observed a 1.18-fold increase in MMP1 and a 1.29-fold elevation in MMP2 (Figure 5C,D), demonstrating enhanced proteolytic capacity for ECM degradation and tissue remodeling essential for keratinocyte migration. Treatment also induced 1.63-fold-higher total MAPK levels with 1.35-fold-increased phosphorylation (Figure 5E,F), confirming activation of this central signaling cascade that coordinates cellular proliferation and migration through ERK1/2-dependent mechanisms [44].

Further analysis identified additional therapeutic mechanisms through which D-Trp enhances diabetic wound repair. The treated groups showed pronounced upregulation of collagen I (Figure 5H), reinforcing ECM stabilization through enhanced structural protein deposition, consistent with the observed TGF-β-mediated fibrogenic response. The concurrent elevation of PI3K (Figure 5I) and SMAD2 phosphorylation (Figure 5J,K) revealed dual activation of both immediate pro-survivals signaling and sustained TGF-β/SMAD transcriptional regulation [45], creating an optimal microenvironment for fibroblast proliferation and matrix production. Particularly significant was the dramatic suppression of Caspase-3 (Figure 5L), demonstrating D-Trp's ability to counteract the excessive apoptosis that characterizes diabetic wounds [46].

These molecular changes work synergistically with D-Trp's previously demonstrated upregulation of TGF-β, MMPs, and MAPK pathways, creating a comprehensive therapeutic effect that enhances wound healing through multiple coordinated mechanisms. The combined action of increased collagen I production and MMP-mediated ECM degradation establishes balanced tissue remodeling, while concurrent activation of PI3K, MAPK, and SMAD2 signaling pathways drives cellular repair processes. Furthermore, the significant reduction in Caspase-3 levels demonstrates potent inhibition of pathological apoptosis, preserving cellular viability in the wound environment. Together, these effects address the key pathological deficits in diabetic wound healing by simultaneously promoting ECM restructuring, activating cellular repair programs, and maintaining tissue viability through integrated molecular pathways.

2.6. D-Tryptophan Boosts Gene Expression to Promote Diabetic Wound Healing

Our investigation demonstrated that D-Trp administration significantly modulates gene expression profiles in diabetic wound models, exerting a potent anti-inflammatory effect. Notably, D-Trp treatment markedly downregulated key pro-inflammatory cytokines, including *IL-1β* (Figure 6A), *IL-6* (Figure 6B), *TNF-α* (Figure 6C), and *MCP-1* (Figure 6D). This suppression of inflammatory mediators highlights the therapeutic potential of D-Trp in attenuating excessive inflammation, a critical factor in facilitating efficient diabetic wound healing [47]. Concurrently, D-Trp treatment significantly upregulated *MMP-9* expression (Figure 6E), a critical mediator of extracellular matrix (ECM) degradation and remodeling, thereby promoting enhanced cellular migration and tissue regeneration [48].

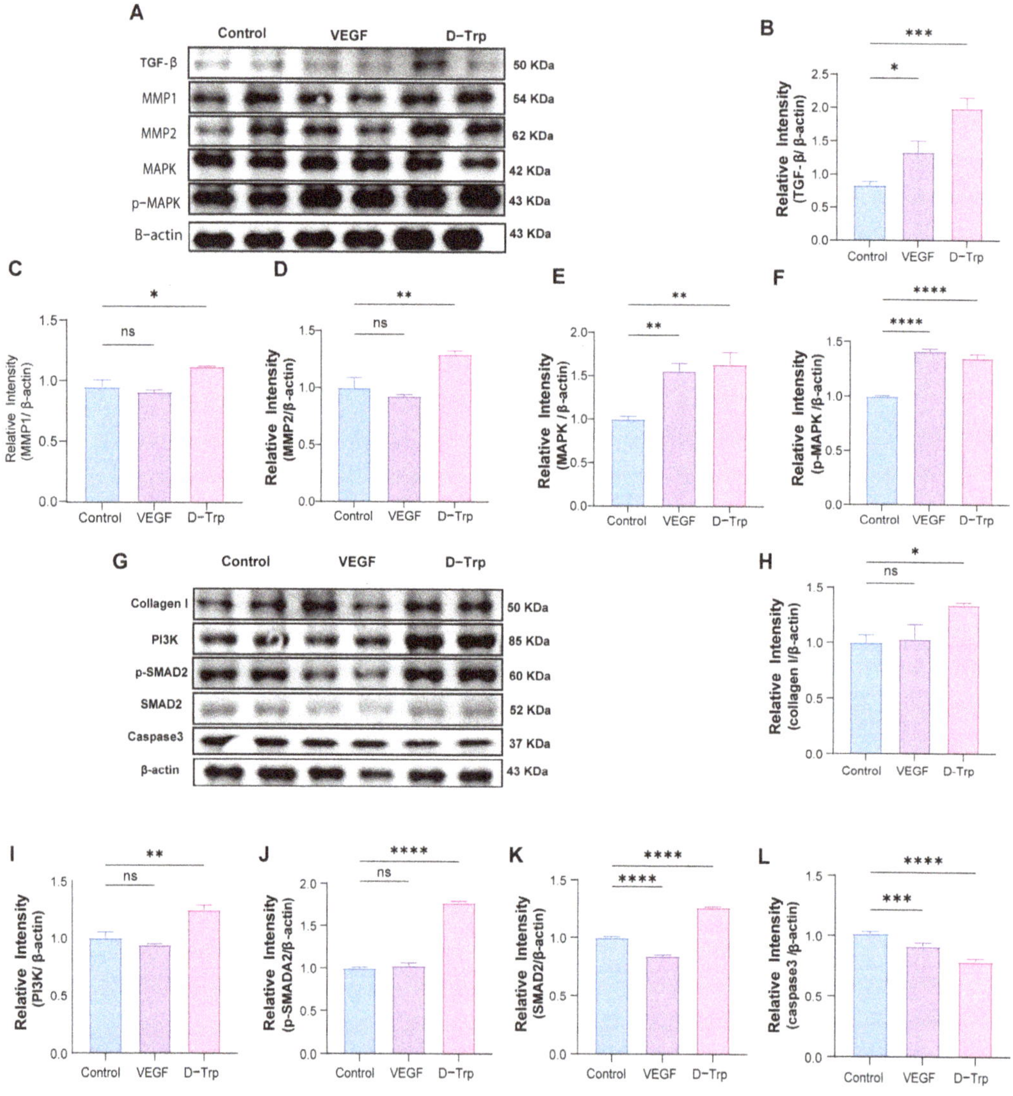

Figure 5. Western blotting analysis conducted in the diabetic wound model, 14 days post-surgery. (**A**) The intensity of Western blotting in skin wound samples. (**B–L**) A quantification of the results of Western blotting for various proteins, including TGF-β, MMP2, MAPK, phosphorylated MAPK (p-MAPK), JAK1, Collagen I, PI3K, phosphorylated SMAD2 (p-SMAD2), SMAD2, and Caspase-3, in skin samples from mice treated with PBS, VEGF, and D-Trp. The data are expressed as the mean ± standard error (SE) with a sample size of $n = 4$, and statistical significance is indicated as follows: ns, $p > 0.05$; * $p < 0.05$; ** $p < 0.01$; *** $p < 0.001$; **** $p < 0.0001$.

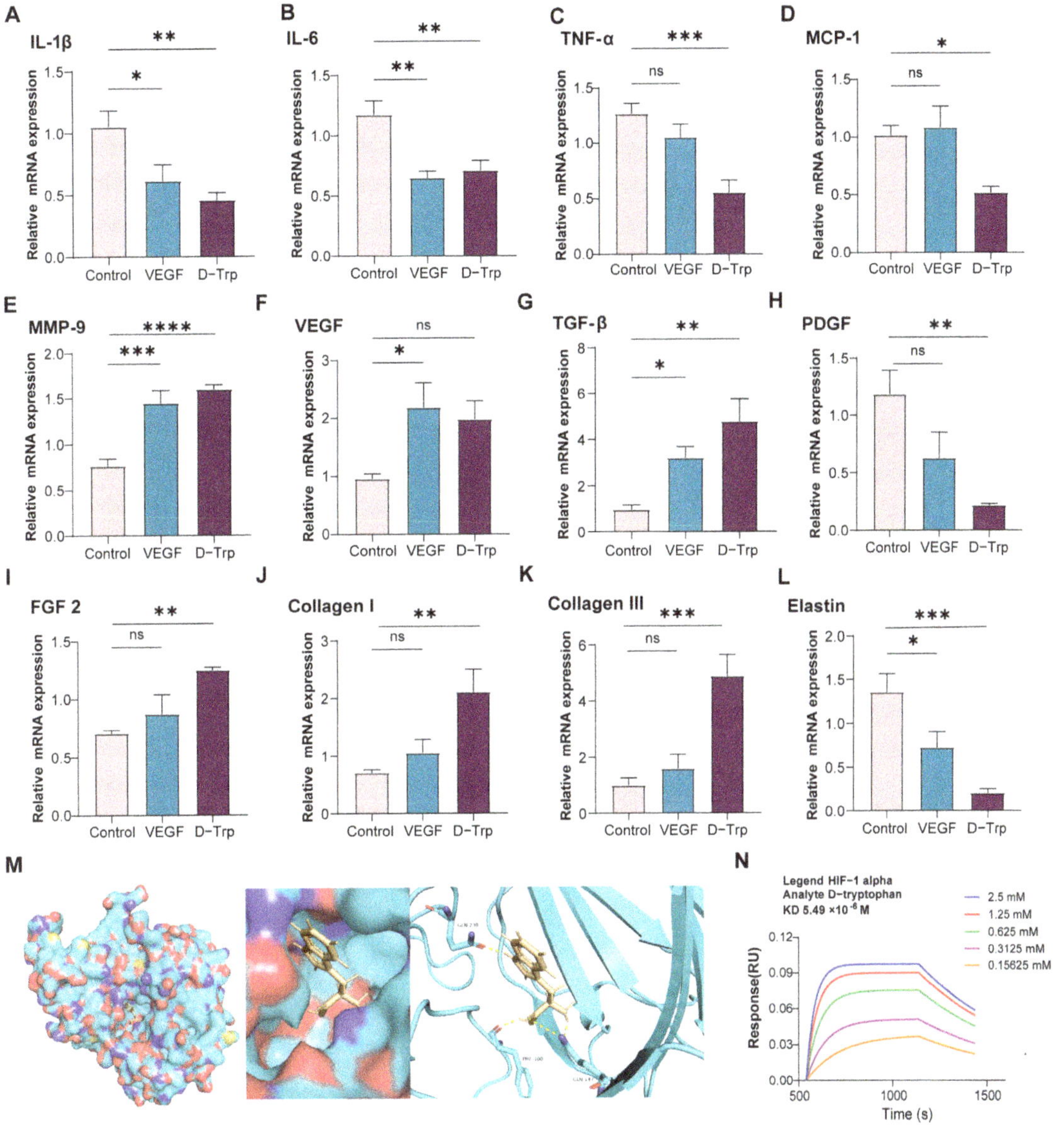

Figure 6. D-Trp stimulates the production of essential growth factors and ECM components, promoting fibroblast activity and ECM synthesis. (**A–L**) mRNA expression levels of *IL-1β*, *IL-6*, *TNF-α*, *MCP-1*, *MMP-9*, *VEGF*, *TGF-β*, *PDGF*, *FGF2*, *collagen I*, *collagen III* and *elastin*, respectively, in wound tissues were determined by RT-PCR. (**M**) Molecular docking analysis predicted a favorable binding interaction between D-Trp and hypoxia-inducible factor 1-alpha (HIF-1α). (**N**) The binding affinity between D-Trp and HIF-1α was assessed using the Octet® BLI system. The data are expressed as the mean ± standard error (SE) with a sample size of $n = 6$, and statistical significance is indicated as follows: ns, $p > 0.05$; * $p < 0.05$; ** $p < 0.01$; *** $p < 0.001$; **** $p < 0.0001$.

Furthermore, D-Trp administration induced a marked increase in *VEGF* levels (Figure 6F), suggesting a robust pro-angiogenic response. This upregulation of *VEGF* underscores the potential of D-Trp to enhance neovascularization, a vital process for restoring nutrient and oxygen supply to the wound microenvironment and supporting effective tissue repair [49,50]. The significant upregulation of *TGF-β* (Figure 6G), *PDGF* (Figure 6H), and *FGF2* (Figure 6I), along with increased expression of *collagen I* (Figure 6J), *collagen III* (Figure 6K), and *elastin* (Figure 6L), demonstrates that D-Trp enhances the production of key growth factors and extracellular matrix (ECM) components. This coordinated upregulation suggests D-Trp promotes fibroblast activation, stimulates ECM synthesis, and drives tissue remodeling, ultimately accelerating wound closure and improving regenerative outcomes in diabetic wounds [51,52].

To elucidate the mechanistic basis of D-Trp-mediated wound healing, we investigated its interaction with hypoxia-inducible factor 1-alpha (HIF-1α), a master regulator of cellular responses to hypoxia [53]. Molecular docking analysis revealed a strong binding affinity between D-Trp and HIF-1α (Figure 6M), a finding subsequently validated by Bio-Layer Interferometry (BLI) using the Octet® system (Figure 6N), which measured a binding affinity of 5.49 μM. Given HIF-1α's pivotal role in orchestrating angiogenesis, cell proliferation, and tissue regeneration under hypoxic conditions, a hall-mark of wound microenvironments, these results suggest that D-Trp exerts its therapeutic effects through direct modulation of HIF-1α activity [54,55]. This mechanistic framework is further supported by our observation of upregulated *VEGF* expression and other wound-healing genes, collectively indicating that D-Trp enhances tissue repair, at least partially, via HIF-1α-dependent pathways.

3. Discussion

This study establishes D-Trp as a promising therapeutic candidate for diabetic wound healing by addressing multiple pathological deficits characteristic of impaired healing. Diabetic wounds typically demonstrate compromised repair due to diminished fibroblast activity, insufficient collagen production, and chronic inflammation [10,56]. The enhanced wound healing observed in the D-Trp group suggested a multifaceted role for D-Trp in promoting tissue repair. The faster wound area shrinkage and thicker wound tissue observed in the D-Trp group pointed towards increased fibroblast activity and collagen deposition. This was further supported by the early development of a newborn epidermis and structures resembling hair follicles in the D-Trp group, suggesting that D-Trp may promote epithelial cell proliferation and differentiation, thereby contributing to re-epithelialization and the restoration of skin integrity. Although hair follicle regeneration on mature scars is generally limited, the presence of hair follicle-like structures in our model highlights the potential of D-Trp to stimulate aspects of skin repair beyond simple wound closure. Further studies are needed to confirm whether these structures can fully mature and restore hair growth in scar tissue.

The coordinated upregulation of MMP1 and MMP3 demonstrates D-Trp's capacity to enhance ECM proteolytic activity, facilitating the tissue remodeling essential for proper wound regeneration [57–60]. The concurrent activation of MAPK signaling is critical in modulating cellular processes, including proliferation, differentiation, and inflammatory responses, all of which are vital for tissue repair [44], suggesting that D-Trp may contribute to enhanced cell signaling and potentially improved tissue repair. The upregulation of SMAD2, a key transcription factor involved in TGF-β signaling, supports the notion that D-Trp may enhance TGF-β signaling, a crucial pathway for wound healing and ECM synthesis [61]. The downregulation of caspase-3, a critical executioner caspase in apoptosis, indicated that D-Trp may protect cells from programmed cell death, contributing to a more favorable healing environment [62].

The intricate relationship between gene expression and wound healing is pivotal in understanding the mechanisms that govern tissue repair [63]. The observed gene expression changes, combined with our binding affinity data, provide compelling evidence for the complicated influence of D-Trp on wound healing, highlighting its potential as a therapeutic agent. The downregulation of pro-inflammatory cytokines indicated a potent anti-inflammatory effect of D-Trp, mitigating the detrimental effects of excessive inflammation that can hinder tissue repair and increase infection susceptibility. Furthermore, the upregulation of MMP-9 and VEGF, coupled with the direct interaction between D-Trp and HIF-1α, suggested that D-Trp promoted efficient ECM remodeling, angiogenesis, and likely played a role in activating HIF-1α, enhancing its pro-angiogenic effects. The increased expression of key growth factors and ECM components further supported its ability to stimulate fibroblast activity, enhance ECM synthesis, and ultimately contribute to improved tissue regeneration and wound closure. These findings, particularly the observed effects in both diabetic and non-diabetic models, suggest that D-Trp might offer a promising therapeutic strategy for promoting wound healing, especially in individuals with impaired healing, such as those with diabetes, who often experience delayed healing and complications.

While VEGF is a well-characterized pro-angiogenic factor that directly stimulates vascular endothelial cell proliferation and new blood vessel formation, D-Trp also modulates epithelial cell proliferation and inflammatory responses, potentially enhancing multiple facets of tissue repair. Moreover, as a small molecule amino acid, D-Trp offers advantages in stability, ease of synthesis, and cost-effectiveness compared to protein-based therapeutics like VEGF, which require complex production and storage. These features suggest that D-Trp could serve as a more practical and accessible therapeutic agent for wound healing applications.

In conclusion, D-Trp influences protein expression related to wound healing by upregulating matrix metalloproteinases (MMPs), activating MAPK and JAK signaling pathways, and downregulating caspase-3. Additionally, it reduces inflammation, enhances extracellular matrix (ECM) turnover, stimulated angiogenesis via HIF-1α activation, and promoted ECM synthesis. Our study offers a promising strategy, especially in diabetic wound care, as well as enlarging the clinical approaches and documental research on amino acid biology in chronic diseases.

Limitation of This Study

One limitation of this study is the absence of pharmacokinetic or tissue distribution data to evaluate the local retention and potential systemic absorption of D-Trp following topical application. Although no overt signs of local irritation or adverse skin reactions were observed during treatment, a formal assessment of skin tolerance particularly in diabetic models was not performed. Additionally, only a single concentration of D-Trp was tested, selected for comparison with VEGF, and a dose–response evaluation was not conducted in order to minimize animal use. Moreover, the study focused on early to mid-stage wound healing and did not assess long-term outcomes such as scar formation, tissue remodeling, or mechanical integrity of the regenerated skin. Future studies should address these aspects to comprehensively evaluate the safety, efficacy, and clinical translational potential of D-Trp in wound healing.

4. Materials and Methods

4.1. In Vitro Wound Healing Assay

We carried out a wound healing assay as previously described [64]. Briefly, HaCaT human keratinocyte cells (catalog number M-C1056) were obtained from the Kunming Cell Bank, Kunming Institute of Zoology, Chinese Academy of Sciences, Kunming, Yunnan, China, and were seeded in 12-well plates at a density of 2×10^5 cells/well. Once the cells reached 80–90% confluency, the medium (DMEM, Corning, Corning, NY, USA; #10-092-CVRC) was replaced with fresh medium containing either no supplement (Control), VEGF (20 ng/mL, MedChemExpress LLC, Monmouth Junction, NJ, USA, HY-P70458A), or D-tryptophan (D-Trp, 20 ng/mL, MedChemExpress LLC, Monmouth Junction, NJ, USA; #HY-W012479). Two hours after treatment, a scratch wound was introduced by scraping the cell monolayer with the tip of a sterile pipette. Cell migration was assessed by measuring wound closure distances using ImageJ software (version 1.47v; National Institutes of Health, Bethesda, MD, USA). Images were captured at 6, 12, 24, and 36 h using an Olympus CX41 microscope (Olympus, Tokyo, Japan). All experiments were performed in triplicate.

4.2. Hemolysis Analysis

Whole blood was collected from healthy mice and immediately mixed with an anticoagulant solution. The anticoagulated blood was subsequently washed twice with sterile normal saline (0.9% NaCl) under aseptic conditions. Following centrifugation ($300\times g$, 10 min), the cell pellet was resuspended in the appropriate buffer to achieve a final concentration of 1×10^7 to 1×10^8 cells/mL. The diluted red blood cell suspension is mixed with a D-Trp sample dissolved in normal saline. The mixture was incubated at 37 °C for a duration of 30 min, subsequent to which it was subjected to centrifugation at 1000 rpm for 5 min. The resultant supernatant was then analyzed at a wavelength of 540 nm. The negative control was treated with normal saline, and the positive control was treated with Triton X-100, and the percentage of hemolysis was calculated as follows: percentage of hemolysis H% = (sample A − negative control)/100% × positive control.

4.3. Cytotoxicity Analysis

The cell counting kit 8 (CCK-8 MedChemExpress LLC, Monmouth Junction, NJ, USA; #HY-KO301) assay was used to determine cytotoxicity on the HaCaT cell line (Kunming cell bank). The cells were maintained in an appropriate growth medium until they reached 70% to 80% confluency within the tissue culture flask. Subsequently, the cells were detached using trypsin and enumerated with a hemocytometer. There were approximately 1×10^4 to 1×10^5 cells per well. After 24 h of incubation to allow cell adherence, the culture medium was replaced with serial dilutions of D-Trp prepared in growth medium, ranging from 0.195 to 100 μg/mL. After cell adhesion, D-Trp-treated serial dilutions in growth medium were prepared and the medium in the wells was replaced with these dilutions. Control wells containing untreated cells were set up for comparison with control wells containing known cytotoxic agents. The plates were incubated at 37 °C for 24 h, with 5% CO_2. Following the incubation period, the CCK-8 solution was introduced to each well according to the manufacturer's guidelines and further incubated for 1~4 h to promote the colorimetric reaction. Subsequently, the absorbance at the wavelength of 450 nm was determined utilizing a microplate reader. Cell viability is quantified as a percentage, employing the following formula: cell viability (%) = (A sample/A control) × 100%, where sample A represents the absorbance of treated cells and control A represents the absorbance of untreated control cells. The cytotoxic effect of D-Trp on HaCaT cells was evaluated by the analysis results.

4.4. Binding Affinity

We used the Octet® Bio-Layer Interferometry (BLI) system (Sartorius AG, Göttingen, Germany) to detect the binding affinity of D-Trp to HIF-1α protein, MCE #HY-P74888. We prepared the assay buffer and the reference buffer. Initially, D-Trp solutions were prepared at concentrations ranging from 2.5 mM to 0.15625 mM. Subsequently, recombinant HIF-1α protein (MedChemExpress, Monmouth Junction, NJ, USA) was biotinylated using a reagent from Thermo Fisher Scientific (Waltham, MA, USA) and diluted to a concentration of 10 μg/mL to facilitate its immobilization. The biosensors were then loaded with streptavidin (Sartorius AG, Göttingen, Germany). Subsequently, the Octet® BLI system was utilized to load streptavidin onto SA biosensors according to the manufacturer's protocol. The biosensors, loaded with streptavidin, were then allowed to incubate in the HIF-1α solution for a sufficient duration to facilitate receptor binding to streptavidin. Following binding, the biosensors were rinsed using assay buffer. Subsequently, the biosensors were placed in the reference buffer for a predetermined duration to establish a baseline signal. Thereafter, the biosensors were transferred to solutions containing varying concentrations of D-Trp to facilitate binding of the ligand to the immobilized HIF-1α for a period of 300 s. The biosensors were then returned to the reference buffer to observe the dissociation of D-Trp from the HIF-1α over a span of 150 s. The experiment incorporated appropriate controls, including biosensors containing only streptavidin (without the receptor) or biosensors with a receptor but without the ligand.

4.5. Molecular Docking

Molecular docking simulations were conducted using AutoDock Vina (version 1.1.2). The three-dimensional structure of human HIF-1α was retrieved from the Protein Data Bank (PDB ID: 4H6J), while the 3D structure of D-Trp was obtained from the PubChem database (CID: 9060). Docking was performed to predict the binding interactions between HIF-1α and D-Trp, and the corresponding binding energies were calculated by AutoDock Vina. Visualization and structural analysis of the docking results were carried out using PyMOL (version 2.5).

4.6. Ethical Considerations

This study adhered to regulatory compliance that involved ethical regulations and guidelines related to human and animal research. This included protecting the confidentiality of the subjects, obtaining informed consent, and ensuring ethical and humane treatment of the animals throughout the study.

4.7. Animal Subjects

The experimental animal protocols received approval from the Institutional Animal Care and Use Committee (IACUC) of the Kunming Institute of Zoology, Chinese Academy of Sciences, under the approval number IACUC-RE-2024-08-013. The handling and housing of the animals adhered to the guidelines established by the Animal Experimental Center of the Kunming Institute of Zoology. Mice were purchased at the age of 8 weeks from Charles River Laboratory Animal Technology Co., Ltd. (Wuhan, China), and were subsequently housed in groups under controlled environmental conditions. The mice were kept on a 12 h light–dark cycle in a temperature range of 22 °C to 24 °C. They received ad libitum access to standard laboratory food and water, except during specified periods for glucose level testing, during which their food access was restricted.

4.8. Induction of Diabetes

Experimental diabetes was established in mice via a single intraperitoneal injection of streptozotocin (STZ, YEASEN Biotechnology, Shanghai, China; #60256ES80) at 80 mg/kg body weight. Glycemic status was assessed hourly using a glucometer. Mice were classified as diabetic if their blood glucose concentrations consistently exceeded 11.1 mmol/L for at least 3 consecutive days.

4.9. Treatment Groups and Administrations

To evaluate the therapeutic potential of D-Trp in promoting wound healing, full-thickness dorsal wounds of 6 mm diameter were created in C57BL/6J mice. A total of 30 male mice aged 10 to 12 weeks were used, including 15 healthy controls and 15 with experimentally induced diabetes. The mice were randomly assigned into three groups of five animals each: control (topical PBS, Corning #21-040-CV), VEGF, or D-Trp treatment. Following a one-week acclimatization period, the mice were anesthetized by intraperitoneal injection of 2% pentobarbital (Cayman Chemical Company, Ann Arbor, MI, USA; #76-74-4) at a dose of 4 to 4.5 μL per 20 g of body weight. The fur on the dorsal region, specifically between the tail and back, was removed using Veet® Pure Hair Removal Cream (Reckitt Benckiser, Rueil-Malmaison, France) to prepare the surgical site under aseptic conditions. Full-thickness wounds were then generated using a 6 mm biopsy punch. After wounding, treatments were applied topically once daily using a pipette. The control group received 30 μL of PBS, while the VEGF and D-Trp groups each received 30 μL of their respective treatments at a concentration of 5 μg/mL. No occlusive dressing was applied to the wounds. Wound area measurements and tissue sample collections were conducted at different time points to account for the distinct healing rates in the two mouse models. For normal mice, wounds were assessed on days 1, 3, and 7 post-wounding, with tissue samples collected on day 7. For diabetic mice, wounds were evaluated on days 1, 7, and 14, with tissue harvested on day 14. This extended timeline for diabetic mice reflects their delayed wound healing compared to normal mice, allowing for a more accurate assessment of treatment effects. Throughout the study, wounds were monitored daily for signs of infection or other complications. Wound sites were photographed using an Olympus E-M10 digital camera, and wound boundaries were traced and measured using ImageJ software.

4.10. Wound Healing Assessment

The percentage of wound closure was calculated at each time point using the following formula: wound closure (%) = [(baseline wound area − current wound area)/baseline wound area] × 100.

4.11. Sample Collection for Histology, Western Blot, and RNA Analysis

At the designated endpoints, skin tissue surrounding the wound site was collected from five mice per group. Approximately 5–10 mm^2 of full-thickness skin was harvested from each mouse. The collected tissue was divided into three portions for downstream analysis. One portion was fixed in 4% paraformaldehyde and processed for histological evaluation. A second portion was snap-frozen in liquid nitrogen and stored at −80 °C for subsequent protein extraction and Western blot analysis. The remaining portion was immediately processed for total RNA extraction and gene expression analysis.

4.12. Histological Analysis

At the experimental endpoint, the mice were euthanized and wound tissues collected for histological examination. Standard tissue processing included fixation in 4% neutral-buffered formalin, paraffin embedding, and sectioning at 5 μm thickness. Hematoxylin and eosin (H&E) staining was performed to assess general tissue morphology.

4.13. Western Blot

Western blot analysis was used to assess protein expression in wound tissues. Proteins were extracted from the wound sites, separated on 12% SDS-PAGE gels, and transferred onto polyvinylidene fluoride (PVDF) membranes. Membranes were blocked with 5% bovine serum albumin (BSA) for 1 h at room temperature, followed by overnight incubation at 4 °C with primary antibodies against collagen I abcam #ab6586, TGF-β Zen-Bio.cn #R22797, MMP1 abcam #ab51074, MMP2 abcam #ab92536, MMP3 abcam #ab52915, MAPK Zen-Bio.cn #R10024, phosphorylated MAPK (P-MAPK) Zen-Bio.cn #310289, JAK2 Zen-Bio.cn #R24775, PI3K abcam #ab283852, phosphorylated SMAD2 (P-SMAD2) abcam #ab2808888, SMAD2 abcam #ab63176, and Caspase-3 abcam (all at 1:1000 dilution), while GAPDH affinity #AF7021 and β-actin Zen-Bio.cn #250132, 1:10,000. After washing, membranes were incubated with HRP-conjugated secondary antibodies for 1 h at room temperature. Immunoreactive bands were detected using the ImageQuant LAS 4000 mini-imaging system (GE Healthcare, Little Chalfont, Buckinghamshire, UK).

4.14. RNA Extraction and Quantitative Real-Time PCR Analysis

Total RNA was extracted from wound tissues using the TRIzol™ Reagent (Invitrogen, Thermo Fisher Scientific, Waltham, MA, USA; Cat# 15596018) according to the manufacturer's protocol. To quantify the expression levels of genes such as TNF-α, IL-1β, IL-6, MCP-1, MMP-9, VEGF, TGF-β, PDGF, FGF 2, collagen I, collagen III, and elastin, quantitative real-time PCR (qPCR) was performed. A total of 500 ng of RNA was reverse transcribed into cDNA using the 5× All-in-One MasterMix Kit (ABM, Applied Biological Materials Inc., Richmond, BC, Canada; Cat# G492), following the manufacturer's instructions. The qPCR reactions were carried out in 20 μL volume using the BlasTaq 2× qPCR MasterMix (Abm, Canada; q#G892) in conjunction with the Step One Plus Real-Time PCR system (Thermo, Waltham, MA, USA). The expression levels of the target genes were normalized to those of GAPDH mRNA, which served as an internal control. The primer sequences used for qPCR are provided in Table 1.

Table 1. Primer sequences devised for RT-PCR analysis.

Gene	Forward (5′ → 3′)	Reverse (5′ → 3′)
GAPDH	GAAGGTCGGTGTGAACGGAT	AATCTCCACTTTGCCACTGC
TNF-α	TCTTCTCATTCCTGCTTGTGG	GGTCTGGGCCATAGAACTGA
IL-1β	GCAACTGTTCCTGAACTCAACT	ATCTTTTGGGGTCCGTCAACT
IL-6	TAGTCCTTCCTACCCCAATTTCC	TTGGTCCTTAGCCACTCCTTC
MCP-1	TTAAAAACCTGGATCGGAACCAA	GCATTAGCTTCAGATTTACGGGT
MMP-9	CGTCGTGATCCCCACTTACT	AACACACAGGGTTTGCCTTC
VEGF	GCACATAGAGAGAATGAGCTTCC	CTCCGCTCTGAACAAGGCT
TGF-β	CTCCCGTGGCTTCTAGTGC	GCCTTAGTTTGGACAGGATCTG
PDGF	CGGCCTGTGACTAGAAGAGG	GGGTCACTTCACACTTGCAT
FGF 2	CACCCTCACATCAAGCTACAACTTCA	TCAGCTCTTAGCAGACATTGGAAGA
Collagen I	GCTCCTCTTAGGGGCCACT	CCACGTCTCACCATTGGGG
Collagen III	CTGTAACATGGAAACTGGGGAAA	CCATAGCTGAACTGAAAACCACC
Elastin	TGCCTGGAGACATTTCCCTAG	GGTGCTCCAACATTTCCCAT

4.15. Statistical Analysis

Data were analyzed using GraphPad Prism version 10.1.2. All results are expressed as mean ± standard deviation (SD). Sample sizes varied between assays due to differences in tissue availability and specific methodological requirements; however, all sample sizes were sufficient to ensure robust statistical analysis. For comparisons between two groups, an unpaired two-tailed Student's *t*-test was used. For comparisons involving more than two groups, one-way analysis of variance (ANOVA) followed by Tukey's post hoc test was performed. A *p*-value of <0.05 was considered statistically significant.

Author Contributions: Conceptualization, R.L.; investigation and methodology, D.A.T., J.M., Y.W., K.C., B.B.M., Z.W., M.K. and M.Y.; validation and formal analysis, D.A.T.; writing—original draft preparation, D.A.T.; writing—review and editing, Q.L.; supervision, R.L.; funding acquisition, R.L. All authors have read and agreed to the published version of the manuscript.

Funding: This work was supported by Key-Area Research and Development Program of Guangdong Province (2023B1111050008); the Ministry of Science and Technology of China (2022YFC2602500 and 2023YFF1304900); Guangxi Key Research and Development Program (AB24010109), the National Science Foundation of China (32400769 and 32370538); Chinese Academy of Sciences (SAJC202402 and KFJ-BRP-008), Yunnan Province (202502AA310005, 202305AH340007 and 202403AC100010); Kunming Science and Technology Bureau (2022SCP007) and Shenzhen New Cornerstone Science Foundation (NCI202238); the Yunnan Characteristic Plant Extraction Laboratory (2025YKZY002); and the University Chinese Academy of Sciences and Alliance of International Science Organization (ANSO) through 2022A8010415002.

Institutional Review Board Statement: The experimental animal protocols received approval from the Institutional Animal Care and Use Committee (IACUC) of the Kunming Institute of Zoology, Chinese Academy of Sciences, under approval number IACUC-RE-2024-08-013, with an approved date of 2024-08-30. The handling and housing of the animals adhered to the guidelines established by the Animal Experimental Center of the Kunming Institute of Zoology.

Informed Consent Statement: Not applicable.

Data Availability Statement: The data that support the findings of this study are available from the corresponding author upon reasonable request.

Acknowledgments: The authors would like to acknowledge Shengwen Zhou for his contribution to the molecular docking analysis.

Conflicts of Interest: The authors declare no conflicts of interest.

Abbreviations

The following abbreviations are used in this manuscript:

DFUs	diabetic foot ulcers
PBS	phosphate-buffered Saline
VEGF	vascular endothelial growth factor
SMAD2	mothers against decapentaplegic homolog 2
MMP	matrix metalloproteinases
ECM	extracellular matrix
MAPK	mitogen-activated protein kinase
HIF-1α	hypoxia-inducible factor 1-alpha
RT-PCR	reverse transcription-polymerase chain reaction
GAPDH	glyceraldehyde 3-phosphate dehydrogenase
TNF-α	tumor necrosis factor-alpha
IL-1β	interleukin-1 beta
IL-6	interleukin-6
MCP-1	monocyte chemoattractant protein-1
TGF-β	transforming growth factor-beta
PDGF	platelet-derived growth factor
FGF 2	fibroblast growth factor

References

1. Sorg, H.; Tilkorn, D.J.; Hager, S.; Hauser, J.; Mirastschijski, U. Skin Wound Healing: An Update on the Current Knowledge and Concepts. *Eur. Surg. Res.* **2017**, *58*, 81–94. [CrossRef]
2. Fallah, N.; Rasouli, M.; Amini, M.R. The current and advanced therapeutic modalities for wound healing management. *J. Diabetes Metab. Disord.* **2021**, *20*, 1883–1899. [CrossRef] [PubMed]
3. Landen, N.X.; Li, D.; Stahle, M. Transition from inflammation to proliferation: A critical step during wound healing. *Cell Mol. Life Sci.* **2016**, *73*, 3861–3885. [CrossRef] [PubMed]

4. Zhang, Y.; Wang, S.; Yang, Y.; Zhao, S.; You, J.; Wang, J.; Cai, J.; Wang, H.; Wang, J.; Zhang, W.; et al. Scarless wound healing programmed by core-shell microneedles. *Nat. Commun.* **2023**, *14*, 3431. [CrossRef] [PubMed]
5. Dangwal, S.; Stratmann, B.; Bang, C.; Lorenzen, J.M.; Kumarswamy, R.; Fiedler, J.; Falk, C.S.; Scholz, C.J.; Thum, T.; Tschoepe, D. Impairment of Wound Healing in Patients with Type 2 Diabetes Mellitus Influences Circulating MicroRNA Patterns via Inflammatory Cytokines. *Arter. Thromb. Vasc. Biol.* **2015**, *35*, 1480–1488. [CrossRef]
6. Spampinato, S.F.; Caruso, G.I.; De Pasquale, R.; Sortino, M.A.; Merlo, S. The Treatment of Impaired Wound Healing in Diabetes: Looking Among Old Drugs. *Pharmaceuticals* **2020**, *13*, 60. [CrossRef]
7. Cai, Y.; Chen, K.; Liu, C.; Qu, X. Harnessing strategies for enhancing diabetic wound healing from the perspective of spatial inflammation patterns. *Bioact. Mater.* **2023**, *28*, 243–254. [CrossRef]
8. Dasari, N.; Jiang, A.; Skochdopole, A.; Chung, J.; Reece, E.M.; Vorstenbosch, J.; Winocour, S. Updates in Diabetic Wound Healing, Inflammation, and Scarring. *Semin. Plast. Surg.* **2021**, *35*, 153–158. [CrossRef]
9. Avishai, E.; Yeghiazaryan, K.; Golubnitschaja, O. Impaired wound healing: Facts and hypotheses for multi-professional considerations in predictive, preventive and personalised medicine. *EPMA J.* **2017**, *8*, 23–33. [CrossRef]
10. Patel, S.; Srivastava, S.; Singh, M.R.; Singh, D. Mechanistic insight into diabetic wounds: Pathogenesis, molecular targets and treatment strategies to pace wound healing. *Biomed. Pharmacother.* **2019**, *112*, 108615. [CrossRef]
11. Samarawickrama, P.N.; Zhang, G.; Zhu, E.; Dong, X.; Nisar, A.; Zhu, H.; Ma, Y.; Zhou, Z.; Yang, H.; Gui, L.; et al. Clearance of senescent cells enhances skin wound healing in type 2 diabetic mice. *Theranostics* **2024**, *14*, 5429–5442. [CrossRef] [PubMed]
12. Zhang, R.; Mao, D.; Fu, Y.; Ju, R.; Wei, G. A self-assembled and H_2O_2-activatable hybrid nanoprodrug for lung infection and wound healing therapy. *Theranostics* **2025**, *15*, 5953–5968. [CrossRef] [PubMed]
13. Ridiandries, A.; Tan, J.T.M.; Bursill, C.A. The Role of Chemokines in Wound Healing. *Int. J. Mol. Sci.* **2018**, *19*, 3217. [CrossRef] [PubMed]
14. Deng, Y.; Zheng, J.; Li, J.; Liu, B.; Chen, K.; Xu, Y.; Deng, L.; Liu, H.; Liu, Y.N. NIR light-driven nanomotor with cascade photodynamic therapy for MRSA biofilm eradication and diabetic wound healing. *Theranostics* **2025**, *15*, 3474–3489. [CrossRef]
15. Farabi, B.; Roster, K.; Hirani, R.; Tepper, K.; Atak, M.F.; Safai, B. The Efficacy of Stem Cells in Wound Healing: A Systematic Review. *Int. J. Mol. Sci.* **2024**, *25*, 3006. [CrossRef]
16. Freedman, B.R.; Hwang, C.; Talbot, S.; Hibler, B.; Matoori, S.; Mooney, D.J. Breakthrough treatments for accelerated wound healing. *Sci. Adv.* **2023**, *9*, eade7007. [CrossRef]
17. Challapalli, R.S.; Dwyer, R.M.; McInerney, N.; Kerin, M.J.; Lowery, A.J. Effect of Breast Cancer and Adjuvant Therapy on Adipose-Derived Stromal Cells: Implications for the Role of ADSCs in Regenerative Strategies for Breast Reconstruction. *Stem Cell Rev. Rep.* **2021**, *17*, 523–538. [CrossRef]
18. Vogel, F.; Luth, A.; Charpentier, A. The impact of microbiological diagnostics on the antimicrobial treatment of hospitalised patients with infectious disease. *Eur. J. Med. Res.* **1996**, *1*, 312–314.
19. Kolimi, P.; Narala, S.; Nyavanandi, D.; Youssef, A.A.A.; Dudhipala, N. Innovative Treatment Strategies to Accelerate Wound Healing: Trajectory and Recent Advancements. *Cells* **2022**, *11*, 2439. [CrossRef]
20. Nourian Dehkordi, A.; Mirahmadi Babaheydari, F.; Chehelgerdi, M.; Raeisi Dehkordi, S. Skin tissue engineering: Wound healing based on stem-cell-based therapeutic strategies. *Stem Cell Res. Ther.* **2019**, *10*, 111. [CrossRef]
21. Wang, C.; Shirzaei Sani, E.; Shih, C.D.; Lim, C.T.; Wang, J.; Armstrong, D.G.; Gao, W. Wound management materials and technologies from bench to bedside and beyond. *Nat. Rev. Mater.* **2024**, *9*, 550–566. [CrossRef]
22. Liu, C.; Cheng, C.; Cheng, K.; Gao, A.S.; Li, Q.; Atala, A.; Zhang, Y. Precision exosome engineering for enhanced wound healing and scar revision. *J. Transl. Med.* **2025**, *23*, 578. [CrossRef]
23. Fani, N.; Moradi, M.; Zavari, R.; Parvizpour, F.; Soltani, A.; Arabpour, Z.; Jafarian, A. Current Advances in Wound Healing and Regenerative Medicine. *Curr. Stem Cell Res. Ther.* **2024**, *19*, 277–291. [CrossRef]
24. Sharma, Y.; Ghatak, S.; Sen, C.K.; Mohanty, S. Emerging technologies in regenerative medicine: The future of wound care and therapy. *J. Mol. Med.* **2024**, *102*, 1425–1450. [CrossRef]
25. Zheng, R.; Geng, T.; Wu, D.Y.; Zhang, T.; He, H.N.; Du, H.N.; Zhang, D.; Miao, Y.L.; Jiang, W. Derivation of feeder-free human extended pluripotent stem cells. *Stem Cell Rep.* **2021**, *16*, 1686–1696. [CrossRef] [PubMed]
26. Pollegioni, L.; Kustrimovic, N.; Piubelli, L.; Rosini, E.; Rabattoni, V.; Sacchi, S. d-amino acids: New functional insights. *FEBS J.* **2025**. *Epub ahead of print*. [CrossRef] [PubMed]
27. Seki, N.; Kimizuka, T.; Gondo, M.; Yamaguchi, G.; Sugiura, Y.; Akiyama, M.; Yakabe, K.; Uchiyama, J.; Higashi, S.; Haneda, T.; et al. (D)-Tryptophan suppresses enteric pathogen and pathobionts and prevents colitis by modulating microbial tryptophan metabolism. *iScience* **2022**, *25*, 104838. [CrossRef]
28. Moghimani, M.; Noori, S.M.A.; Afshari, A.; Hashemi, M. D-tryptophan, an eco-friendly natural, safe, and healthy compound with antimicrobial activity against food-borne pathogens: A systematic review. *Food Sci. Nutr.* **2024**, *12*, 3068–3079. [CrossRef]

29. Elafify, M.; Darwish, W.S.; Al-Ashmawy, M.; Elsherbini, M.; Koseki, S.; Kawamura, S.; Abdelkhalek, A. Prevalence of Salmonella spp. in Egyptian dairy products: Molecular, antimicrobial profiles and a reduction trial using d-tryptophan. *J. Consum. Prot. Food Saf.* **2019**, *14*, 399–407. [CrossRef]
30. Elafify, M.; Sadoma, N.M.; Abd El Aal, S.F.A.; Bayoumi, M.A.; Ahmed Ismail, T. Occurrence and D-Tryptophan Application for Controlling the Growth of Multidrug-Resistant Non-O157 Shiga Toxin-Producing Escherichia coli in Dairy Products. *Animals* **2022**, *12*, 922. [CrossRef]
31. Gumede, D.B.; Abrahamse, H.; Houreld, N.N. Targeting Wnt/β-catenin signaling and its interplay with TGF-β and Notch signaling pathways for the treatment of chronic wounds. *Cell Commun. Signal.* **2024**, *22*, 244. [CrossRef]
32. Glady, A.; Vandebroek, A.; Yasui, M. Human keratinocyte-derived extracellular vesicles activate the MAPKinase pathway and promote cell migration and proliferation in vitro. *Inflamm. Regen.* **2021**, *41*, 4. [CrossRef]
33. Peña, O.A.; Martin, P. Cellular and molecular mechanisms of skin wound healing. *Nat. Rev. Mol. Cell Biol.* **2024**, *25*, 599–616. [CrossRef]
34. Wilkinson, H.N.; Hardman, M.J. Wound healing: Cellular mechanisms and pathological outcomes. *Open Biol.* **2020**, *10*, 200223. [CrossRef]
35. Sorgdrager, F.J.H.; Naude, P.J.W.; Kema, I.P.; Nollen, E.A.; Deyn, P.P. Tryptophan Metabolism in Inflammaging: From Biomarker to Therapeutic Target. *Front. Immunol.* **2019**, *10*, 2565. [CrossRef]
36. Duran-Padilla, M.; Serrano-Loyola, R.; Perez-Garcia, A.; Carrillo-Betancourt, R.; Campos-García Rojas, C.; Reyes-Alberto, M.; Franco, W.; Hernandez-Ruiz, J.; Gutierrez-Herrera, E. Tryptophan intrinsic fluorescence from wound healing correlates with re-epithelialization in a rabbit model. *Skin. Res. Technol.* **2024**, *30*, e13834. [CrossRef]
37. Barouti, N.; Mainetti, C.; Fontao, L.; Sorg, O. L-Tryptophan as a Novel Potential Pharmacological Treatment for Wound Healing via Aryl Hydrocarbon Receptor Activation. *Dermatology* **2015**, *230*, 332–339. [CrossRef]
38. Krishnaswamy, V.R.; Benbenishty, A.; Blinder, P.; Sagi, I. Demystifying the extracellular matrix and its proteolytic remodeling in the brain: Structural and functional insights. *Cell. Mol. Life Sci.* **2019**, *76*, 3229–3248. [CrossRef]
39. Sarapultsev, A.; Gusev, E.; Komelkova, M.; Utepova, I.; Luo, S.; Hu, D. JAK-STAT signaling in inflammation and stress-related diseases: Implications for therapeutic interventions. *Mol. Biomed.* **2023**, *4*, 40. [CrossRef]
40. Das, A.; Ganesh, K.; Khanna, S.; Sen, C.K.; Roy, S. Engulfment of Apoptotic Cells by Macrophages: A Role of MicroRNA-21 in the Resolution of Wound Inflammation. *J. Immunol.* **2014**, *192*, 1120–1129. [CrossRef]
41. Poormasjedi-Meibod, M.-S.; Hartwell, R.; Taghi Kilani, R.; Ghahary, A. Anti-Scarring Properties of Different Tryptophan Derivatives. *PLoS ONE* **2014**, *9*, e91955. [CrossRef]
42. Arribas-Lopez, E.; Zand, N.; Ojo, O.; Snowden, M.J.; Kochhar, T. The Effect of Amino Acids on Wound Healing: A Systematic Review and Meta-Analysis on Arginine and Glutamine. *Nutrients* **2021**, *13*, 2498. [CrossRef]
43. Md Fadilah, N.I.; Shahabudin, N.A.; Mohd Razif, R.A.; Sanyal, A.; Ghosh, A.; Baharin, K.I.; Ahmad, H.; Maarof, M.; Motta, A.; Fauzi, M.B. Discovery of bioactive peptides as therapeutic agents for skin wound repair. *J. Tissue Eng.* **2024**, *15*, 20417314241280359. [CrossRef] [PubMed]
44. Zhang, W.; Liu, H.T. MAPK signal pathways in the regulation of cell proliferation in mammalian cells. *Cell Res.* **2002**, *12*, 9–18. [CrossRef] [PubMed]
45. Liu, L.; Liu, B.; Li, L.; Zhou, X.; Li, Q. Effects of Resveratrol on Pulmonary Fibrosis via TGF-β/Smad/ERK Signaling Pathway. *Am. J. Chin. Med.* **2023**, *51*, 651–676. [CrossRef]
46. Tothova, Z.; Semelakova, M.; Solarova, Z.; Tomc, J.; Debeljak, N.; Solar, P. The Role of PI3K/AKT and MAPK Signaling Pathways in Erythropoietin Signalization. *Int. J. Mol. Sci.* **2021**, *22*, 7682. [CrossRef]
47. Zheng, D.; Liwinski, T.; Elinav, E. Inflammasome activation and regulation: Toward a better understanding of complex mechanisms. *Cell Discov.* **2020**, *6*, 36. [CrossRef]
48. Winkler, J.; Abisoye-Ogunniyan, A.; Metcalf, K.J.; Werb, Z. Concepts of extracellular matrix remodelling in tumour progression and metastasis. *Nat. Commun.* **2020**, *11*, 5120. [CrossRef]
49. Pérez-Gutiérrez, L.; Ferrara, N. Biology and therapeutic targeting of vascular endothelial growth factor A. *Nat. Rev. Mol. Cell Biol.* **2023**, *24*, 816–834. [CrossRef]
50. Zhou, P.; Zhang, S.; Li, L.; Zhang, R.; Guo, G.; Zhang, Y.; Wang, R.; Liu, M.; Wang, Z.; Zhao, H.; et al. Targeted degradation of VEGF with bispecific aptamer-based LYTACs ameliorates pathological retinal angiogenesis. *Theranostics* **2024**, *14*, 4983–5000. [CrossRef]
51. Diller, R.B.; Tabor, A.J. The Role of the Extracellular Matrix (ECM) in Wound Healing: A Review. *Biomimetics* **2022**, *7*, 87. [CrossRef]
52. Kusindarta, D.L.; Wihadmadyatami, H. The Role of Extracellular Matrix in Tissue Regeneration. In *Tissue Regeneration*; Essayed Kaoud, H.A., Ed.; IntechOpen: Rijeka, Croatia, 2018.
53. Xu, M.; Pang, Q.; Xu, S.; Ye, C.; Lei, R.; Shen, Y.; Xu, J. Hypoxia-inducible factor-1α activates transforming growth factor-β1/Smad signaling and increases collagen deposition in dermal fibroblasts. *Oncotarget* **2017**, *9*, 3188–3197.

54. Shi, Y.H.; Fang, W.G. Hypoxia-inducible factor-1 in tumour angiogenesis. *World J. Gastroenterol.* **2004**, *10*, 1082–1087. [CrossRef] [PubMed]
55. Pitsouli, C.; Tamamouna, V. The Hypoxia-Inducible Factor-1α in Angiogenesis and Cancer: Insights from the Drosophila Model. In *Gene Expression and Regulation in Mammalian Cells-Transcription Toward the Establishment of Novel Therapeutics*; Uchiumi, F., Ed.; IntechOpen: Rijeka, Croatia, 2018.
56. Ding, X.; Yang, C.; Li, Y.; He, T.; Xu, Y.; Cheng, X.; Song, J.; Xue, N.; Min, W.; Feng, W.; et al. Reshaped commensal wound microbiome via topical application of Calvatia gigantea extract contributes to faster diabetic wound healing. *Burns Trauma* **2024**, *12*, tkae037. [CrossRef] [PubMed]
57. Chen, K.; Xu, M.; Lu, F.; He, Y. Development of Matrix Metalloproteinases-Mediated Extracellular Matrix Remodeling in Regenerative Medicine: A Mini Review. *Tissue Eng. Regen. Med.* **2023**, *20*, 661–670. [CrossRef]
58. Fernandez-Guarino, M.; Bacci, S.; Perez Gonzalez, L.A.; Bermejo-Martinez, M.; Cecilia-Matilla, A.; Hernandez-Bule, M.L. The Role of Physical Therapies in Wound Healing and Assisted Scarring. *Int. J. Mol. Sci.* **2023**, *24*, 7487. [CrossRef]
59. Laronha, H.; Caldeira, J. Structure and Function of Human Matrix Metalloproteinases. *Cells* **2020**, *9*, 1076. [CrossRef]
60. Levi, N.; Papismadov, N.; Solomonov, I.; Sagi, I.; Krizhanovsky, V. The ECM path of senescence in aging: Components and modifiers. *FEBS J.* **2020**, *287*, 2636–2646. [CrossRef]
61. Tzavlaki, K.; Moustakas, A. TGF-beta Signaling. *Biomolecules* **2020**, *10*, 487. [CrossRef]
62. Nichani, K.; Li, J.; Suzuki, M.; Houston, J.P. Evaluation of Caspase-3 Activity During Apoptosis with Fluorescence Lifetime-Based Cytometry Measurements and Phasor Analyses. *Cytom. A* **2020**, *97*, 1265–1275. [CrossRef]
63. Yu, X.; Zhou, J.; Ye, W.; Xu, J.; Li, R.; Huang, L.; Chai, Y.; Wen, M.; Xu, S.; Zhou, Y. Time-course swRNA-seq uncovers a hierarchical gene regulatory network in controlling the response-repair-remodeling after wounding. *Commun. Biol.* **2024**, *7*, 694. [CrossRef]
64. Pinto, B.I.; Cruz, N.D.; Lujan, O.R.; Propper, C.R.; Kellar, R.S. In Vitro Scratch Assay to Demonstrate Effects of Arsenic on Skin Cell Migration. *J. Vis. Exp.* **2019**, e58838. [CrossRef]

International Journal of
Molecular Sciences

MDPI

Article

Identification of a Novel Antagonist of BRS-3 from Natural Products and Its Protective Effects Against H_2O_2-Induced Cardiomyocyte Injury

Jihong Lu [1,†], Lehao Wu [1,†], Jianzheng Zhu [2], Han Zhou [3], Mingzhu Fang [1], Hongshuo Liang [1], Miao Guo [2], Mo Chen [1], Yuhang Zhu [4], Jixia Wang [3], Hua Xiao [2] and Yan Zhang [1,*]

1 Shanghai Frontiers Science Center of Drug Target Identification and Delivery, School of Pharmacy, Shanghai Jiao Tong University, Shanghai 200240, China; lujihong@sjtu.edu.cn (J.L.); wulehaogo@sjtu.edu.cn (L.W.); cris08@sjtu.edu.cn (M.F.); dlutlhs@163.com (H.L.); chen-mo@sjtu.edu.cn (M.C.)

2 State Key Laboratory of Microbial Metabolism, Joint International Research Laboratory Metabolic & Developmental Sciences, School of Life Sciences and Biotechnology, Shanghai Jiao Tong University, Shanghai 200240, China; zhujzh9@sjtu.edu.cn (J.Z.); guoniao@sjtu.edu.cn (M.G.); huaxiao@sjtu.edu.cn (H.X.)

3 Key Laboratory of Phytochemistry and Natural Medicines, Dalian Institute of Chemical Physics, Chinese Academy of Sciences, Dalian 116023, China; zhouhan418@dicp.ac.cn (H.Z.); jxwang@dicp.ac.cn (J.W.)

4 Zhiyuan College, Shanghai Jiao Tong University, Shanghai 200240, China; zhuyhang18119625017@sjtu.edu.cn

* Correspondence: zhangyan_sjtu@sjtu.edu.cn

† These authors contributed equally to this work.

Abstract: The identification of exogenous ligands from natural products is an alternative strategy to explore the unrevealed physiological functions of orphan G-protein-coupled receptors (GPCRs). In this study, we have successfully identified and pharmacologically characterized licoisoflavone A (LIA) as a novel selective antagonist of BRS-3, an orphan GPCR. Functional studies showed that pretreatment with LIA ameliorated hydrogen peroxide (H_2O_2)-induced cardiomyocyte injury. Furthermore, LIA pretreatment significantly restored the activities of malondialdehyde (MDA), superoxide dismutase (SOD), and catalase (CAT), as well as lactate dehydrogenase (LDH) levels, in H9c2 cells following H_2O_2 exposure. The protective effect of LIA was also evident in primary cardiomyocytes from rats and mice against H_2O_2-induced cell injury but was absent in primary cardiomyocytes derived from bombesin receptor subtype-3 knockout ($Brs3^{-/y}$) mice, strongly confirming the mechanism of LIA's action through BRS-3 antagonism. Proteomics studies further revealed that LIA exerted its protective effects via activating the integrin/ILK/AKT and ERK/MAPK signaling pathways. Complementary findings from Bantag-1, a well-recognized antagonist of BRS-3, in human embryonic kidney 293 mBRS-3 (HEK293-mBRS-3) stable cells and B16 cell lines, which demonstrated resistance to H_2O_2-induced damage, further supported the pivotal role of BRS-3 in oxidative stress-induced cell injury. Our study contributes to expanding our understanding of the potential pharmacological functions of BRS-3, unveiling previously unknown pharmacological functionality of this orphan receptor.

Keywords: BRS-3; orphan GPCR; licoisoflavone A; antagonist; cardiomyocyte

1. Introduction

G protein-coupled receptors (GPCRs), known as seven transmembrane receptors, are the therapeutic targets of approximately 40% of modern medicinal drugs [1]. Of all the validated GPCRs, 25% have no known endogenous ligands and are, thus, classified as orphan receptors [2]. The orphan receptor bombesin receptor subtype-3 (BRS-3, BB_3) was identified based on its high degree of homology to neuromedin B receptor (NMBR, BB_1)

and the gastrin-releasing peptide receptor (GRPR, BB_2), both of which are recognized as mammalian bombesin receptors [3].

BRS-3 couples with a guanine nucleotide-binding protein (G_q), decomposing the 4,5-diphosphatidylinositol biphosphate (PIP2) into inositol triphosphate (IP3) and diacylglycerol. This process leads to an elevation of cytosolic calcium (Ca^{2+}) and the subsequent activation of protein kinase C (PKC) [4]. Given its implications for energy homeostasis, feeding behavior, and insulin regulation [5], BRS-3 continues to attract interest in both academia and industry. With its broad expression in both central and peripheral tissues, the role of BRS-3 in various human diseases is gradually being elucidated [6]. However, the exploration of BRS-3's physiological and pathological roles is hindered not only by the absence of endogenous ligands, but also by the scarcity of specific pharmacological tools. Currently, there are only a few exogenous ligands available, notably the agonist MK-5046 and the antagonist Bantag-1 [7], with antagonists being particularly scarce and in limited supply.

As part of our ongoing project to identify new ligands capable of interacting with BRS-3, this study has successfully unveiled a novel antagonist, licoisoflavone A (LIA), derived from natural products. LIA is a natural flavonoid and the main component of Glycyrrhiza [8]. LIA is also an effective ingredient in numerous traditional Chinese medicine prescriptions, notably the Tongmaiyangxin pill [9] and Baoyuan decoction [10]. Previous studies have shown the therapeutic efficacy of the Tongmaiyangxin pill and Baoyuan decoction in treating cardiovascular diseases, particularly myocardial injury [9,10]. LIA has also been reported to exhibit anti-myocardial hypertrophy activity [9]. We have, for the first time, demonstrated that LIA exerts a protective effect against H_2O_2-induced cardiomyocyte injury in a BRS-3-dependent manner. This study holds significant importance as it marks the novel pharmacological functionality of the orphan receptor BRS-3, paving the way for potential therapeutic applications.

2. Results

2.1. LIA Is a Novel Antagonist with a High Selectivity of BRS-3

For G_q-coupled GPCRs, intracellular Ca^{2+} signaling is initiated upon agonist binding by converting $G_{\alpha q}$-GDP to $G_{\alpha q}$-GTP, which triggers phospholipase C (PLC)-dependent IP3 generation and Ca^{2+} release. As shown in Figure 1A, LIA, as well as Bantag-1, a well-characterized BRS-3 antagonist, dose-dependently reversed 50 nM MK-5046-induced intracellular Ca^{2+} mobilization in HEK293-mBRS-3 stable cells. The IC_{50} value of LIA was 7.063×10^{-7} M (4.250×10^{-7} M to 1.167×10^{-6} M, 95% CI), while that of Bantag-1 was 2.316×10^{-9} M (1.301×10^{-9} M to 4.172×10^{-9} M, 95% CI).

To further verify the antagonistic effect of LIA, we examined the IP1 accumulation, i.e., the stable degradation product of IP3, in H1299-hBRS-3 stable cells. As shown in Figure 1B, the accumulation caused by 50 nM MK-5046 could be inhibited by LIA with an IC_{50} value of 8.410×10^{-7} M (1.564×10^{-7} M to 3.482×10^{-6} M, 95% CI). We also established the dynamic mass redistribution (DMR) assay in HEK293-mBRS-3 and found that LIA reversed the DMR response of 50 nM MK-5046 in a dose-dependent manner with an IC_{50} value of 1.185×10^{-6} M (8.387×10^{-7} M to 1.674×10^{-6} M, 95% CI) (Figure 1C).

In docking studies, the calculated binding energy was −53.38 kJ/mol and the docking score was −7.847, indicating the tight binding between LIA and the receptor. The amino acid residues Thr^{106}, Arg^{316}, and Trp^{113} contributed to the binding interactions. The predicted binding site Arg^{316} is the same as for the previously studied agonists [11,12]. The binding between LIA and BRS-3 was visualized in Figure 1D.

To assess the receptor subtype selectivity of LIA, HEK293 cells were transiently transfected with either BB_1 and BB_2. [D-Phe^6, β-Ala^{11}, Phe^{13}, Nle^{14}] Bombesin (6-14) (Bomb), a full agonist capable of activating both BB_1 and BB_2, substantially triggered Ca^{2+} mobilization. Notably, LIA, even at a concentration of 50 μM (Figure 1E), failed to reverse this Ca^{2+} mobilization, therefore highlighting the subtype selectivity of LIA.

Collectively, these results indicate that LIA, a natural product, is a novel selective antagonist of BRS-3.

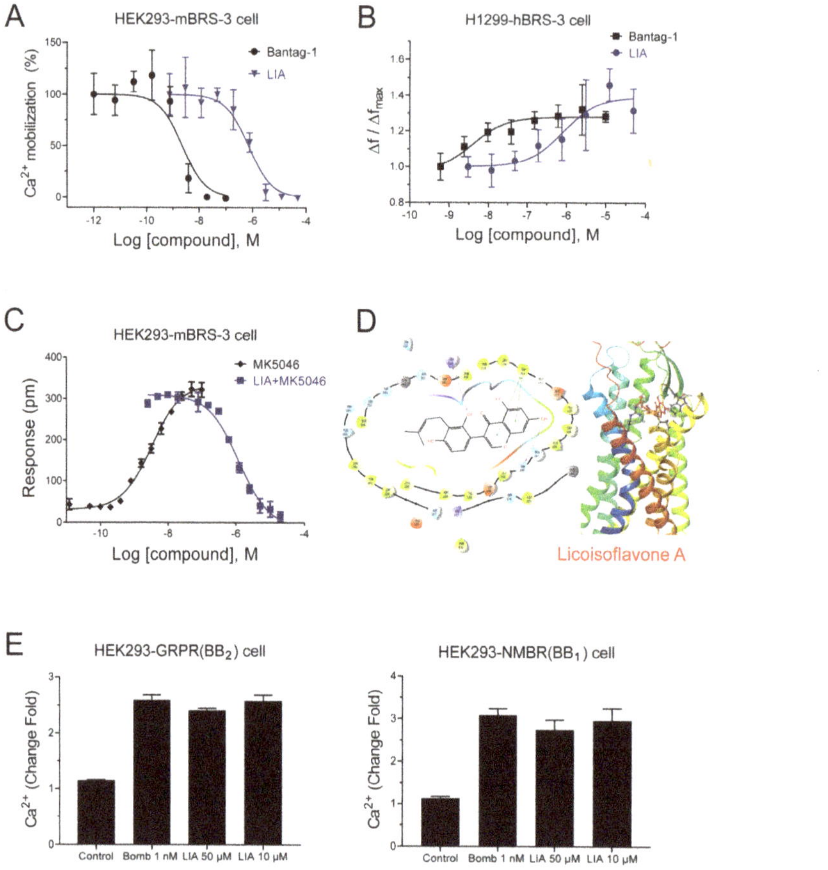

Figure 1. Pharmacological characterization of LIA as a novel antagonist of BRS-3. Dose–inhibition curves of LIA for 50 nM MK-5046-induced calcium mobilization in HEK293-mBRS-3 cells (**A**), IP-1 accumulation in H1299-hBRS-3 cell (**B**), and DMR responses in HEK293-mBRS-3 cells (**C**). The predicted binding model of LIA with human BRS-3 (**D**). The ability of LIA to reverse 1 nM Bomb-induced calcium mobilization in HEK293 expressing GRPR and NMBR cells (**E**). Data shown are means ± SD, *n* = 3. Dose–response curves were fitted using the log(agonist) vs. response using GraphPad Prism 8.0 software.

2.2. LIA Protects Cardiomyocytes Against H_2O_2-Induced Injury in H9c2 Cells

H9c2 cells were incubated with various concentrations of H_2O_2 for 24 h to mimic myocardial injury in vitro. H_2O_2 caused a dose-dependent decrease in cell viability (Figure 2A). It was found that when 200 μM H_2O_2 was used, the inhibition rate of H9c2 cells reached approximately 50%. Subsequently, 200 μM H_2O_2 was applied for modeling. As shown in Figure 2B, the H_2O_2-induced decrease in cell viability was significantly increased after 20 and 50 μM LIA treatment. Since 50 μM LIA was observed to affect cell viability (Figure 2B), 20 μM LIA was used as the highest concentration in the following experiments.

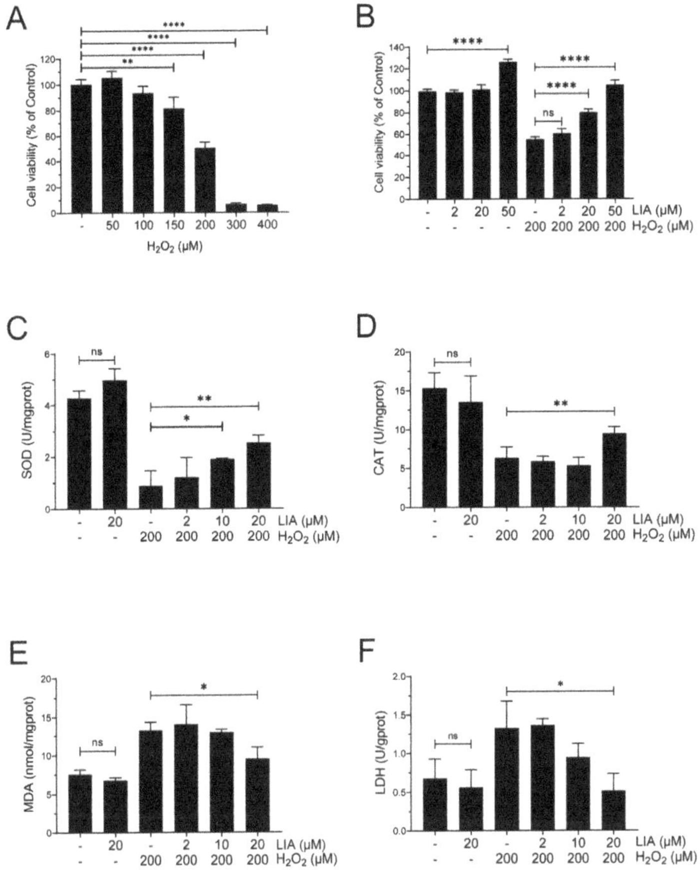

Figure 2. The protective effects of LIA on H_2O_2-induced injury in H9c2 cells. Cell viability at different concentrations of H_2O_2 (**A**). Protective effect of LIA on 200 μM H_2O_2-induced cell injury (**B**). Protective effect of LIA on the activities of SOD (**C**), CAT (**D**), and the levels of MDA (**E**) and LDH (**F**). The values are represented as means ± SD, $n = 3$. * $p < 0.05$, ** $p < 0.01$, **** $p < 0.001$. "ns" stands for "not significant", indicating $p > 0.05$.

As shown in Figure 2C,D, exposure of H9c2 cells to H_2O_2 induced oxidative damage, and resulted in a marked decrease in superoxide dismutase (SOD) activities compared to the control group. These reductions were effectively reversed by LIA treatment. Furthermore, as depicted in Figure 2E,F, the levels of malondialdehyde (MDA) and lactate dehydrogenase (LDH) were increased by H_2O_2 stimulation; however, these increases were significantly counteracted following LIA treatment.

Taken together, the above results suggest that LIA confers a protective effect against H_2O_2-induced cellular damage in H9c2 cells.

2.3. Proteomics Analysis of the Effect of LIA in H9c2 Cells Induced by H_2O_2

Quantitative proteomics analysis was performed to further explore the mechanism of the protective effect of LIA in H9c2 cells. LC-MS/MS was used to analyze three biological repeats for each group (C: Control, H: H_2O_2, LH: LIA + H_2O_2), and the Pearson correlation of these raw data was calculated (Figure 3A). In total, 1853 proteins were identified. Volcano plots (Figure 3B) showed that 140 proteins were downregulated and 10 were upregulated after H_2O_2 stimulation. Moreover, compared with solvent control, LIA pretreatment resulted in 55 proteins being upregulated and 22 being downregulated.

To obtain a deep insight into LIA-induced signaling pathways, the canonical pathway enrichment was analyzed by IPA (Figure S1). Several major pathways, including integrin signaling, PI3K/AKT signaling, ERK/MAPK signaling, ILK signaling, and p70S6K signaling, were revealed (Figure 3C) based on their $-\log(p\text{-value})$. Interestingly, these pathways are closely related to cardiac biology and pathophysiology. Therefore, we focused on these pathways during subsequent pathway verification.

Western blot analysis showed that in H9c2 cells, ILK, *p*-AKT (Ser^{473}), and *p*-ERK1/2 were downregulated in the H_2O_2 group, although this downregulation was reversed upon LIA treatment (Figure 3D). However, treatment with LIA alone exhibited no significant impact on the levels of ILK, *p*-AKT (Ser^{473}), and *p*-ERK1/2 levels in normal control cells. Furthermore, MK-5046 has the trend to diminish the protective effects of LIA against H_2O_2-induced cell damage. In addition, the phosphorylation of p70S6K1 at Thr^{389} showed no significant change with LIA treatment (Figure S3). These findings were in line with the proteomics analysis results.

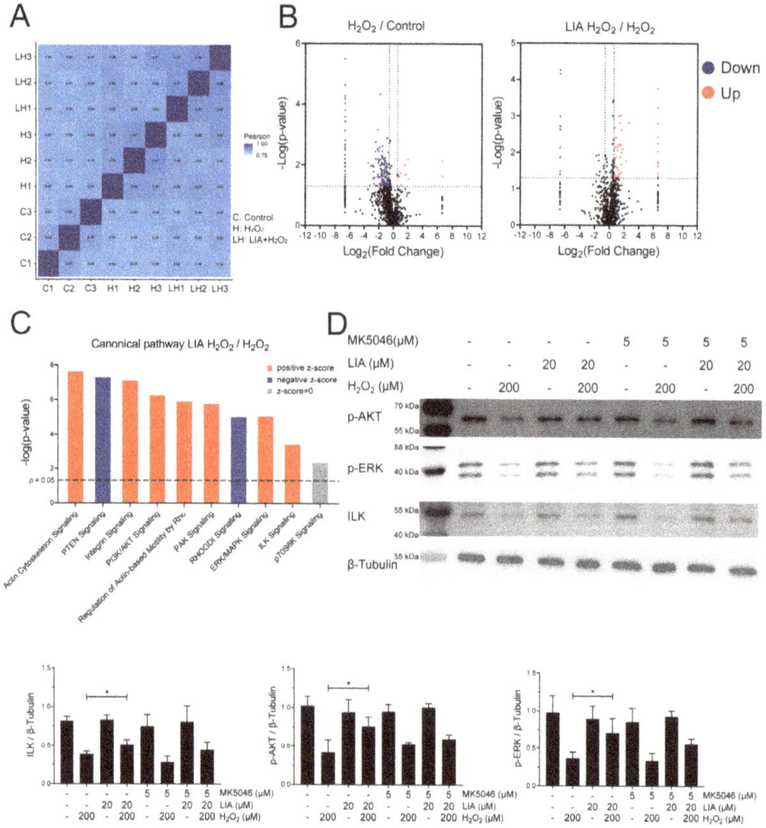

Figure 3. Proteomics analysis of the effect of LIA in H9c2 cells induced by H_2O_2. Pearson correlation (**A**) and Volcano plots (**B**). The red dots represent proteins that were upregulated ($p < 0.05$), and the blue dots represent proteins that were downregulated ($p < 0.05$). The black dots represent proteins with fold change < 1.5 or without significant changes ($p > 0.05$). "Canonical pathway" enrichment (**C**), the dash line represents $p = 0.05$. The effects of LIA on the protein expressions of ILK, *p*-P70S6K, *p*-ERK1/2, and *p*-AKT (Ser^{473}) were detected by Western blot (**D**). The values are represented as means ± SD, $n = 3$. * $p < 0.05$.

2.4. LIA Protects H_2O_2-Induced H9c2 Cell Injury via ILK/AKT and ILK/ERK Signaling Pathways

The roles of ILK, PI3K/AKT, and ERK/MAPK signaling pathways in the protective effects of LIA were further investigated. H_2O_2-induced H9c2 cells were pretreated with specific inhibitors of ILK (OSU-T315, 0.6 μM), AKT (Capivasertib, 0.75 μM), and ERK1/2 (FR 180204, 1 μM), respectively, before LIA treatment. The results showed that the three inhibitors—OSU, CAP, and FR—each significantly diminished the protective effects of LIA, as illustrated in Figure 4A. Interestingly, OSU, an ILK inhibitor, was observed to counteract the LIA-induced elevation of *p*-AKT (Ser^{473}) and *p*-ERK1/2, as shown in Figure 4B. This finding implies that ILK may be upstream in the PI3K/AKT and ERK/MAPK signaling cascades.

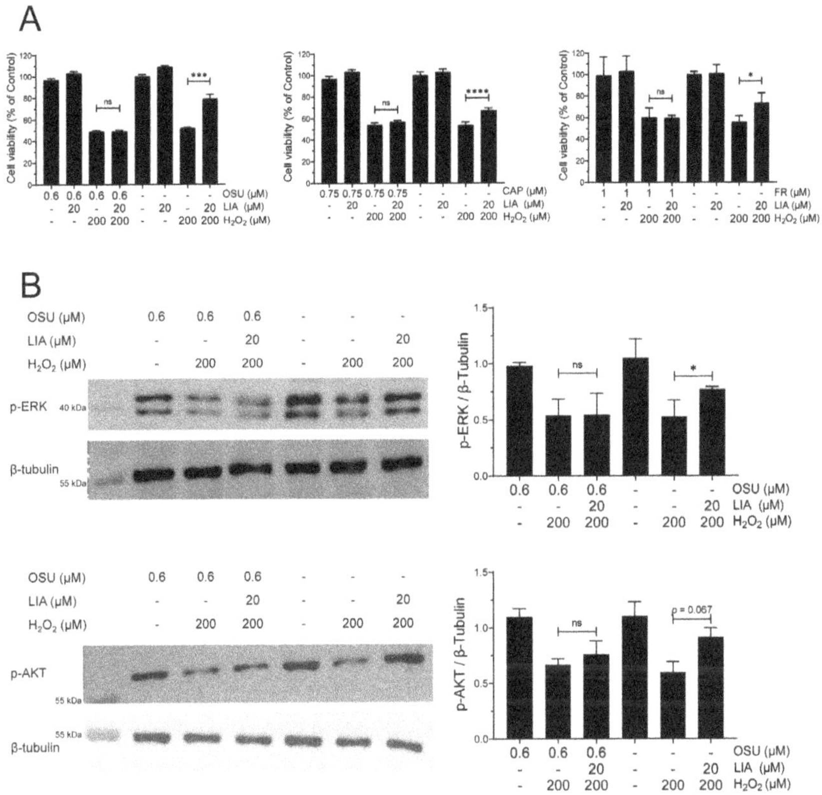

Figure 4. Roles of ILK/AKT and ILK/ERK signaling pathways in the protective effects of LIA against H_2O_2-induced H9c2 cell injury. Effects of LIA on cell viability in H9c2 cells pretreated with specific inhibitors (**A**). The protein expressions of *p*-ERK1/2 and *p*-AKT (Ser^{473}) after treatment with ILK inhibitor (**B**). The values are represented as means ± SD, $n = 3$. * $p < 0.05$, *** $p < 0.005$, **** $p < 0.001$. "ns" stands for "not significant", indicating $p > 0.05$.

2.5. LIA Exerts Protective Effects Against H_2O_2-Induced Injury in Primary Rat and Mouse Cardiomyocytes

To substantiate the protective effects of LIA against H_2O_2-induced cellular damage, we isolated and cultured primary cardiomyocytes from rats and mice. Due to the limited proliferative capacity of primary cardiomyocytes [13], a concentration of 50 μM LIA was chosen in the experiment to achieve a better signal. Notably, LIA exhibited pronounced protective effects in both primary rat and mouse cardiomyocytes compared to the solvent control group (Figure 5A).

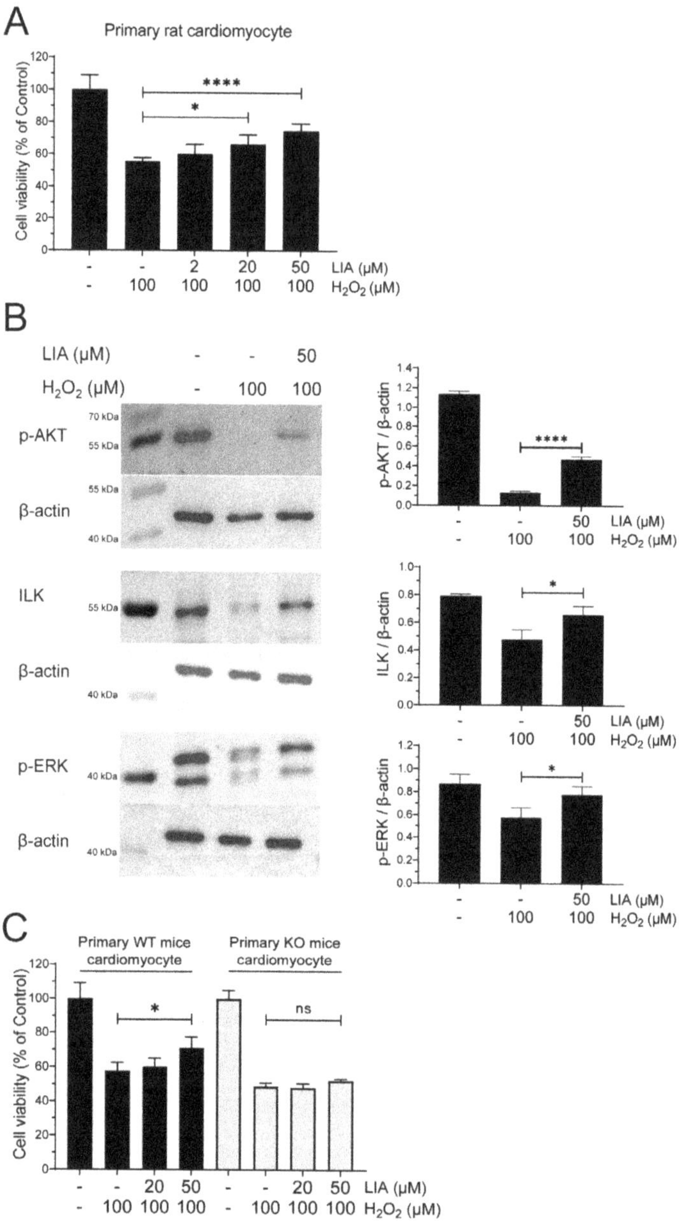

Figure 5. The protective effects of LIA against H_2O_2-induced injury in primary rat and mouse cardiomyocytes. The effect of LIA on cell viability in primary rat and mouse cardiomyocytes (**A**). LIA alters the protein expressions of ILK, *p*-P70S6K, *p*-ERK1/2, and *p*-AKT (Ser^{473}) against H_2O_2-induced injury in primary rat cardiomyocytes (**B**). The comparative analysis of the effects of LIA on cell viability against H_2O_2-induced injury in primary cardiomyocytes from wild-type (WT) and $Brs3^{-/y}$ (KO) mice (**C**). The values are represented as means ± SD, $n = 6$ for (**A**), $n = 3$ for (**B**), and $n = 5$ for (**C**). * $p < 0.05$, **** $p < 0.001$. "ns" stands for "not significant", indicating $p > 0.05$.

Consistent with the observations in H9c2 cells, treatment with LIA effectively activated the ILK, AKT (Ser^{473}), and ERK1/2 signaling pathways in primary rat cardiomyocytes subjected to H_2O_2-induced injury, as depicted in Figure 5B.

To further validate the target, we acquired primary cardiomyocytes from $Brs3^{-/y}$ mice along with their littermates, for a comparative analysis. As shown in Figure 5C, our results demonstrated that 50 μM LIA specifically increased the viability of cardiomyocytes derived from wild-type mice, whereas it had no such effect on those from $Brs3^{-/y}$ mice. This distinct response provides compelling evidence that the protective effect of LIA is mediated by its interaction with BRS-3.

2.6. BRS-3 Inhibition Regulates H_2O_2-Induced Cell Injury

Our findings demonstrate that LIA, a novel selective antagonist of BRS-3, effectively protects against H_2O_2-induced cardiomyocyte injury. This strongly suggests the role of BRS-3 in the pathogenesis of oxidative stress-induced cell injury. To validate our hypothesis, we performed quantitative real-time reverse transcription polymerase chain reaction (qRT-PCR) tests to detect and quantify the expression of BRS-3 in cardiomyocytes as shown in Figure S4A. The data confirm that BRS-3 is endogenously expressed in both H9c2 and rat primary cardiomyocytes. Furthermore, we assessed the phosphorylation of ERK in H9c2 cells in response to MK-5046, a well-characterized BRS-3 agonist. The results revealed that MK-5046 promoted ERK phosphorylation, while both Bantag-1 and LIA inhibited this agonist-induced phosphorylation, indicating that BRS-3 is not only expressed but also functional in H9c2 cells, as depicted in Figure S4B.

We then proceeded to investigate the impact of Bantag-1, a recognized antagonist of BRS-3 on H9c2 cells, instead of LIA. Interestingly, our observations revealed that Bantag-1 had no influence on cell viability in the control group; however, in the presence of H_2O_2, Bantag-1, but not MK-5046 (Figure S2), significantly increased cell viability (Figure 6A), demonstrating a protective effect similar to that of LIA.

To rule out the potential influence of cell line specificity and for target validation, we conducted a comparative analysis of the H_2O_2-induced response in both HEK293 and HEK293-mBRS-3 stable cells. Our findings indicated that 250 μM H_2O_2 reduced the cell viability of HEK293-mBRS-3 cells to 47.86%, whereas a higher concentration of 300 μM H_2O_2 was required to achieve a comparable reduction in HEK293 cells (Figure 6B). In addition, when pretreated with 10 μM Bantag-1, the cell viability of HEK293-mBRS-3 cells was markedly improved compared to the H_2O_2 model group (Figure 6C). However, such enhancements were absent in HEK293 cells.

We recently found that BRS-3 was endogenously expressed in B16 cells, the murine melanoma cell line, and generated BRS-3 knockout B16 cells (B16-KO) using CRISPR-Cas9 techniques [12]. A comparison of cell viability between H_2O_2-induced B16 cells and B16-KO cells (Figure 6D) revealed that BRS-3 may be involved in the cellular tolerance to hydrogen peroxide-induced injury. Additionally, Bantag-1 demonstrated efficacy in mitigating injury in B16 cells but not in B16-KO cells (Figure 6E), further supporting the role of BRS-3 in the process of cell injury.

Taken together, these findings indicated a prominent role of BRS-3 against oxidative stress-induced cell injury.

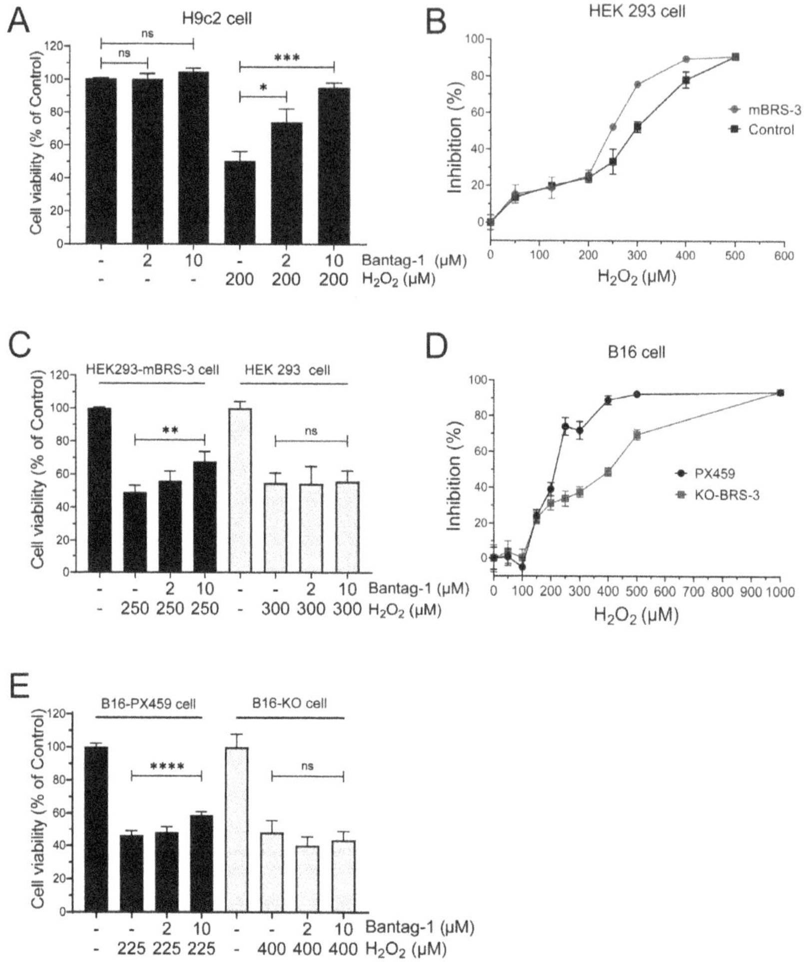

Figure 6. Validation of target for the BRS-3 regulation of H_2O_2-induced cell injury. Effect of Bantag-1 on cell viability in H9c2 cells (**A**). Cell viability in different concentrations of H_2O_2 in HEK293 and HEK293-mBRS-3 stable cells (**B**). Effect of Bantag-1 and LIA on cell viability against H_2O_2-induced injury in HEK293 and HEK293-mBRS-3 stable cells (**C**). Cell viability in different concentrations of H_2O_2 in B16 (B16-PX459) cells and B16-KO cells (**D**). Comparative analysis of the effects of Bantag-1 on cell viability against H_2O_2-induced injury in B16 cells and B16-KO cells (**E**). The values are represented as means ± SD, $n = 3$ for (**A–D**) and $n = 4$ for (**E**). * $p < 0.05$, ** $p < 0.01$, *** $p < 0.005$, **** $p < 0.001$. "ns" stands for "not significant", indicating $p > 0.05$.

3. Discussion

In this study, we have identified a novel antagonist of the orphan receptor BRS-3. Our studies on BRS-3 overexpression in HEK293 and H1299 cells revealed that LIA was able to specifically inhibit MK-5046, a specific agonist of BRS-3, and induced calcium mobilization and IP1 accumulation with a sub-micromolar concentration range. To the best of our knowledge, LIA represents the first natural compound recognized as an antagonist for BRS-3.

Our team has been dedicated to the identification of exogenous ligands for GPCRs from natural products and their derivatives [11,12,14,15]. The screening of potential target compounds from natural products offers unique advantages. The vast diversity inherent in the structures of natural products is expected to significantly enhance the probability of identifying receptor ligands that possess high affinity and selectivity. Moreover, the well-documented pharmacological activities of natural product ligands facilitate the establishment of possible correlations with their corresponding receptors. This, in turn, provides invaluable insights and clues for exploring the unknown biological functions of receptors.

Despite numerous studies highlighting the important roles of BRS-3 in various physiological processes [5], thus far, no therapeutic agents have yet been developed for clinical use. This can largely be attributed to the limited research on BRS-3's function and the lack of druggable ligands. MK-5046, for instance, has been demonstrated to cause cardiovascular complications, thereby restricting its clinical applicability [16]. Bantag-1, a high affinity and specific peptide antagonist of BRS-3, faces challenges in in vivo studies due to its short half-life [17]. ML-18, the first non-peptide BRS-3 antagonist, exhibited a moderate affinity for BRS-3, with an IC_{50} value of 4.8 μM, while maintain an affinity for GRPR (BB_2), with an IC_{50} value of 16 μM [18]. The identification of the natural compound LIA as an antagonist of BRS-3 has significantly enhanced the feasibility of in vivo studies of BRS-3 and have improved its specificity. Given the known pharmacological effects, it may provide valuable insights into the novel function of BRS-3. Future work will focus on structural modifications of LIA to enhance its activity.

In our current study, we induced myocardial injury in vitro by treating cardiac cells with H_2O_2 and found that LIA exerted a protective effect on the injured cardiac cells. Furthermore, LIA was found to enhance the expression of the antioxidant enzymes SOD and CAT in H9c2 cells, while simultaneously inhibiting the production of lipid peroxidation end products, i.e., MDA and LDH. LIA also demonstrated protective properties in primary cardiomyocytes derived from rats and mice against H_2O_2-induced cell injury. Importantly, the protective effect of LIA was absent in primary cardiomyocytes from $Brs3^{-/y}$ mice, strongly confirming that LIA acts by antagonizing BRS-3.

ILK is highly expressed in cardiac muscle, where it plays a key role in cell migration and the progression of cardiac diseases related to integrin function [19]. ILK binds to integrins and links integrins and receptor tyrosine kinases to the actin cytoskeleton, facilitating downstream signaling cascades, in particular the activation of AKT [20]. Additionally, the ERK/MAPK signaling pathway is recognized for its critical role in cardiac cell migration, survival, and cardiac repair [21]. Our proteomics studies illuminated the significance of ILK, AKT, and ERK in the protective mechanism of LIA against oxidative stress-induced cardiac injury. Subsequent immunoblotting analyses support that LIA reversed the H_2O_2-induced down-regulation of ILK, *p*-ERK1/2, and *p*-AKT (Ser^{473}) in both H9c2 cells and primary rat cardiomyocytes. In H9c2 cells, all these beneficial effects of LIA were abolished when co-treated with the inhibitors targeting ILK, AKT, and ERK, respectively, further reinforcing the notion that LIA acts through these three pathways. Furthermore, ILK inhibitors reversed LIA-induced upregulation of *p*-AKT and *p*-ERK1/2, indicating that ILK is located upstream of AKT and ERK1/2. Taken together, by blocking BRS-3, LIA activates ILK, which in turn induces AKT and ERK1/2 phosphorylation, thereby exerting a protective effect against oxidative stress-induced cardiomyocyte injury.

The pharmacological profile of LIA points to the possibility that BRS-3 may play a crucial role in oxidative stress-induced cell injury, a previously unrecognized biological function of this orphan receptor. Therefore, to obtain further insight into the novel aspect of this receptor function, we utilized a variety of cell lines, as previously established in our research. HEK293 cells were stably transfected with murine-derived BRS-3 to facilitate high expression levels and functional studied in a well-characterized system. H1299 cells were stably transfected with human-derived BRS-3 to model the human receptor in a human cellular context, which is important for translational relevance. Additionally, B16 cells naturally expressed high levels of BRS-3 and were used to assess antagonizing effects in an endogenous receptor setting, mimicking physiological conditions. Our findings demonstrated that cell lines with elevated BRS-3 expression exhibited greater sensitivity to H_2O_2, rendering them more vulnerable to H_2O_2-induced cellular injury. Furthermore, Bantag-1, a well-known specific antagonist of BRS-3, also showed protective effects in both HEK293-BRS-3 and B16 cell lines exposed to H_2O_2. These results provide additional evidence supporting the involvement of BRS-3 in oxidative stress-induced cell injury.

4. Materials and Methods

4.1. Chemicals and Materials

LIA (cat#66056-19-7, purity $\geq$ 98%) was supplied by ChemFaces (Wuhan, China). MK-5046 was purchased from Shanghai Macklin Biochemical Technology Co., Ltd. (Shanghai, China). Bantag-1 was purchased from Sigma-Aldrich (St. Louis, MO, USA). OSU-T315, Capivasertib, FR 180,204 were purchased from MedChemExpress (Shanghai, China).

4.2. Cell Culture and Experiment Design

The H9c2 rat cardiomyocyte cell line was purchased from the Cell Bank of the Chinese Academy of Sciences (Shanghai, China) and cultivated in Dulbecco's modified Eagle's medium (DMEM, Gibco, Waltham, MA, USA) supplemented with 10% (*v/v*) fetal bovine serum (FBS, Gibco, Waltham, MA, USA) and 1% antibiotic penicillin/streptomycin solution ($\times$100; Gibco, Waltham, MA, USA). Human embryonic kidney 293 mBRS-3 stable cell samples were kindly provided by Professor Olivier Civelli from the University of California, Irvine. All cells were placed in a humid incubator with 5% CO_2 at 37 °C. B16 cells with BRS-3 knockout and H1299 stably expressing human BRS-3 cells were cultivated as described previously [8]. Primary cardiomyocytes were obtained by isolations from the ventricles of 1-day-old Wistar rats or C57BL/6J mice (SLAC Laboratory Animal Company, Shanghai, China). All animal experiments protocols were approved by the Animal Ethics Committee of SJTU (The registration number: O_A2021020-3).

Hydrogen peroxide (H_2O_2, Adamas, Shanghai, China) was used to induce oxidative injury in H9c2 cells. Cells were divided into four groups, namely the vehicle control group (Control), H_2O_2 treated group (H_2O_2), drug and H_2O_2-treated group (Drug + H_2O_2), and drug-treated group (Drug + Control). Cells were pretreated with tested compounds for one hour prior to exposure to H_2O_2. All experiments were repeated at least three times.

4.3. Calcium Mobilization

Intracellular Ca^{2+} mobilization was detected by fluorometric imaging plate reader (FLIPR) assay (Molecular Devices, Sunnyvale, CA, USA). HEK293-mBRS-3 stable cells were seeded into a black 96-well plate (Corning, Corning, NY, USA) with 8 $\times$ 10^4 cells per well. For antagonist assays, antagonists were preincubated with the cells for 10 min before the addition of agonists. Calcium mobilization was monitored from the addition of the antagonist until 5 min after the addition of agonists. The maximum calcium response was determined by the peak calcium level elicited by the agonist. Normalization was performed by calculating the average calcium concentration over 20 s before agonist addition, which was set as the baseline (0%). The assay was conducted in a buffer solution containing 20 mM HEPES and 1$\times$ Hank's balanced salt solution (HBSS) (Gibco, Waltham, MA, USA), pH 7.4, as the manual. The intracellular Ca^{2+} concentration was measured by FLIPR assay (Molecular Devices, Sunnyvale, CA, USA).

4.4. Dynamic Mass Redistribution (DMR) Assay

The DMR assay was performed using the Epic BT system (Corning, New York, NY, USA) as previously described [11]. HEK293-mBRS-3 stable cells were seeded into Epic 384-well biosensor microplates overnight; after 2 min baseline, compounds were added, and the DMR signals were recorded for 1 h. In the antagonistic assay, cells were incubated with LIA for 1 h before adding 25 nM of MK-5046, and then the DMR signals were monitored for 1 h.

4.5. Measurement of Inositol Phosphates (IP1) Accumulation

Following the IP-One-Gq KIT manufacturer's protocol (Molecular Devices, Sunnyvale, CA, USA), IP1 accumulation was detected in H1299 and H1299-hBRS-3 stable cells based on the manufacturer's protocol [22]. Briefly, cells were diluted in stimulation buffer at 5×10^5 cells/mL in 384-well plates and were immediately treated with tested compounds for 1 h. Then, the IP1 d2 reagent (acceptor) was added to each well followed by the IP1 Tb cryptate antibody (donor), and the plate was sealed and incubated for another 1 h at room temperature. A Tecan Spark multimode microplate reader (Tecan, Männedorf, Switzerland) was used to determine the IP1 accumulation by using a homogeneous time-resolved fluorescence (HTRF) protocol (Ex 350 nm, Em 665/620 nm HTRF). Data were expressed as the percentage of the non-stimulation group.

4.6. Cell Viability Analysis

Cell viability was assessed by MTT assay. Cells were seeded into a 96-well plate at a density of 3×10^3 cells per well, and the marginal wells were filled with PBS. Cells were incubated with various concentrations of LIA or Bantag-1 1 h before exposure to H_2O_2. For signaling pathway inhibitors, cells were pretreated 15 min before adding LIA or Bantag-1. Twenty-four hours after the H_2O_2 treatment, MTT solution (0.5 mg/mL, Sigma-Aldrich, St. Louis, MO, USA) was added to each well for another 4 h at 37 °C. After the incubation, the MTT reagent was removed and replaced with 100 μL of DMSO to dissolve the formazan for 15 min. Cell viability was analyzed by measuring the optical density at 570 nm with a microplate reader. All experiments were repeated in triplicate.

4.7. Malondialdehyde (MDA), Superoxide Dismutase (SOD), Lactate Dehydrogenase (LDH), and Catalase (CAT) Determination

The LDH level in the culture medium was tested using the LDH assay kit (Nanjing Jiancheng Bioengineering Institute, Nanjing, China). The intracellular MDA, SOD, and CAT activities were determined by using commercial kits (Nanjing Jiancheng Bioengineering Institute). Briefly, cells were lysed via the freeze—thaw method and the reagents were mixed according to the manufacturer's instructions. Protein concentrations were determined by using a Pierce™ BCA protein Assay kit (BCA, Thermo Scientific, Waltham, MA, USA).

4.8. Quantitative Real-Time Reverse Transcription Polymerase Chain Reaction (qRT-PCR)

Total RNA was extracted from cells by using TRIzol reagent (Beyotime, Shanghai, China), following the manufacturer's instructions. Reverse transcription was performed by Rever Tra Ace® qPCR RT Kit (Toroivd, Shanghai, China) and the fluorescence real-time PCR was performed using Bestar® Sybr Green qPCR Master Mix (DBI® Bioscience, Ludwigshafen, Germany). The sequences of the oligonucleotide primers are listed in Table 1. The parameter crossing point (Cp) values were normalized to those of glyceraldehyde 3-phosphate dehydrogenase (GAPDH), which was applied as the internal reference.

Table 1. Primer sequences for qRT-PCR.

Brs3	Forward	5′-GAAACATCAAGCTCTGCCGTCT-3′
	Reverse	5′-CCACTGAAATGATCACAGCAT-3′
Actb	Forward	5′-CGAGTACAACCTTCTTGCAGC-3′
	Reverse	5′-TATCGTCATCCATGGCGAACTG-3′

4.9. Proteomics Analysis by LC-MS/MS

H9c2 cells were seeded in 6-well plates at a density of 3×10^5 per well. After 20 μM of LIA treatment for 1 h, cells were incubated for an additional hour with H_2O_2. The cells were harvested, and the proteins were extracted and measured by BCA. Then, cellular proteins (30 μg/each sample) were digested in trypsin (Promega, Madison, WI, USA) at 37 °C overnight. After desalting, digested peptides were dissolved in 10 μL of buffer A (0.1% formic acid in water), and equivalent peptides measured by Nanodrop were analyzed with EasynanoLC1000 coupled with a TripleTOF™ 5600+ system Mass spectrometer (AB Sciex, Framingham, MA, USA). Peptides were separated by a 150×0.3 mm reverse-phase column (ChromXP C18, 3 μm 120 Å, AB Sciex, Framingham, MA, USA). Buffer A and buffer B (80% acetonitrile with 0.1% formic acid) were applied as the mobile phases. The flow rate was 300 nL/min and the gradient was listed as follows: 2–20% buffer B (98 min), 20–30% buffer B (10 min), 30–95% buffer B (2 min), 95% buffer B (8 min), 95–2% buffer B (1 min), and kept in 2% buffer B for 1 min. The MS scan with a resolution of 70,000 ranged from 350 to 1500 m/z, and the scan for MS/MS ranged from 200 to 2000 m/z with a resolution of 17,500. MaxQuant (version 1.6.1.0, Martinsried, Germany) software was used for database search, and label-free quantification was performed using intensity-based absolute quantification (iBAQ). Differently expressed proteins were screened following the criteria of at least two unique peptides, 1.5-fold change (FC), and p value < 0.05.

4.10. Western Blot Assay

Protein preparation was the same as described in Section 4.8, and cells were pretreated with signaling inhibitors 15 min before the addition of LIA or Bantag-1. After 10 min of denaturing at 99 °C, proteins were loaded and separated by sodium dodecyl sulfate–polyacrylamide gel electrophoresis (SDS-PAGE), then transferred to nitrocellulose membranes. The membranes were blocked for 1 h in 5% BSA dissolved in Tris-buffered saline with 0.1% Tween (TBST) and incubated with primary antibodies with a dilution of 1:1000. The primary antibodies targeting phospho (*p*)-AKT, *p*-ERK, *p*-P70S6K, and β-Tubulin were purchased from cell signaling technology (Danvers, MA, USA), while the anti-ILK was purchased from Servicebio® (Wuhan, China). The membranes were washed with TBST, incubated with secondary antibodies (Teyebio, Shanghai, China) for 1 h, then detected with a Tanon 5200 multi-imaging system (Tanon, Shanghai, China).

4.11. Molecular Docking

The predicted 3D structure of human BRS-3 was downloaded from the AlphaFold Protein Structure Database (P32247). The human BRS-3 structure was prepared in Protein Preparation Wizard of Maestro (Schrödinger LLC, New York, NY, USA): hydrogen atoms were added; bond orders were assigned; H-bond assignment was optimized at pH 7.0 using PROPKA. All atoms were energy-minimized to reach the convergent RMSD of 0.3 Å with the OPLS4 force field. The three-dimensional conformation of LIA was prepared in LigPrep with the OPLS4 force field. The Induced Fit Docking module was used to fine-tune the receptor structure. The centroid of residues Arg^{127}, His^{294}, and His^{107} was defined as the box center. Then, a protein grid box for docking was generated by enclosing the residues in the box at a size of 10 Å × 10 Å × 10 Å centered on the LIA using the receptor grid generation module with no constraints. To probe the possible binding mode of LIA with the BRS-3 prepared structure, we conducted molecular docking using the Extra Precision (XP) Glide module. The docking parameters were set to default. Molecular mechanics/generalized Born and surface area solvation (MM-GBSA) were calculated using the Prime module.

4.12. Statistical Analysis

All values were displayed as mean ± standard deviation (SD), and the statistical analysis was performed by using GraphPad Prism 8.0 software. Statistical difference was analyzed by one-way analysis of variance followed by the Dunnett post hoc test. The difference was significant if $p < 0.05$.

5. Conclusions

The present study elucidates that the natural compound LIA serves as an exogenous antagonist of BRS-3. LIA was effective in preventing H_2O_2-induced cell injury in both H9c2 cells and primary cardiomyocytes. This protective effect was associated with the activation of integrin/ILK/AKT and ERK/MAPK signaling pathways. This study also contributes significantly to expanding the understanding of the potential roles of BRS-3, uncovering previously unknown pharmacological functionality of this orphan receptor. These findings provide promising insights into potential therapeutic interventions for cardiac injuries and may pave the way for future drug development in this area.

Supplementary Materials: The following supporting information can be downloaded at: https://www.mdpi.com/article/10.3390/ijms26062745/s1.

Author Contributions: Conceptualization Y.Z. (Yan Zhang) and H.X.; Methodology, J.L., L.W. and J.Z.; software, H.Z. and J.W.; Validation and formal analysis, M.F., H.L. and Y.Z. (Yuhang Zhu); Investigation, J.L., L.W., J.Z., M.G. and M.C.; Data curation, J.L.; Writing—original draft preparation, J.L.; Writing—review and editing, Y.Z. (Yan Zhang); Supervision, Y.Z. (Yan Zhang) and H.X.; Project administration, Y.Z. (Yan Zhang); Funding acquisition, Y.Z. (Yan Zhang) and H.X. All authors have read and agreed to the published version of the manuscript.

Funding: This research was funded by the National Natural Science Foundation of China (No. 22374098), the Natural Science Foundation of Shanghai (No. 23ZR1434200, No. 19ZR1427800), and the Key Scientific Project of Shanghai Jiao Tong University (No. TMSK-2020-130).

Institutional Review Board Statement: All animal experiments protocols were approved by the Animal Ethics Committee of SJTU (The registration number: O_A2021020-3).

Informed Consent Statement: Not applicable.

Data Availability Statement: All datasets generated or analyzed during this study are available from the corresponding author on reasonable request.

Conflicts of Interest: The authors declare no conflicts of interest. The funders had no role in the design of the study; in the collection, analyses, or interpretation of data; in the writing of the manuscript; or in the decision to publish the results.

Abbreviations

The following abbreviations are used in this manuscript:

BRS-3/BB_3	Bombesin receptor subtype-3
DMR	Dynamic mass redistribution
FLIPR	Fluorometric imaging plate reader
GPCR	G protein-coupled receptor

GRPR/BB_2	Gastrin-releasing peptide receptor
H_2O_2	Hydrogen peroxide
HTRF	Homogeneous time-resolved fluorescence
iBAQ	Intensity-based absolute quantification
ILK	Integrin-linked kinase
LIA	Licoisoflavone A
MAPK	Mitogen-activated protein kinases
NMBR/BB_1	Neuromedin B receptor

References

1. Insel, P.A.; Snead, A.; Murray, F.; Zhang, L.; Yokouchi, H.; Katakia, T.; Kwon, O.; Dimucci, D.; Wilderman, A. GPCR expression in tissues and cells: Are the optimal receptors being used as drug targets? *Br. J. Pharmacol.* **2012**, *165*, 1613–1616. [CrossRef] [PubMed]
2. Romanova, E.V.; Sweedler, J.V. Peptidomics for the discovery and characterization of neuropeptides and hormones. *Trends Pharmacol. Sci.* **2015**, *36*, 579–586. [CrossRef] [PubMed]
3. Guo, M.; Qu, X.; Qin, X.Q. Bombesin-like peptides and their receptors: Recent findings in pharmacology and physiology. *Curr. Opin. Endocrinol. Diabetes Obes.* **2015**, *22*, 3–8. [CrossRef]
4. Xiao, C.; Reitman, M.L. Bombesin-Like Receptor 3: Physiology of a Functional Orphan. *Trends Endocrinol. Metab.* **2016**, *27*, 603–605. [CrossRef]
5. Li, M.; Liang, P.; Liu, D.; Yuan, F.; Chen, G.C.; Zhang, L.; Liu, Y.; Liu, H. Bombesin Receptor Subtype-3 in Human Diseases. *Arch. Med. Res.* **2019**, *50*, 463–467. [CrossRef]
6. Sano, H.; Feighner, S.D.; Hreniuk, D.L.; Iwaasa, H.; Sailer, A.W.; Pan, J.; Reitman, M.L.; Kanatani, A.; Howard, A.D.; Tan, C.P. Characterization of the bombesin-like peptide receptor family in primates. *Genomics* **2004**, *84*, 139–146. [CrossRef]
7. Moreno, P.; Mantey, S.A.; Nuche-Berenguer, B.; Reitman, M.L.; González, N.; Coy, D.H.; Jensen, R.T. Comparative pharmacology of bombesin receptor subtype-3, nonpeptide agonist MK-5046, a universal peptide agonist, and peptide antagonist Bantag-1 for human bombesin receptors. *J. Pharmacol. Exp. Ther.* **2013**, *347*, 100–116. [CrossRef]
8. Bobrowska-Hägerstrand, M.; Wróbel, A.; Rychlik, B.; Bartosz, G.; Söderström, T.; Shirataki, Y.; Motohashi, N.; Molnár, J.; Michalak, K.; Hägerstrand, H. Monitoring of MRP-like activity in human erythrocytes: Inhibitory effect of isoflavones. *Blood Cells Mol. Dis.* **2001**, *27*, 894–900. [CrossRef]
9. Guo, R.; Liu, N.; Liu, H.; Zhang, J.; Zhang, H.; Wang, Y.; Baruscotti, M.; Zhao, L.; Wang, Y. High content screening identifies licoisoflavone A as a bioactive compound of Tongmaiyangxin Pills to restrain cardiomyocyte hypertrophy via activating Sirt3. *Phytomedicine* **2020**, *68*, 153171. [CrossRef]
10. Zhang, X.; Chen, Q.; Zhao, J.; Zhao, W.; Fan, N.; Wang, Y.; Chen, H.; Rong, J. A four-compound remedy AGILe protected H9c2 cardiomyocytes against oxygen glucose deprivation via targeting the TNF-α/NF-κB pathway: Implications for the therapy of myocardial infarction. *Front. Pharmacol.* **2023**, *14*, 1050970. [CrossRef]
11. Zhu, Y.; Wu, L.; Zhao, Y.; Wang, Z.; Lu, J.; Yu, Y.; Xiao, H.; Zhang, Y. Discovery of oridonin as a novel agonist for BRS-3. *Phytomedicine* **2022**, *100*, 154085. [CrossRef] [PubMed]
12. Wu, L.; Cui, J.; Zhao, C.; Wang, Z.; Lu, J.; Li, S.; Jia, J.; Xiao, H.; Zhang, Y. Discovery of Dimethyl Shikonin Oxime 5a, a Potent, Selective Bombesin Receptor Subtype-3 Agonist for the Treatment of Type 2 Diabetes Mellitus. *J. Med. Chem.* **2023**, *66*, 8011–8029. [CrossRef] [PubMed]
13. Simpson, P.; Savion, S. Differentiation of rat myocytes in single cell cultures with and without proliferating nonmyocardial cells. Cross-striations, ultrastructure, and chronotropic response to isoproterenol. *Circ. Res.* **1982**, *50*, 101–116. [CrossRef] [PubMed]
14. Qiu, X.; Wu, L.H.; Yu, Y.; Jin, Y.; Wang, J.X.; Wang, C.R.; Zhang, Y. Discovery of exogenous ligands for orphan receptor BRS-3 from Chinese herbs. *Zhongguo Zhong Yao Za Zhi* **2022**, *47*, 1595–1602.
15. Wu, L.; Zhang, W.; Qiu, X.; Wang, C.; Liu, Y.; Wang, Z.; Yu, Y.; Ye, R.D.; Zhang, Y. Identification of Alkaloids from Corydalis yanhusuo W. T. Wang as Dopamine D_1 Receptor Antagonists by Using CRE-Luciferase Reporter Gene Assay. *Molecules* **2018**, *23*, 2585. [CrossRef]
16. Reitman, M.L.; Dishy, V.; Moreau, A.; Denney, W.S.; Liu, C.; Kraft, W.K.; Mejia, A.V.; Matson, M.A.; Stoch, S.A.; Wagner, J.A.; et al. Pharmacokinetics and pharmacodynamics of MK-5046, a bombesin receptor subtype-3 (BRS-3) agonist, in healthy patients. *J. Clin. Pharmacol.* **2012**, *52*, 1306–1316. [CrossRef]
17. Guan, X.M.; Chen, H.; Dobbelaar, P.H.; Dong, Y.; Fong, T.M.; Gagen, K.; Gorski, J.; He, S.; Howard, A.D.; Jian, T.; et al. Regulation of energy homeostasis by bombesin receptor subtype-3: Selective receptor agonists for the treatment of obesity. *Cell Metab.* **2010**, *11*, 101–112. [CrossRef]

18. Moody, T.W.; Mantey, S.A.; Moreno, P.; Nakamura, T.; Lacivita, E.; Leopoldo, M.; Jensen, R.T. ML-18 is a non-peptide bombesin receptor subtype-3 antagonist which inhibits lung cancer growth. *Peptides* **2015**, *64*, 55–61. [CrossRef]
19. Quang, K.L.; Maguy, A.; Qi, X.Y.; Naud, P.; Xiong, F.; Tadevosyan, A.; Shi, Y.F.; Chartier, D.; Tardif, J.C.; Dobrev, D.; et al. Loss of cardiomyocyte integrin-linked kinase produces an arrhythmogenic cardiomyopathy in mice. *Circ. Arrhythm. Electrophysiol.* **2015**, *8*, 921–932. [CrossRef]
20. Li, G.; Li, Y.Y.; Sun, J.E.; Lin, W.H.; Zhou, R.X. ILK-PI3K/AKT pathway participates in cutaneous wound contraction by regulating fibroblast migration and differentiation to myofibroblast. *Lab. Investig.* **2016**, *96*, 741–751. [CrossRef]
21. Song, Y.; Zhang, C.; Zhang, J.; Sun, N.; Huang, K.; Li, H.; Wang, Z.; Huang, K.; Wang, L. An injectable silk sericin hydrogel promotes cardiac functional recovery after ischemic myocardial infarction. *Acta. Biomater.* **2016**, *41*, 210–223. [CrossRef] [PubMed]
22. Zawilska, J.B.; Urbańska, A.; Sokołowska, P. Orexins/hypocretins stimulate accumulation of inositol phosphate in primary cultures of rat cortical neurons. *Pharmacol. Rep.* **2013**, *65*, 513–516. [CrossRef] [PubMed]

International Journal of
Molecular Sciences

MDPI

Article

Identification and Screening of Novel Antioxidant Peptides from Yak Skin and Their Protective Effect on H_2O_2-Induced HepG2 Cells Oxidation

Yan Jin [1], Nan Zhang [1], Yurong Huang [1], Ziyao Zhang [1], Enhui Jin [1], Yu Kong [1], Wenjie Sui [1], Tao Wu [1,*] and Min Zhang [1,2,*]

[1] College of Food Science and Engineering, Tianjin University of Science & Technology, Tianjin 300457, China; jinyan@tust.edu.cn (Y.J.); nzhanga@126.com (N.Z.); hyrong0420@163.com (Y.H.); ziyaozhang123@163.com (Z.Z.); 15641660860@163.com (E.J.); yu_kong@tust.edu.cn (Y.K.); wjsui@tust.edu.cn (W.S.)

[2] College of Food Science and Bioengineering, Tianjin Agricultural University, Tianjin 300384, China

* Correspondence: wutao@tust.edu.cn (T.W.); zm0102@tust.edu.cn (M.Z.)

Abstract

To improve the bioavailability of yak by-products, novel antioxidant peptides were prepared and identified from yak skin hydrolysate. The results showed that the ultrafiltration fraction of a molecular weight of less than 1 kDa had the strongest free radical scavenging activity. A total of 219 novel peptides were identified by mass spectrometry and five antioxidant peptides were screened based on molecular docking with Keap1 (LMGPR, GFDGD, FGFDGDF, GHNGLDGL, and GPAGPQGPR). These peptides may bind with Keap1 competitively and exert antioxidant effects by activating the Nrf2/ARE pathway. After synthesis, FGFDGDF showed a better free radical scavenging ability and protective effect on H_2O_2-induced oxidative damage of HepG2 cells among these peptides. The pretreatment of peptides could enhance the activity of intracellular antioxidant enzymes and reduce the level of malondialdehyde and IL-8. This study provides a scientific basis for the application of yak skin peptide as a novel antioxidant in functional food.

Keywords: yak skin; antioxidant peptide; molecular docking; oxidative stress; Keap1-Nrf2/ARE pathway

1. Introduction

Free radicals are inevitable products of metabolism and are usually eliminated by the antioxidant defense system, which is an ongoing balancing process. When stimulated by external factors, this balance might be disrupted, leading to an excessive production of free radicals and initiating oxidative stress to damage health [1]. Although synthetic antioxidants such as butylated hydroxyanisole, tert-butylhydroquinone, and propyl gallate have the ability to reduce free radicals, their potential toxicity and carcinogenicity limit their use in food application [2]. Therefore, the development of natural and non-toxic antioxidants has emerged as a key research focus. In recent years, an increasing number of antioxidant peptides have been identified and derived primarily from legumes, animal-based foods, and marine proteins, which were typically released from proteins through enzymatic hydrolysis [3–5]. Protease hydrolysis is the preferred technique for producing antioxidant peptides due to its high selectivity, mild reaction conditions, minimal side reactions and high efficiency selectivity [6]. Studies have shown that the antioxidant activity of peptides is affected by molecular weight, the type of amino acid, and their sequence. It was found that the molecular weight of the optimal antioxidant activity of rice, walnut, and other protein hydrolysates prepared by enzymatic hydrolysis was concentrated below 1 kDa [7,8]. Overall, a reasonable combination of molecular weight and amino acid sequence is crucial for enhancing the activity of antioxidant peptides. Therefore, the extensive use of different food resources and high-efficiency preparation progress on antioxidant peptides with strong antioxidant activity and stability still need to be investigated.

Yak (*Bos grunniens*) is a rare breed mainly found in the Qinghai–Tibet Plateau. China harbors the largest population and diversity of yaks accounting for more than 94% worldwide [9]. Because of the harsh environment of extreme coldness, hypoxia, and strong radiation, yak products may have unique properties regarding nutrition, sensory characteristics, and processing properties [10]. As the main by-product of yak processing in meat and milk, yak skin (YS) is rich in protein, accounting for 30–40%, which can be a good resource of bioactive peptides [11]. Nowadays, YS is usually discarded or is traditionally utilized in low-quality leather production, leading to a wastage of this valuable resource. Studies show that YS gelatin has great potential to replace commercial gelatin because of its superior properties including oxidation resistance, emulsification, and heat resistance [12]. Its collagen has whitening, moisturizing, and wrinkle-eliminating effects, and is easy to absorb, with good biocompatibility and degradability. Xiaotong Ma et al. [13] identified three novel antifreeze peptides from YS hydrolysates which could increase the survival rate of probiotics. Moreover, YS boasts outstanding nutritional values, which may apply in the mitigation of myocardial ischemia–reperfusion injury [14]. The potential angiotensin I-converting enzyme (ACE)-inhibitory activities of YS gelatin peptides were also investigated, but the sequences were not clarified [15]. Only two novel peptides (GADGAPGKDGVRG and GPRGDQGPVGR) with a good ability for iron-chelating were identified and studied from YS [16]. Overall, current research mainly focuses on gelatin, while the development of bioactive peptides derived from YS is limited. Further research on the preparation, identification, and healthy functions of yak skin peptides (YSP) needs to be clarified, as well as the specific mechanism of its bioactivity.

The molecular docking techniques based on machine learning and artificial intelligence are widely used in the area of functional food and drug development [17]. Molecular docking is derived from Fisher E's lock and key model, which posits that the spatial alignment of receptors and ligands is crucial for their recognition [18]. The interaction between peptides and biological targets can be directly observed through molecular docking to elucidate the biological mechanism of food-derived peptides [19]. Nuclear factor erythrocyte-associated factor 2 (Nrf2) is recognized as a neurotherapeutic target, and the Keap1-Nrf2/ARE pathway serves as a pivotal signaling pathway for endogenous oxidant damage [20]. In recent years, research has focused on exploring the mechanism of antioxidant peptides through Keap1-Nrf2/ARE pathway [21], such as the milk-derived peptide KVLPVPEK (K-8-K), which inhibits lipid peroxidation and prevents the interaction between Nrf2 and Keap1 [22]. These comprehensive methods offer robust support for the research and application of antioxidant peptides.

Therefore, in this study, antioxidant peptides were prepared from YS through enzymatic hydrolysis and ultrafiltration separation, identified by mass spectrometry and screened based on molecular docking with Keap1. Then the cell protection effect of YSP was determined by using the H_2O_2-induced oxidative damage model of HepG2 cells. This study provides a theoretical foundation for the development and application of YSP, aiming to improve the high-value utilization of YS and broaden the source of antioxidants.

2. Results and Discussion

2.1. Structural Characterization of YSP

The structural characterization on YS and YSP are shown in Figure 1A. Both YS and YSP had a special absorption peak in the range of 3500–3300 cm^{-1}, which was mainly associated with N-H and O-H stretching. It reflected the presence of hydrogen bonds in the peptide chain skeleton [23]. Both samples exhibited an absorption peak at 2926 cm^{-1}, which was caused by asymmetric C-H stretching vibration, indicating that both samples contained aliphatic amino acids side chain [24]. In addition, absorption was observed at 1641 cm^{-1} in the range of amide I band (1600–1700 cm^{-1}), indicating that the protein had a more complete triple helix structure. Moreover, the absorption peak at 1453 cm^{-1} and 1081 cm^{-1} also appeared in both YS and YSP. It indicated that the secondary structures of both samples were relatively stable. Notably, YSP showed several different groups compared with YS, including C=O bond vibration in carboxylic acid (1404 cm^{-1}) and C-H bond stretching vibration (1335 cm^{-1}). These differences indicated that the protein structure was unfolded, and its secondary structure might be destroyed by enzymatic hydrolysis. These changes might expose the originally tightly wrapped active groups, making the enzyme hydrolysis product more likely to capture electrons, thereby exhibiting improved antioxidant capacity.

A

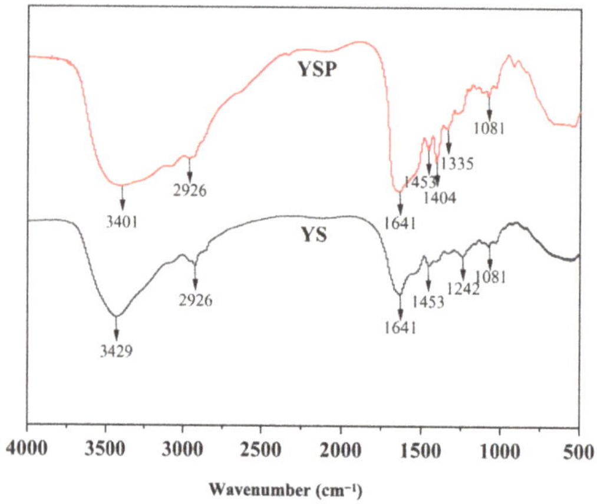

B

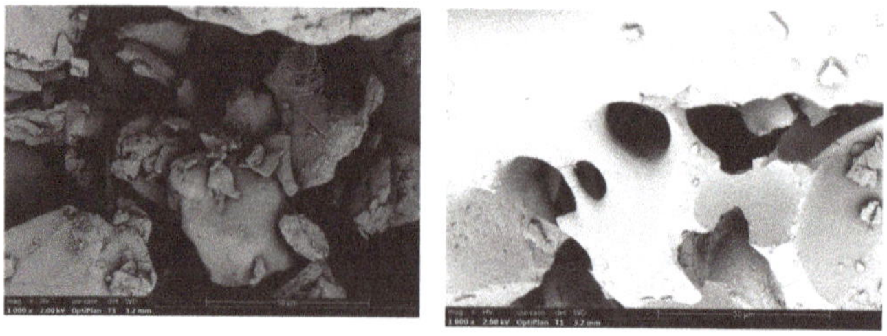

(a) (b)

Figure 1. Fourier transform infrared spectroscopy of YS and YSP (**A**) and the SEM images of YS (**B-a**) and YSP (**B-b**) (1000×).

In addition, the microstructure of YS and YSP had great differences, as shown in Figure 1B. YS was observed to be a dense block with an irregular concave surface. After enzymatic hydrolysis, YSP exhibited a large number of porous and fracture microstructures with a relatively loose texture. This may be due to the enzymatic hydrolysis-induced destruction of hydrogen bonds and van der Waals forces between protein molecules, which weaken the intermolecular interaction, resulting in a certain degree of stretching of protein molecules [25]. This result was consistent with previous reports indicating that proteins were broken down into small fragments, and they adopt a loosely structured conformation after enzyme treatment [26,27]. These structural changes might result in more hydrophobic amino acids being exposed to the surface, which could enhance the antioxidant capacity of the sample.

2.2. Free Radical Scavenging Activities of Ultrafiltrate Fractions

To accurately screen and identify its components, YSP was separated by ultrafiltration membranes. The free radical scavenging activities of four fractions are shown in Figure 2. Obviously, the components with a molecular weight of less than 1 kDa (YSP-I) exhibited the highest scavenging rate on the DPPH radical (75.80 ± 0.76%), ABTS radical (77.42 ± 2.77%), and hydroxyl radical (41.96 ± 0.67%) compared to other fractions, with no significant difference from YSP. It indicated that the components with lower molecular weight might have a stronger antioxidant ability. Similarly, Yawen Kong et al. [28] reported that the molecular weight of less than 3 kDa demonstrated excellent antioxidant activity. Long He et al. [29] reported that the radical scavenging ability of DPPH and the ABTS scavenging activity of yak skin gelatin hydrolysates were significantly increased after ultrasonic treatment. Overall, the components with a molecular weight of <1 kDa were selected for further analysis.

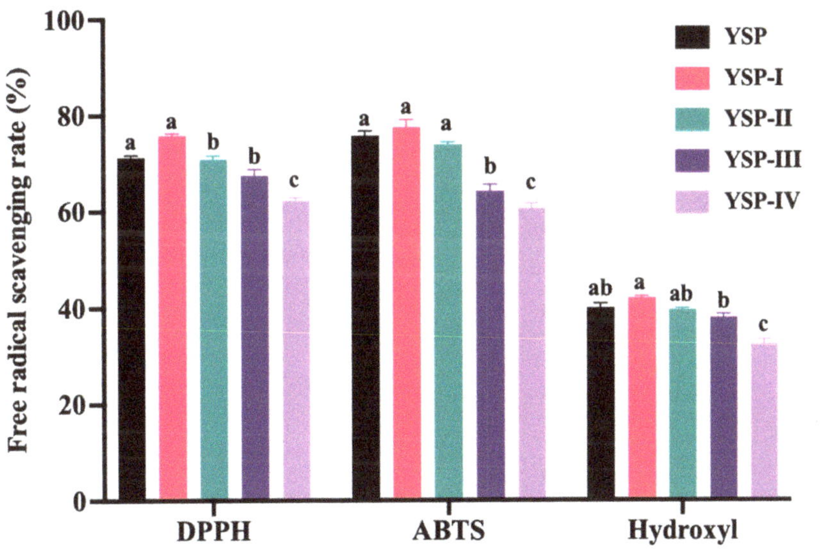

Figure 2. Free radical scavenging activity of YSP and ultrafiltration fractions, including YSP-I (<1 kDa), YSP-II (1–3 kDa), YSP-III (3–10 kDa), and YSP-IV (>1 kDa). Different letters represent significant differences ($p < 0.05$).

2.3. Identification and Screening of Antioxidant Peptide from YSP by LC-MS/MS

YSP-I samples were further analyzed by LC-MC/MC, and 219 peptides were identified (Supplementary Table S1). This showed that YSP-I was mainly composed of short peptides, where 138 peptides were tripeptides and the molecular weight of peptides were mainly concentrated between 200 and 600 Da, accounting for 86% (Figure 3). It indicated that YS protein was effectively hydrolyzed by trypsin. Furthermore, an online bioinformatics analysis was performed to predict the potential biological activities of these peptides. The peptides with a probability threshold where the peptide ranker exceeded 0.5 were selected for further analysis [30]. Meanwhile, non-toxicity, non-hypersusceptibility, and good water solubility were the key factors for screening highly specific peptides [31–33]. The good solubility of small molecular weight peptides could react with lipid free radicals, and they had strong antioxidant activity [4]. Based on these, 29 peptides with potential antioxidant activity, good water solubility, and non-toxicity were selected as shown in Table 1. The molecular weight of these peptides ranged from 259.11682 to 950.40937 Da, and the number of amino acid residues ranged from 3 to 11. Meanwhile, the 29 novel peptides have not been reported before.

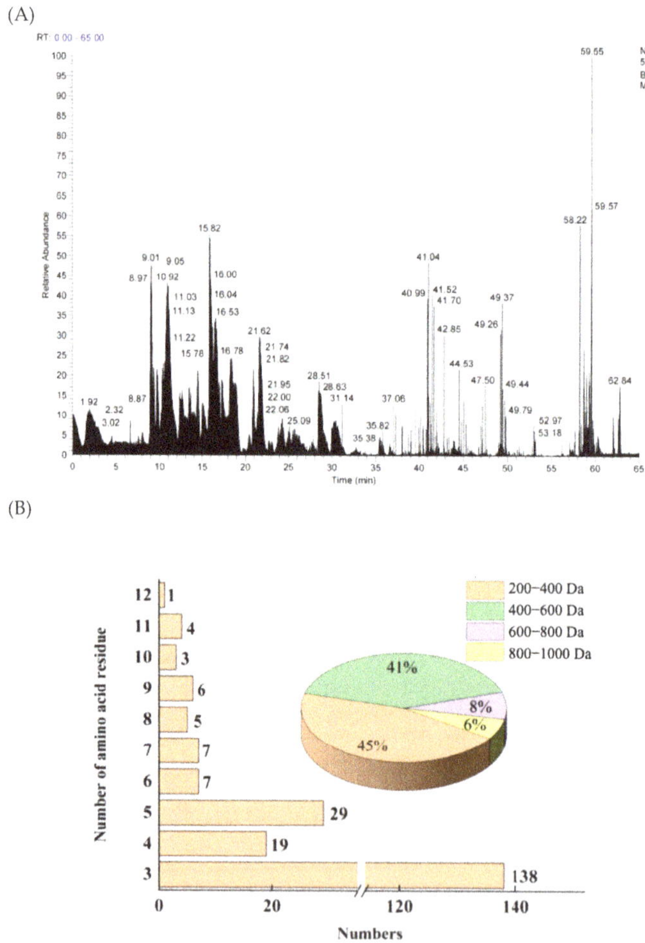

Figure 3. Total ion diagram of mass spectrum (**A**) and the peptides distribution (**B**) by LC-MS/MS.

Table 1. A total of 29 antioxidant peptides screened with good water solubility and non-toxicity and their binding energy of interaction with Keap1.

NO.	Sequence	Length	Mass	Score	Peptide Ranker	CDockerenergy (kcal/mol)
1	FGFDGDF	7	803.31262	187.48	0.93535	−141.445
2	GHNGLDGL	8	781.37187	114.54	0.534861	−136.723
3	GFDGDF	6	656.24421	148.12	0.913215	−119.541
4	FDGDF	5	599.22274	78.156	0.94884	−116.697
5	GFDGD	5	509.17579	118.96	0.637708	−115.582
6	GPSGPPGPDGN	11	950.40937	206.38	0.839787	−90.731
7	FEL	3	407.20564	18.31	0.566207	−84.218
8	GPAGPQGPR	9	835.43005	183.43	0.846699	−83.766
9	GFE	3	351.14304	12.647	0.720069	−83.496
10	GPAGPAGRPG	10	835.43005	131.04	0.873123	−82.544
11	FGE	3	351.14304	65.815	0.675338	−82.025
12	GEGGPQGPR	9	853.40423	142.64	0.536924	−81.513
13	WDT	3	420.1645	25.115	0.593031	−81.295
14	DML	3	377.16206	25.115	0.667515	−80.952
15	MDL	3	377.16206	47.364	0.679551	−78.408
16	WAD	3	390.15393	20.412	0.721933	−77.465
17	GPAGPAGRP	9	778.40859	155.73	0.846699	−71.902
18	PGPAGPAGRP	10	875.46135	107.32	0.871669	−71.298
19	DWP	3	416.16958	20.675	0.948036	−69.084
20	LMGPR	5	572.31045	85.518	0.767987	−68.852
21	PAGPAGPR	8	721.38712	109.7	0.813961	−68.468
22	DPC	3	333.09946	5.9869	0.819965	−68.046
23	FSGLD	5	537.24348	41.717	0.692962	−67.018
24	FCK	3	396.18313	11.715	0.907916	−62.174
25	GPK	3	300.17976	40.566	0.567438	−45.198
26	PQPPQ	5	565.28601	104.03	0.631168	−42.868
27	PAGRP	5	496.27578	91.004	0.74837	−42.160
28	PGS	3	259.11682	11.355	0.526816	−38.120
29	PGR	3	328.1859	32.689	0.839926	−34.945

2.4. Molecular Docking Simulation of Antioxidant Peptides Interacting with Keap1

The Keap1-Nrf2 signaling pathway was recognized as an important mechanism, in which bioactive compounds exerted antioxidant effects [34]. When oxidative damage occurred, excessive free radicals interacted with Keap1, and then Nrf2 was released and up-regulated the antioxidant response element (ARE) to reduce oxidative damage [20]. A large number of studies have shown that peptides might change the Keap1/Nrf2-ARE signaling pathway and thereby affect the activity of antioxidant enzymes in cells [35]. In this study, YSP has exhibited good free radical scavenging activity, but its molecular mechanisms remained unclear. Therefore, the molecular docking analysis was conducted to explore the molecular mechanism of antioxidant peptides by interacting with Keap1.

It was found that peptides with lower binding energy exhibited stronger binding affinity with Keap1 protein [36]. In the present study, the binding energy values of 29 peptides were ranked (Table 1) and five novel peptides were selected, including Phe-Gly-Phe-Asp-Gly-Asp-Phe (FGFDGDF), Gly-His-Asn-Gly-Leu-Asp-Gly-Leu (GHNGLDGL), Gly-Phe-Asp-Gly-Asp (GFDGD), Gly-Pro-Ala-Gly Pro-Gln-Gly-Pro-Arg (GPAGPQGPR), Leu-Met-Gly-Pro-Arg(LMGPR) (secondary mass spectrum shown in Figure 4), where the CDocker energy binding with Keap1 were −141.445 kcal/mol, −136.723 kcal/mol, −115.582 kcal/mol, −83.766 kcal/mol and −68.852 kcal/mol, respectively. FGFDGDF showed the strongest affinity with Keap1 protein. Studies showed that hydrophobic amino acids (such as Leu, Val, Ala, Pro, and Phe), aromatic amino acids, and basic amino acids contributed to the antioxidant activity of peptides [37]. Hydrophobic amino acids at the N-terminal or C-terminal could enhance the interaction of peptides with lipid components to promote their antioxidant function in vivo [38]. Pro, Ala, and Tyr were the key amino acids for free radical scavenging, while His exhibited strong antioxidant activity due to its imidazole group structure with a closed π bond arrangement and unpaired electrons [39]. These amino acids existed in the five screened peptides; it can be inferred that these peptides may possess good antioxidant activities.

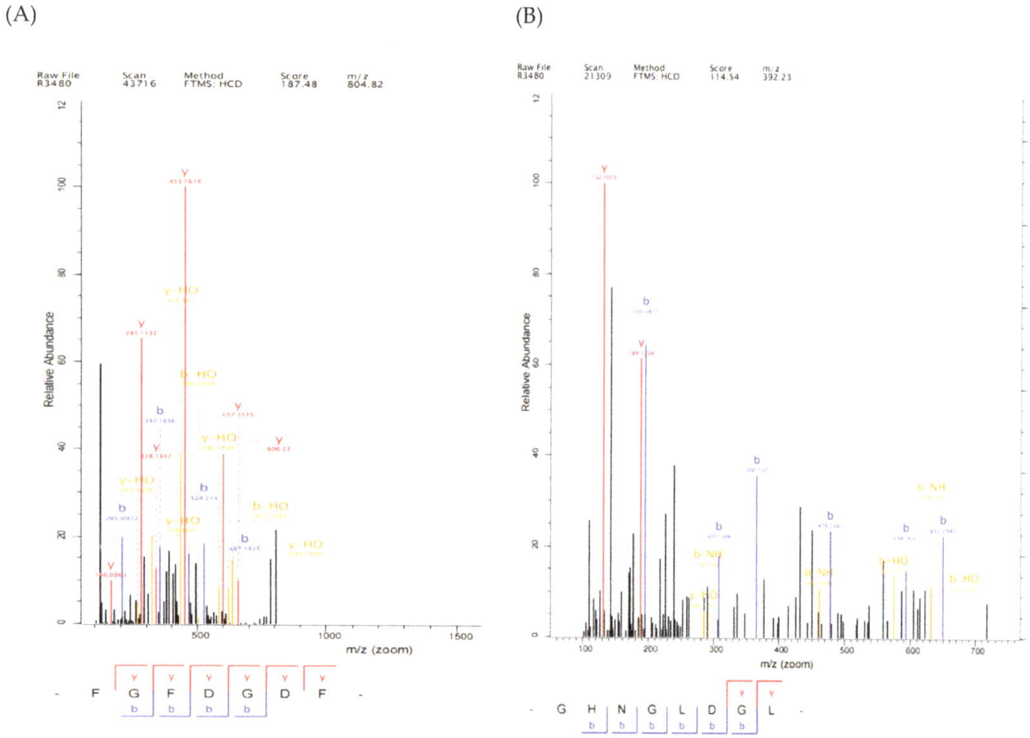

Figure 4. *Cont.*

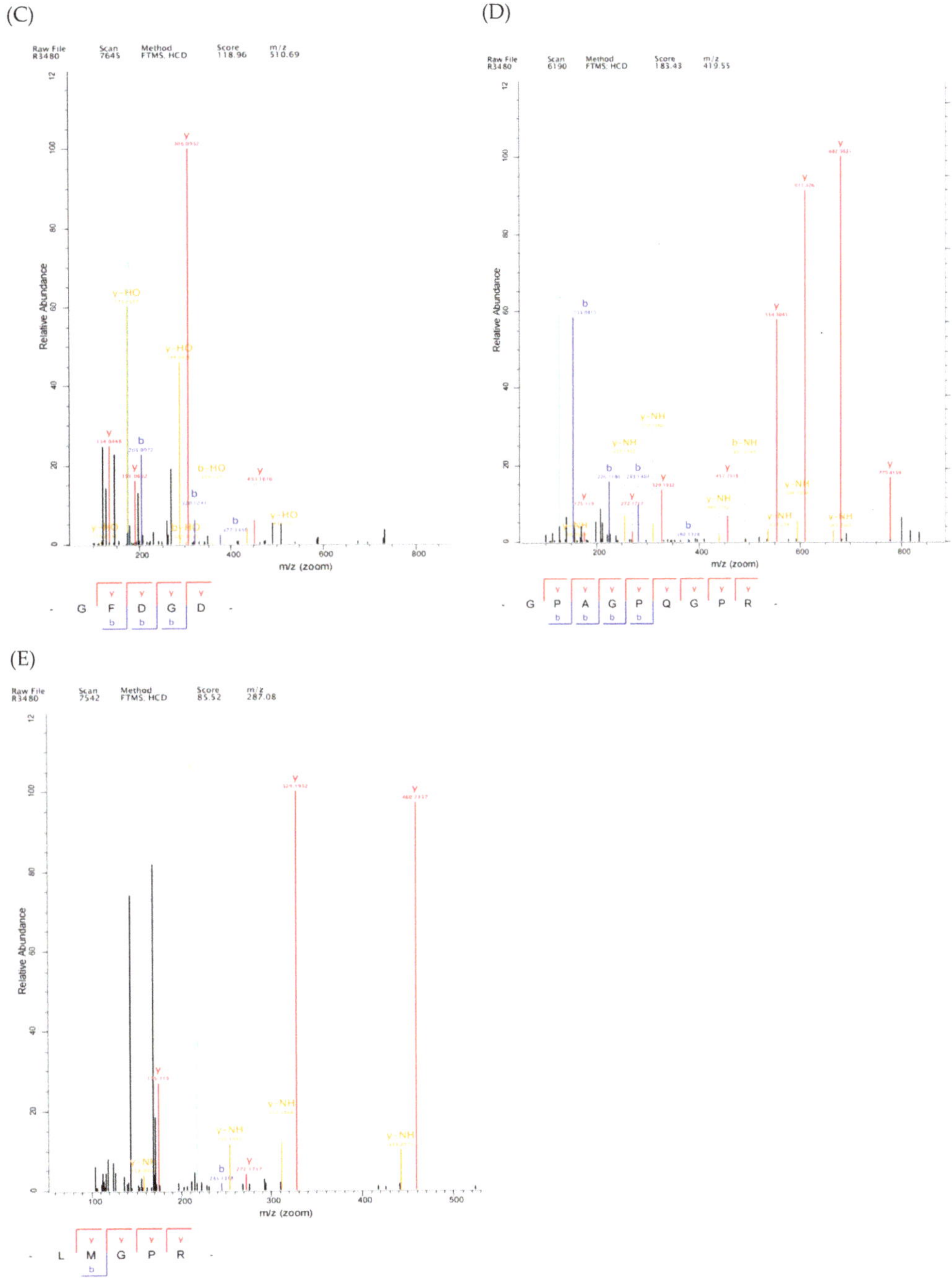

Figure 4. Secondary mass spectrum of five peptides, FGFDGDF (**A**), GHNGLDGL (**B**), GFDGD (**C**), GPAGPQGPR (**D**), and LMGPR (**E**).

According to the crystal structure of the Keap1-Nrf2 complex, the binding pocket can be divided into six bags, namely P1-P6 [40]. The interaction sites of peptides binding with Keap1 are shown in Figure 5. GFDGD was a part of FGFDGDF, and they shared the same active sites to bind with Keap1 (Figure 5A,C), including ASN-382, ARG-380, and ARG-415, which was basically consistent with the previous study [41]. Both FGFDGDF and GFDGD were bound to the P1, P2, P3, and P5 pockets of Keap1. Additionally, the GFDGD peptide was bound to the P4 and P6 active pockets because of its shorter chain. Considering the strong binding energy of FGFDGDF, further research is needed to in order to clarify whether adding or removing a specific amino acid could improve the antioxidant activity of the small peptide. As for the other three peptides, LMGPR was bound to the TYR-572 and ARG-483 residues in the P1 and P4 pockets of the Keap1 protein; GHNGLDGL was bound at the ARG-380, ASN-382 in P1-P4 and P6 pockets; and GPAGPQGPR interacted with SER-431, ASN-414, and other residues in the P1, P2, P4, and P6 pockets to form varieties of interactions. Keap1 mainly recognizes Nrf2 through ETGE and DLG motifs, which mainly binds to ARG-380, ARG-483, and other sites [42], which overlaps with the binding sites of identified peptides. Meanwhile, intermolecular hydrogen bonds, electrostatic interactions, and hydrophobic interactions were the main interactions observed between the five peptides and Keap1, which were previously reported [43]. These results showed that the five peptides could form novel and stable complexes with Keap1. In a word, all five peptides interacted with the six active pockets where Keap1 interacted with ligand Nrf2, and they were in contact with the key residues of Keap1-Nrf2 interactions. The results suggested that these peptides might induce Nrf2 release by competitively binding to Keap1, and they may exert antioxidant activity through activating the Keap1-Nrf2/ARE pathway. These peptides potentially facilitate the release of Nrf2 by competitively binding to Keap1. Subsequently, the activated Nrf2 is translocated into the nucleus, where it forms a heterodimer that binds to the antioxidant response element (ARE), thereby initiating the expression of target genes and exerting antioxidative effects [44]. Pei et al. [45]. also found that antioxidant peptides CRPCGPTP and ANSCNEPCVR, derived from feather keratin, could bind well to key amino acids in the Keap1 Kelch domain and directly disrupt the Keap1-Nrf2 interaction.

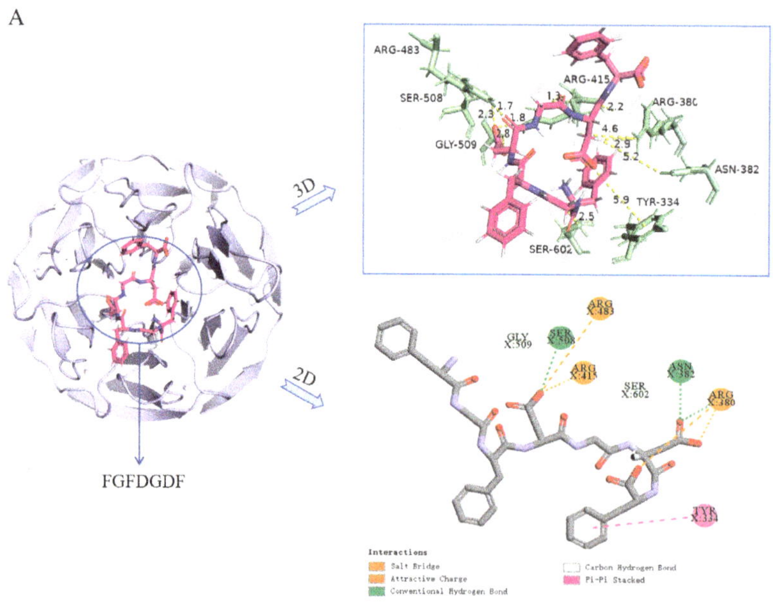

Figure 5. *Cont.*

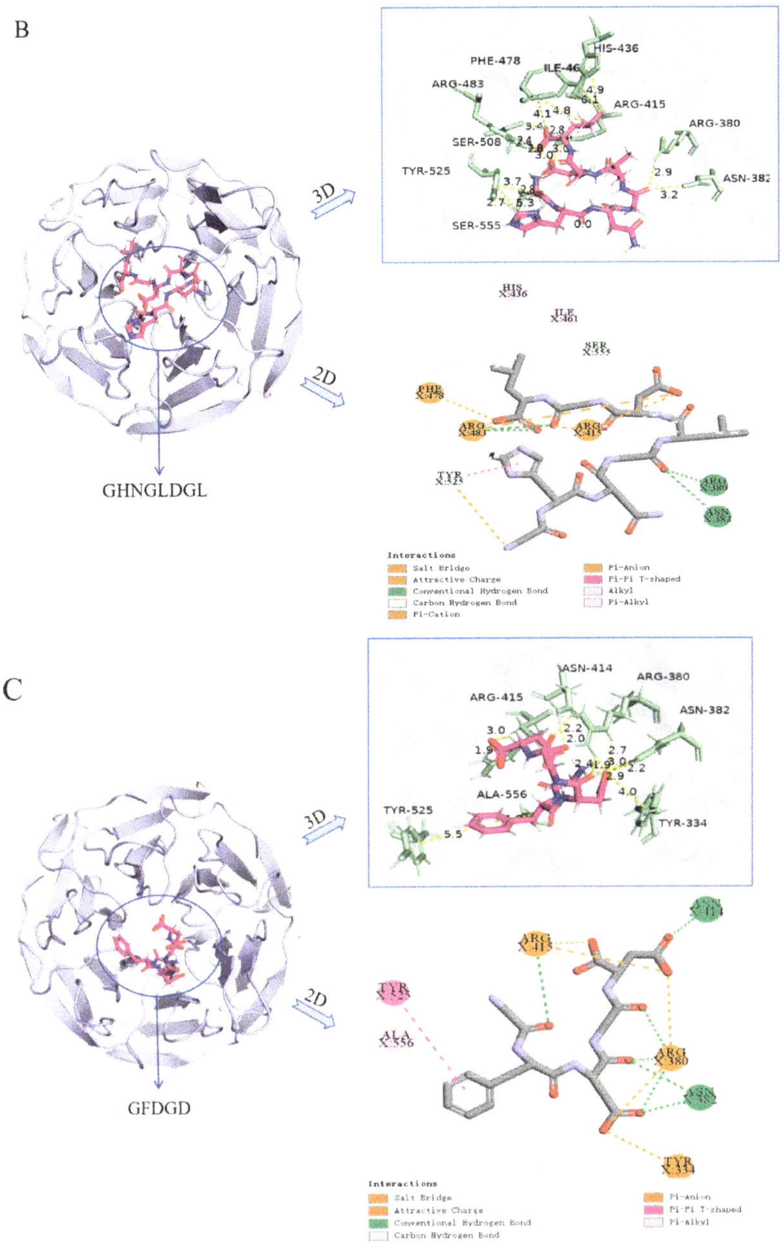

Figure 5. *Cont.*

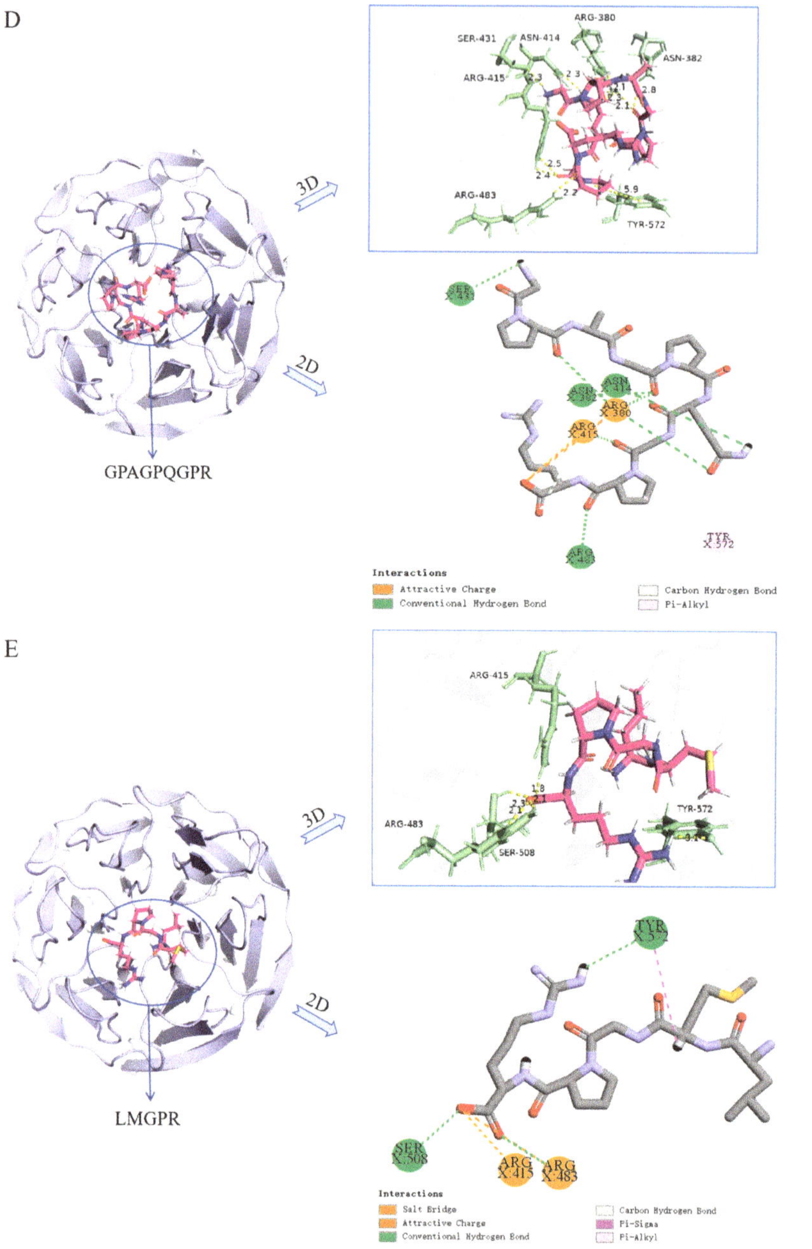

Figure 5. The 2D and 3D simulation diagrams of five antioxidant peptides binding with Keap1 (PDB ID:2FLU) by molecular docking, FGFDGDF (**A**), GHNGLDGL (**B**), GFDGD (**C**), GPAGPQGPR (**D**), and LMGPR (**E**), respectively.

2.5. Free Radical Scavenging Activities of Peptides

The above five screened peptides were synthesized, and their free radical scavenging activities are shown in Figure 6. FGFDGDF and LMGPR showed relatively better DPPH radical scavenging activities compared to the other three peptides, with significant differences ($p < 0.05$). The scavenging rates were 25.80% ± 1.82% and 23.56% ± 2.68%, respectively. FGFDGDF also demonstrated a better ABTS free radical scavenging activity than the other four peptides, with a significant difference ($p < 0.05$), and its clearance rate was 19.15% ± 1.59%. FGFDGDF and GFDGD showed higher hydroxyl radical scavenging activity than the remaining three peptides, with significant differences ($p < 0.05$), and their clearance rates were 44.44% ± 1.94% and 41.61% ± 1.73%, respectively. It can be observed that all five peptides reflected scavenging activity on the free radical. Among them, FGFDGDF had the best free radical scavenging activity, which was coincident with the results of molecular docking. However, the antioxidant activity of peptides may be limited due to the insufficient density of amino acid groups containing phenolic hydroxyl/indole groups such as tyrosine (Tyr) and tryptophan (Trp). Liu et al. [46] believed that the phenolic hydroxyl group of Tyr in the peptide chain was active. The antioxidant activity increased when Tyr was between DD (Asp-Asp) or DQ (Gln-Asp).

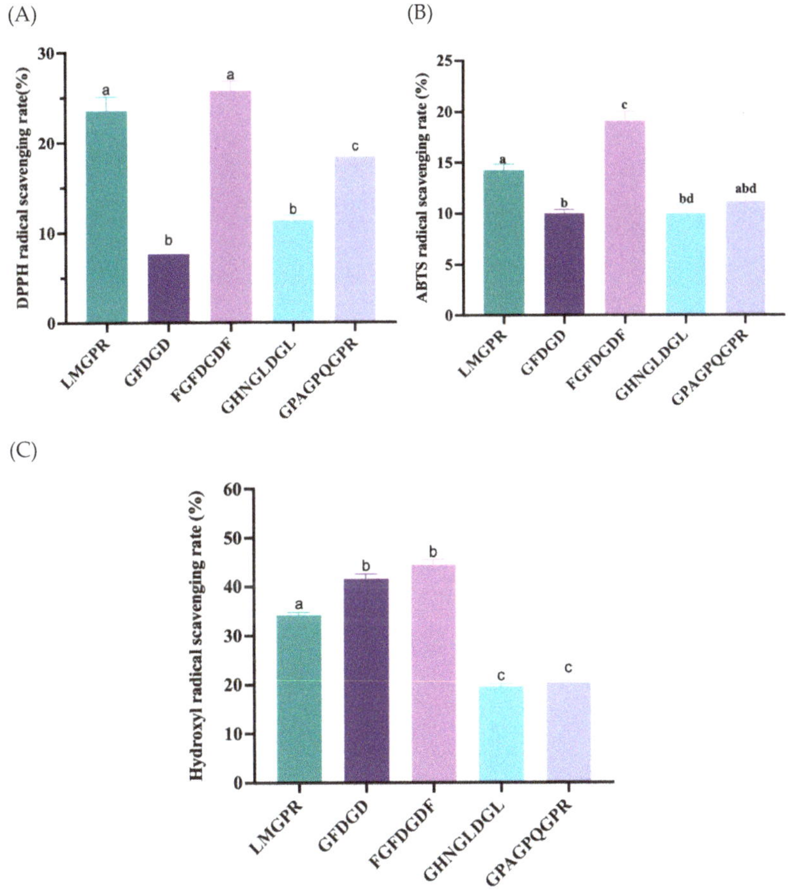

Figure 6. The free radical scavenging activities of synthetic peptides, including the DPPH (**A**), ABTS (**B**), and hydroxyl radical (**C**) scavenging rate of five peptides, respectively. Different letters represent significant differences ($p < 0.05$).

2.6. Protective Effect of Peptides on the Oxidative Damage of HepG2 Cells Induced by H_2O_2

To further verify the biological antioxidant activities of peptides, the protective effect on the oxidative damage of HepG2 cells was investigated. As a hepatotoxic chemical with a long half-life, H_2O_2 can be converted into hydroxyl free radicals and oxygen free radicals, directly causing the oxidative stress damage of cells [47]. As shown in Figure 7A, with the increase of H_2O_2 concentration from 0.8 μM to 1.6 mM, the cell viability significantly decreased from 91.77 ± 4.53% to 37.01 ± 3.20% ($p < 0.05$). Under the treatment with 1.4 mM H_2O_2, the cell survival rate was 51.29 ± 3.06%, which was selected to establish the cell oxidative damage model for subsequent experiments. Meanwhile, the cytotoxicity of peptides was also determined (Figure 7B). With the peptides concentration ranging from 0.1 mg/mL to 2.5 mg/mL, the cell viability was over 90% for all, which indicated that the five peptides had no toxic effect on the normal proliferation of HepG2 cells. When pretreated with antioxidant peptides, the cellular oxidative damage by H_2O_2 was reduced (Figure 7C). Compared with the model group, the survival rate of HepG2 cells significantly increased with the pretreatment concentration of five peptides. It indicated that the five peptides had a good protective effect on the oxidative damage of HepG2 cells.

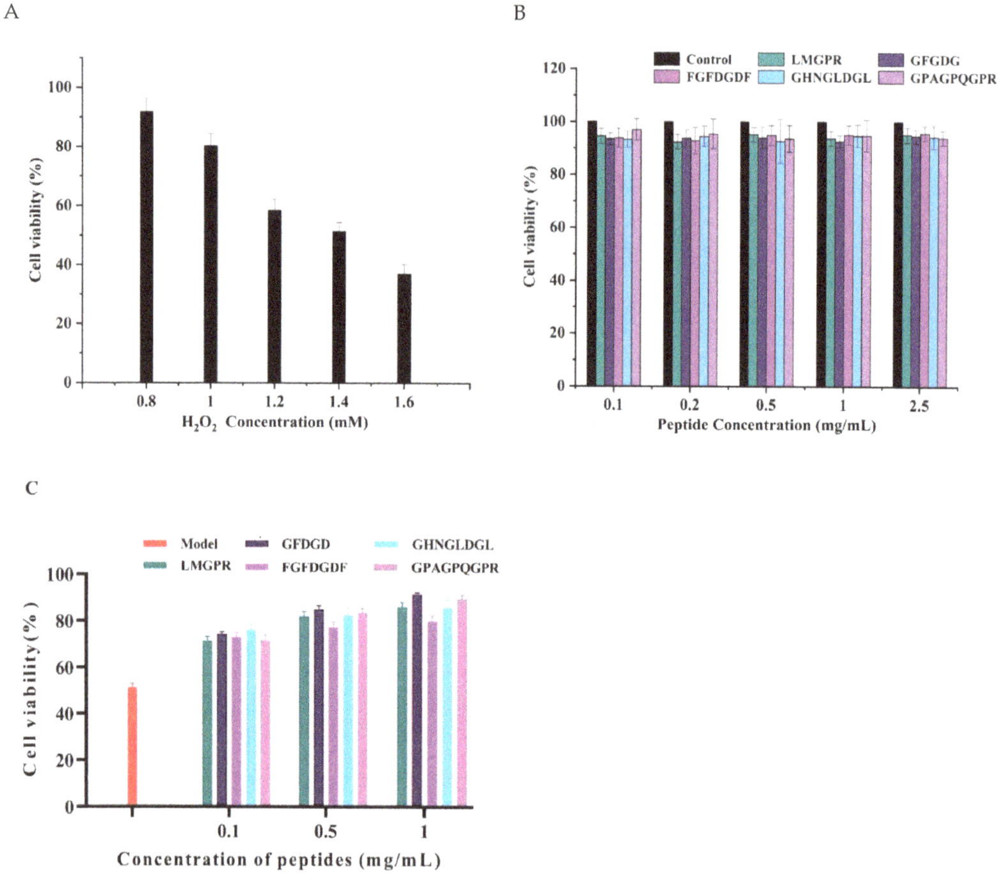

Figure 7. Effect of different concentrations of H_2O_2 (**A**) and five peptides (**B**) on the cell viability of HepG2 cells and the protective effect of peptides against oxidative stress induced by H_2O_2 (**C**).

2.7. Effects of Antioxidant Peptides on Antioxidant Enzymes, Malondialdehyde (MDA), and IL-8 Contents in H_2O_2-Induced HepG2 Cells

As shown in Figure 8A,B, the activities of CAT and SOD in HepG2 cells decreased to 26.01 ± 1.78 U/mg prot and 6.17 ± 0.41 U/mg pro, which were significantly lower than those in the control group (60.80 ± 1.10 U/mg prot, 14.35 ± 0.06 U/mg prot). Under the pretreatment of peptides (LMGPR, GFDGD, FGFDGDF, GHNGLDGL, GPAGPQGPR), the activities of SOD and CAT were significantly more improved than those in the model group ($p < 0.05$). The regulation of endogenous antioxidant enzymes SOD and CAT in cells is one of the main mechanisms of antioxidant peptides [48]. The results indicated that LMGPR, GFDGD, FGFDGDF, GHNGLDGL, and GPAGPQGPR could regulate the activity of antioxidant enzymes and protect the cells from oxidation damage. Specifically, FGFDGDF exhibited the best protective effect, where the activities of SOD and CAT in cells improved, ebing 11.83 ± 0.16 U/mg prot and 47.49 ± 1.14 U/mg prot, which were consistent with the above results.

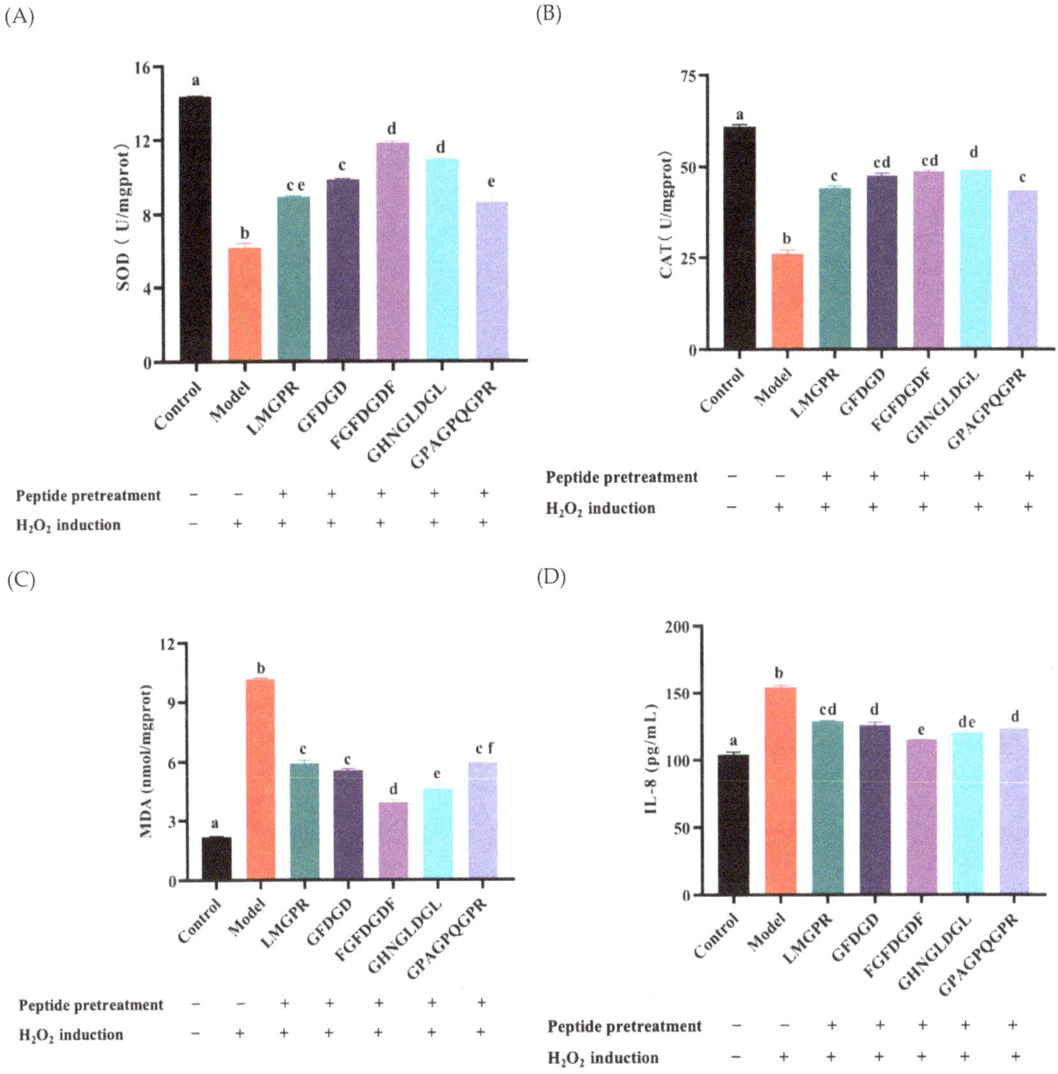

Figure 8. Effects of different peptides on the content of CAT (**A**), SOD (**B**), MDA (**C**), and IL−8 (**D**) in H_2O_2−induced HepG2 cells. Different letters represent significant differences ($p < 0.05$).

Moreover, MDA is one of the key products of cell membrane lipid peroxidation under oxidative stress [49]. Studies have shown that the higher the MDA content, the more serious the cell damage [28]. As shown in Figure 8C, compared with the control group, the MDA content in HepG2 cells of the group significantly increased (10.31 ± 0.23 nmol/mg prot) ($p < 0.05$). It indicated that lipid peroxidation occurred in the HepG2 cells, induced by H_2O_2. After pretreatment with LMGPR, GFDGD, FGFDGDF, GHNGLDGL, and GPAG-PQGPR, the MDA contents were significantly decreased to 5.91 ± 0.27 nmol/mg prot, 5.55 ± 0.16 nmol/mg prot, 3.945 ± 0.21 nmol/mg prot, 4.65 ± 0.39 nmol/mg prot, and 5.96 ± 0.30 nmol/mg prot, respectively ($p < 0.05$).

In addition, IL-8 is a pro-inflammatory cytokine, and H_2O_2-induced oxidative stress can trigger a cellular inflammatory response [50]. After the induction of 1.4 mM H_2O_2, the IL-8 content in cells significantly increased to 154.02 ± 2.80 pg/mL ($p < 0.05$), compared with that in the control group (110.92 ± 2.52 pg/mL) (Figure 8D). Under the pretreatment of five peptides (LMGPR, GFDGD, FGFDGDF, GHNGLDGL, GPAGPQGPR), the intracellular IL-8 content significantly reduced ($p < 0.05$), which indicated potential anti-inflammatory and antioxidant effects.

In general, the five peptides exhibited good cellular antioxidant activities by up-regulating the activity of intracellular antioxidant enzymes in reaction to hydrogen peroxide-induced oxidative stress and decreasing the levels of MDA and IL-8, especially FGFDGDF. The antioxidant peptides screened from YS could reduce oxidative damage and inflammation, further exerting a certain protective effect on cells. However, in vitro cell culture is usually carried out in a highly controlled environment, lacking a complex physiological environment in the body, and cannot accurately simulate the absorption and distribution of drugs in the body. So, further animal experiments and cellular markers needed to be conducted to validate the antioxidant effects and specific mechanism of these synthetic antioxidant peptides.

3. Materials and Methods

3.1. Materials and Reagents

Yak (Maiwa Yak) skins were collected and supplied from Plateau Song Food Co. Ltd. in Hongyuan County, Aba Prefecture, Sichuan Province of China, and transported to the laboratory through refrigeration under 20 °C. Trypsin (100,000 U/g) was purchased from Nanning Donghenghuadao Biotechnology Co., Ltd. (Guangxi, China). 1,1-diphenyl-2-trinitrophenylhydrazine (DPPH) was purchased from Sigma-Aldrich Chemical Co. (St. Louis, MO, USA). 2,2′-diazo-bis (3-ethylbenzothiazolin-6-sulfonic acid) diammonium salt (ABTS) was purchased from Keyuan Biochemical Co., Ltd. (Jinan, China). The HepG2 cell was donated by the laboratory of Tianjin University of Science and Technology (Tianjin, China). DMEM medium, fetal bovine serum (FBS), cell counting kit-8 (CCK-8), penicillin, and streptomycin were purchased from Dalian Meilun Biotechnology Co., Ltd. (Dalian, China). Phosphate-buffer saline (PBS) was purchased from Beijing Lanjieke Biotechnology Co., Ltd. (Beijing, China). The superoxide dismutase (SOD) test kit, hydrogen peroxide (CAT) test kit, malondialdehyde (MDA) test kit, and human interleukin 8 (IL-8) enzyme-linked immunosorbent test kits were all purchased from Nanjing Jiancheng Co., Ltd. (Nanjing, China). All other chemicals used were analytical grade.

3.2. Preparation of YSP by Enzymatic Hydrolysis

Enzymatic hydrolysis was adopted for the preparation of yak skin peptides (YSP). After being thawed at 4 °C, the yak skin (YS) was removed from the hair and subcutaneous connective tissue and cut into small pieces. After homogenization with water, YS powder was prepared under 80 °C, drying for 3 h and then sifted through 60 mesh. Then the YS was hydrolyzed at a ratio of powder-to-water of 1:30 (g/mL) using trypsin at a concentration of 7200 U/g in a water bath at 54 °C for 3.2 h. Subsequently, the enzyme was deactivated under 100 °C for 10 min and the supernatant was collected by centrifugation at 3500 r/min for 20 min. After being freeze-dried, the YSP was obtained and stored at −20 °C for use.

3.3. Fourier Transform Infrared Spectroscopy (FTIR) Analysis

The secondary structure of YS and YSP was analyzed by Fourier transform infrared spectroscopy (FTIR) with some modifications according to the previous scheme [51]. The transparent film was prepared by mixing a 1 mg sample with 150 mg potassium bromide and pressing the tablet. Then the infrared spectrum of the sample was recorded by 32 scans at a resolution of cm^{-1} in the range of 4000–500 cm^{-1}. The data was analyzed using OMNIC 8.2 software.

3.4. Scanning Electron Microscopy Analysis

Solid samples of YS and YSP were fixed on the slide with conductive adhesive. After being coated with gold, the samples were observed using an emission scanning electron microscope at a magnification of 1000× and an acceleration voltage of 20 kV.

3.5. Ultrafiltration Separation

The YSPs were separated by ultrafiltration membranes with molecular weight cutoffs of 1 kDa, 3 kDa, and 10 kDa (TG 16.6, Lu Xiangyi Centrifuge Instrument Co., Ltd., Shanghai, China). Four fractions with different molecular weight were obtained, including YSP-I (<1 kDa), YSP-II (1–3 kDa), YSP-III (3–10 kDa), and YSP-IV (>1 kDa). Each fraction was gathered separately and freeze-dried for further analysis (MODULYOD-230, Thermo Fisher Technology Co. Ltd., Waltham, MA, USA).

3.6. Evaluation on the Free Radical Scavenging Activity

3.6.1. DPPH Radical Scavenging Activity

Slightly modified according to the previous method [52], the 100 μL sample solution was mixed with 100 μL 0.1 mmol/L DPPH anhydrous ethanol solution and reacted in the dark at room temperature for 30 min. Then the absorbance was measured at 517 nm. The DPPH free radical scavenging rate of the sample was calculated by the formula below.

$$DPPH\ radical\ scavenging\ rate(\%) = \left(1 - \frac{A_1 - A_2}{A_0}\right) \times 100, \quad (1)$$

where A1 represented the absorbance value the sample reaction with DPPH anhydrous ethanol solution; A2 represented the absorbance value of the sample's reaction with anhydrous ethanol solution; A0 represented the absorbance of the reaction mixture solution without the sample.

3.6.2. ABTS Radical Scavenging Activity

Slightly modified according to the previous method [53], 7.4 mmol/L ABTS solution was mixed with the 2.6 mmol/L $K_2S_2O_8$ solution in equal volume, and then was kept in the dark for 16 h. The ABTS working solution was obtained when the mixture was subsequently diluted with anhydrous ethanol to adjust the absorbance to a range of 0.70 ± 0.02 at 734 nm. Then, the sample solution of 40 μL was mixed with 160 μL ABTS working solution and reacted at room temperature for 10 min in the dark on 96 microporous plates. Meanwhile, the sample was replaced with distilled water as the blank. The absorbance was measured at 734 nm, and the ABTS radical scavenging rate of samples was calculated using the formula below.

$$ABTS\ radical\ scavenging\ rate(\%) = \frac{A_0 - A_1}{A_1} \times 100, \quad (2)$$

where A_0 and A_1 represented the absorbance values of the blank and sample groups.

3.6.3. Hydroxyl Radical Scavenging Activity

The hydroxyl radical scavenging activities of samples were performed as in the previous method with slight modifications [54]. Briefly, 6.0 mmol/L salicylic acid solution and $FeSO_4$ solution were prepared, and 0.5 mL of each solution was added into a reaction tube, followed by the addition of a 0.5 mL sample solution and a 0.5 mL H_2O_2 solution of 6.0 mmol/L. After mixing, the reaction was conducted in a 37 °C water bath for 0.5 h, and then the absorbance was measured at 510 nm. Meanwhile, the sample was replaced with distilled water and used as the blank group, and H_2O_2 was replaced with distilled water and used as the control group. The hydroxyl radical scavenging rate of the sample was calculated according to the formula below.

$$hydroxyl\ radical\ scavenging\ rate(\%) = \left(1 - \frac{A_0 - A_1}{A}\right) \times 100, \quad (3)$$

where A_0, A_1, and A represented the absorbance values of the sample group, blank group, and control group, respectively.

3.7. Sequence Identification of YSP

The amino acid sequence of an ultrafiltration fraction with relatively high antioxidant activity was identified by LC-MS/MS. The samples were uploaded by an autosampler to Zorbax 300SB-C18 peptide traps (Agilent Technologies, Wilmington, DE, USA) and separated on a liquid chromatography column (0.15 mm × 150 mm, RP-C18, Column Technology luc., Lombard, IL, USA). The mobile phases consisted of phase A (water with 0.1% formic acid) and phase B (84% acetonitrile in water with 0.1% formic acid). The elution gradient of phase B was increased linearly from 4% to 50% in 0–50 min, increased from 50% to 100% in 50–54 min, and then kept at 100% in 54–60 min. Subsequently, the samples and their fragments were analyzed using a Q Exactive HF-X mass spectrometer (Thermo Fisher, USA) with a positive ion for 60 min. The m/z of peptides and their fragments were collected according to the 10 fragmentation profiles after each full scan. The amino acid sequences of peptides were identified and quantified by searching the corresponding database with MaxQuant 1.5.5.1.

3.8. Online Informatics Screening of Potential Antioxidant Peptides

Peptide sequences with high activity, non-toxicity, and good water solubility were screened. The Peptide Ranker (http://distilldeep.ucd.ie/PeptideRanker/) (accessed on 18 November 2023) and ToxinPred (http://crdd.osdd.net/raghava/toxinpred/index.html) (accessed on 18 November 2023) online programs were used to predict peptide bioactivity probability and the toxicity of the peptides [55,56]. In addition, the water solubility of peptides was also determined using the online computational tool Innovagen (http://www.innovagen.com/proteomics-tools) (accessed on 18 November 2023) [31].

3.9. Molecular Docking of YSP with Keap1 Protein

The interactions between peptides and Keap1 protein were investigated by molecular docking. Firstly, the 2D structural formula of peptides were drawn using Kingdraw Offical 4.0 software, and transformed into 3D structures by Kingdraw Offical 4.0 software. MM2 in the calculation function was selected for the energy minimization of peptides, then, the small ligand molecules of peptides were obtained. The crystal structure of the Keap1-Nrf2 complex (PDB ID: 2FLU) was obtained from the PDB protein database. Afterwards, the molecular docking was performed using Autodock 4.2.3. Water molecules, small solvent molecules, Nrf2 16-mer peptides, and other ligands were removed from the protein structure, and the protein was hydrogenated. The molecular docking mode was performed based on semi-flexible molecular docking (CDOCKER), with the receptor protein set to rigid and the ligand small molecule set to flexible [20]. The active pocket center coordinates were x: 5, y: 9, and z: 1, and the radius was 15 Å. The CDOCKER_ENERGY value, interaction sites, and intermolecular action forces were used to analyze the molecular docking results. The 3D and 2D interactions between the peptides and Keap1 were visualized using PyMol 2.5 software.

3.10. *Synthesis of Antioxidant Peptides*

The screened peptides with potential antioxidant activity were further synthesized with purity over 90%, as in the solid phase method explored by Nanjing Yuan-Peptide Biotechnology Co., Ltd. (Nanjing, China). Then the free radical scavenging activities and cellular antioxidant activities of the synthesized peptides were evaluated.

3.11. *Evaluation on the Cellular Antioxidant Activity of YSP*

3.11.1. Cell Culture

HepG2 cells were cultured in DMEM growth medium supplemented with 10% FBS and 50 U/mL penicillin-streptomycin in a culture flask. The flask was incubated in an incubator of 5% CO_2 and kept at 37 °C. The culture medium was refreshed every 2–3 days. Once the HepG2 cells adhered to the flask wall and achieved a coverage rate of over 85%, the cells were digested using a 0.25% trypsin solution for passage.

3.11.2. Determination of Cell Viability

Cell viability was determined by the CCK-8 method. The synthesized peptides were dissolved in DMEM growth medium containing 10% FBS with concentrations of 0.1, 0.2, 0.5, 1.0, and 2.5 mg/mL. HepG2 cell suspensions were seeded in 96-well plates at a density of 5×10^4 cells/mL with 200 μL in each well, and cultured for 24 h until the cell fusion reached about 85%. After washing twice with PBS, 100 μL of each peptide solution was added to the cells and incubated for 24 h. Finally, 90 μL of culture medium along with 10 μL of CCK-8 reagent were added to each well for a further 2 h incubation. Then the absorbance of each well was measured at 450 nm using a microplate reader. Meanwhile, a blank group without cells and a control group without peptides were established. The cell viability was calculated using the formula below.

$$\text{Cell viability} = \left(\frac{As - A_0}{Ac - A_0}\right) \times 100 \tag{4}$$

As represented the absorbance of sample treatment, while A_0 and *Ac* represented the absorbances of the blank and control groups.

3.11.3. Establishment of Oxidative Stress Model in HepG2 Cells

HepG2 cells were cultured in 96-well plates at a density of 5×10^4 cells/mL for 24 h. After the medium was removed, 100 μL of complete medium containing different concentrations of H_2O_2 (800, 1000, 1200, 1400, 1600 μM) was added and incubated for 2 h. After the medium was discarded, cell viability was determined using the CCK-8 assay to establish the optimal H_2O_2 concentration for inducing an oxidative stress model. The cells without H_2O_2 treatment were used as control group.

3.11.4. Treatment of Peptides on H_2O_2-Induced HepG2 Cells

HepG2 cells were cultured in 96-well plates at a density of 5×10^4 cells/mL for 24 h until the cell confluence reached about 85%. After the medium was discarded, 100 μL peptides of different concentrations were added to the wells and cultured for an additional 24 h. After discarding the medium and then being washed, the cells were incubated with 100 μL of an appropriate concentration of H_2O_2. After incubation for 2 h, the supernatant was collected, and the content of IL-8 was determined strictly, according to the kit instructions. Afterwards, the cells were washed twice with PBS, and then the cell viability was measured. Alternatively, the cells were disrupted with 300 μL of cell lysate and centrifuged at 6000× *g* at 4 °C for 15 min. The supernatant was collected and stored at −80 °C. The protein concentration was determined using the Bio-Rad DC analysis with BSA as the standard curve. In addition, the contents of SOD, CAT, and MDA in cell lysates were determined in strict accordance with the kit instructions.

3.12. Statistical Analysis

All experiments were repeated at least three times, and the data were expressed as mean ± standard deviation. Prism 9 and PyMol 2.5 were used for statistical analysis and data plotting. One way ANOVA was performed for significance difference analysis and multiple comparison, where $p < 0.05$ represented a significant difference between the data.

4. Conclusions

In conclusion, YSP was prepared and analyzed using the LC-MS/MS technique in this study. A total of 219 peptides were identified and five peptides with high scores (LMGPR, GFDGD, FGFDGDF, GHNGLDGL, and GPAGPQGPR) were selected for verification through molecular docking prediction with Keap1. Through a H_2O_2-induced oxidative HepG2 cell model, these five antioxidant peptides demonstrated a significant protective effect by increasing the activity of intracellular antioxidant enzymes and reducing the levels of MDA and IL-8. Among them, FGFDGDF exhibited relatively strong free radical scavenging activities and protective effects. These findings provide support for the application of YSP as a new antioxidant in functional foods or in the treatment of diseases related to oxidative stress. Future research may explore the application of YSP as a food additive to enhance the taste, color, and nutritional value of food products. In the cosmetics industry, YSP could serve as an active ingredient to augment the moisturizing, sunscreen, and antioxidant properties of cosmetic formulations.

Supplementary Materials: The following supporting information can be downloaded at: https://www.mdpi.com/article/10.3390/ijms26135976/s1.

Author Contributions: Investigation: Y.H., Y.J. and Z.Z.; data curation: Y.H.; writing—original draft: Y.H. and N.Z.; writing—review and editing: N.Z., Y.J. and E.J.; formal analysis: N.Z.; supervision: Y.J., T.W. and M.Z.; resources: Y.K. and W.S.; funding acquisition: M.Z. All authors have read and agreed to the published version of the manuscript.

Funding: This work was supported by the National Key R&D Program of China (2022YFF1100201) and Tianjin Innovation Platform Project (24PTLYHZ00020).

Institutional Review Board Statement: Not applicable.

Informed Consent Statement: Not applicable.

Data Availability Statement: The data that support the findings of this study are available from the first author upon reasonable request.

Conflicts of Interest: The authors declare no conflicts of interest.

References

1. Wu, J.N.; Yi, Z.W.; Chen, X.T.; Pan, N.; Su, X.; Shi, H.F.; Liu, Z.Y. Screening and mechanistic study of antioxidant peptides from Bangia fusco-purpurea targeting the Keap1–Nrf2 pathway. *Food Biosci.* **2024**, *59*, 104122. [CrossRef]
2. Liang, R.; Xu, L.; Fan, C.; Cao, L.L.; Guo, X.F. Structural Characteristics and Antioxidant Mechanism of Donkey-Hide Gelatin Peptides by Molecular Dynamics Simulation. *Molecules* **2023**, *28*, 7975. [CrossRef]
3. Abdo, A.A.A.; Hou, Y.; Hassan, F.A.; Al-Sheraji, S.H.; Aleryani, H.; Alanazi, A.; Sang, Y. Antioxidant potential and protective effect of modified sea cucumber peptides against H2O2-induced oxidative damage in vitro HepG2 cells and in vivo zebrafish model. *Int. J. Biol. Macromol.* **2024**, *266*, 131090. [CrossRef]
4. Wang, L.; Li, Z.Y.; Fan, X.; Zhang, T.; Wang, H.; Ye, K.P. Novel antioxidant peptides from bovine blood: Purification, identification and mechanism of action. *Lwt Food Sci. Technol.* **2024**, *205*, 116499. [CrossRef]
5. Wang, R.H.; Zhao, Y.; Xue, W.J.; Xia, Y.T.; Liang, G.Z. Novel antioxidant peptides from soybean protein by employ computational and experimental methods and their mechanisms of oxidative stress resistance. *J. Mol. Struct.* **2024**, *1318*, 139284. [CrossRef]
6. Wang, Y.M.; Zhang, Z.; Sheng, Y.; Chi, C.F.; Wang, B. A systematic review on marine umami peptides: Biological sources, preparation methods, structure-umami relationship, mechanism of action and biological activities. *Food Biosci.* **2024**, *57*, 103637. [CrossRef]
7. Chen, H.J.; Dai, F.J.; Chen, C.Y.; Fan, S.L.; Zheng, J.H.; Chau, C.F.; Lin, Y.S.; Chen, C.S. Effects of molecular weight fraction on antioxidation capacity of rice protein hydrolysates. *Sci. Rep Uk* **2023**, *13*, 3464. [CrossRef]
8. Zhang, Z.J.; Shang, Y.T.; Li, S.T.; Chen, Z.; Xia, J.X.; Tian, Y.L.; Jia, Y.M.; Ma, A.J. Molecular Docking Revealed the Potential Anti-Oxidative Stress Mechanism of the Walnut Polypeptide on HT22 Cells. *Foods* **2023**, *12*, 1554. [CrossRef]
9. Feng, R.F.; Zou, X.X.; Wang, K.; Liu, H.G.; Hong, H.; Luo, Y.K.; Tan, Y.Q. Antifatigue and microbiome reshaping effects of yak bone collagen peptides on Balb/c mice. *Food Biosci.* **2023**, *52*, 102447. [CrossRef]
10. Joshi, S.; Shrestha, L.; Bisht, N.; Wu, N.; Ismail, M.; Dorji, T.; Dangol, G.; Long, R. Ethnic and Cultural Diversity amongst Yak Herding Communities in the Asian Highlands. *Sustainability* **2020**, *12*, 957. [CrossRef]
11. Yang, H.; Xue, Y.; Liu, J.; Song, S.; Zhang, L.; Song, Q.; Tian, L.; He, X.; He, S.; Zhu, H. Hydrolysis Process Optimization and Functional Characterization of Yak Skin Gelatin Hydrolysates. *J. Chem.* **2019**, *2019*, 9105605. [CrossRef]
12. Xu, M.Q.; Wei, L.X.; Xiao, Y.C.; Bi, H.T.; Yang, H.X.; Du, Y.Z. Molecular structural properties of extracted gelatin from Yak skin as analysed based on molecular weight. *Int. J. Food Prop.* **2017**, *20*, S543–S555. [CrossRef]
13. Ma, X.; Wang, W.; Shi, H.; Kong, X.; Zhang, L. Identification of novel antifreeze peptides from yak skin gelatin ultrasound-assisted enzymatic hydrolysate. *Ultrason. Sonochemistry* **2024**, *111*, 107102. [CrossRef]
14. Jiang, C.N.; Meng, A.G.; Shi, X.Y.; Fu, Z.P.; Wang, Y.L.; Zhou, J.J.; Zhang, X.W.; Liu, C.Y. Preparation of antioxidant peptides from yak skin gelatin and their protective effect on myocardial ischemia reperfusion injury. *Food Funct.* **2024**, *15*, 7961–7973. [CrossRef] [PubMed]
15. Tian, L.; Liu, J.H.; Ma, L.; Zhang, L.; Wang, S.P.; Yan, E.F.; Zhu, H.J. Isolation and Purification of Antioxidant and ACE-Inhibitory Peptides from Yak Skin. *J. Food Process Pres.* **2017**, *41*, e13123. [CrossRef]
16. Ci, X.M.; Liu, R.; Sun, Y.T.; Rifky, M.; Liu, R.; Jin, Y.; Zhu, Q.M.; Zhang, M.; Wu, T. A novel antioxidant iron-chelating peptide from yak skin: Analysis of the chelating mechanism and digestion stability. *J. Sci. Food Agr.* **2024**, *104*, 7907–7916. [CrossRef]
17. Liu, R.; Xing, L.J.; Fu, Q.Q.; Zhou, G.H.; Zhang, W.G. A Review of Antioxidant Peptides Derived from Meat Muscle and By-Products. *Antioxidants* **2016**, *5*, 32. [CrossRef]
18. Guo, X.J.; Liu, J.X.; Wang, C.; Wen, Z.S.; Zheng, B. The Antioxidant Mechanism of Peptides Extracted from Tuna Protein Revealed Using a Molecular Docking Simulation. *Antioxidants* **2024**, *13*, 166. [CrossRef]
19. Ouertani, A.; Chaabouni, I.; Mosbah, A.; Long, J.; Barakat, M.; Mansuelle, P.; Mghirbi, O.; Najjari, A.; Ouzari, H.I.; Masmoudi, A.S.; et al. Two new secreted proteases generate a casein-derived antimicrobial peptide in Bacillus cereus food born isolate leading to bacterial competition in milk. *Front. Microbiol.* **2018**, *9*, 1148. [CrossRef]
20. Shaw, P.; Chattopadhyay, A. Nrf2–ARE signaling in cellular protection: Mechanism of action and the regulatory mechanisms. *J. Cell. Physiol.* **2020**, *235*, 3119–3130. [CrossRef]
21. Miao, X.; Liu, X.; Chen, H.S.; Wang, C.Y.; Diao, J.J. Restoration and preservation effects of mung bean antioxidant peptides on H2O2-induced WRL-68 cells via Keap1-Nrf2 pathway. *Food Sci. Nutr.* **2023**, *11*, 7130–7144. [CrossRef] [PubMed]
22. Tonolo, F.; Folda, A.; Cesaro, L.; Scalcon, V.; Marin, O.; Ferro, S.; Bindoli, A.; Rigobello, M.P. Milk-derived bioactive peptides exhibit antioxidant activity through the Keap1-Nrf2 signaling pathway. *J. Funct. Foods* **2020**, *64*, 103696. [CrossRef]
23. Sun, C.Z.; Shan, Y.W.; Tang, X.; Han, D.; Wu, X.Y.; Wu, H.; Hosseininezhad, M. Effects of enzymatic hydrolysis on physicochemical property and antioxidant activity of mulberry (Morus atropurpurea Roxb.) leaf protein. *Food Sci. Nutr.* **2021**, *9*, 5379–5390. [CrossRef] [PubMed]
24. Gouda, M.; Zu, L.L.; Ma, S.M.; Sheng, L.; Ma, M.H. Influence of bio-active terpenes on the characteristics and functional properties of egg yolk. *Food Hydrocolloid* **2018**, *80*, 222–230. [CrossRef]

25. Rahimipanah, M.; Mahoonak, A.S.; Ghorbani, M.; Tabarestani, H.S.; Meybodi, M.N. Expeller-pressed pomegranate seed (*Punica granatum* L.) as a protein source for the production of antioxidant peptides. *Int. J. Pept. Res. Ther.* **2022**, *28*, 122. [CrossRef]
26. Serajul, I.; Hongxin, W.; Habtamu, A.; Anwar, N.; Chaoyang, M.; Fu An, W. Degree of hydrolysis, functional and antioxidant properties of protein hydrolysates from Grass Turtle (Chinemys reevesii) as influenced by enzymatic hydrolysis conditions. *Food Sci. Nutr.* **2021**, *9*, 4031–4047. [CrossRef]
27. Fathi, M.; Hosseini, F.S.; Ramezani, R.; Rashidi, L. Optimized Enzymatic Hydrolysis of Olive Pomace Proteins Using Response Surface Methodology. *Appl. Food Biotechnol.* **2022**, *9*, 79–90. [CrossRef]
28. Kong, Y.W.; Feng, M.Q.; Sun, J. Novel antioxidant peptides in fermented pork sausage: Purification, characterization, and cytoprotective functions on Caco-2 cells. *Food Chem.* **2023**, *426*, 136566. [CrossRef]
29. He, L.; Han, L.; Wang, Y.R.; Yu, Q.L. Appropriate ultrasonic treatment improves the production of antioxidant peptides by modifying gelatin extracted from yak skin. *Int. J. Food Sci. Tech.* **2022**, *57*, 5897–5908. [CrossRef]
30. Mooney, C.; Haslam, N.J.; Pollastri, G.; Shields, D.C. Towards the Improved Discovery and Design of Functional Peptides: Common Features of Diverse Classes Permit Generalized Prediction of Bioactivity. *PLoS ONE* **2012**, *7*, e45012. [CrossRef]
31. Zhao, W.Z.; Su, L.J.; Huo, S.T.; Yu, Z.P.; Li, J.R.; Liu, J.B. Virtual screening, molecular docking and identification of umami peptides derived from Oncorhynchus mykiss. *Food Sci. Hum. Well* **2023**, *12*, 89–93. [CrossRef]
32. Zhao, W.Z.; Xu, G.; Yu, Z.P.; Li, J.R.; Liu, J.B. Identification of nut protein-derived peptides against SARS-CoV-2 spike protein and main protease. *Comput. Biol. Med.* **2021**, *138*, 104937. [CrossRef]
33. Gupta, S.; Kapoor, P.; Chaudhary, K.; Gautam, A.; Kumar, R.; Raghava, G.P.S.; Discovery, O.S.D. In silico approach for predicting toxicity of peptides and proteins. *PLoS ONE* **2013**, *8*, e73957. [CrossRef] [PubMed]
34. Hu, G.H.; Dou, L.; Zhang, J.; Su, R.A.; Corazzin, M.; Sun, L.A.; Zhao, L.H.; Jin, Y.; Su, L. Exploring the antioxidant stability of sheep bone protein hydrolysate-identification and molecular docking. *Lwt Food Sci. Technol.* **2024**, *192*, 115682. [CrossRef]
35. Kim, M.T.; Huang, R.L.; Sedykh, A.; Wang, W.Y.; Xia, M.H.; Zhu, H. Mechanism Profiling of Hepatotoxicity Caused by Oxidative Stress Using Antioxidant Response Element Reporter Gene Assay Models and Big Data. *Environ. Health Persp* **2016**, *124*, 634–641. [CrossRef]
36. Naik, J.; Kulkarni, D.; Kadu, P.; Pandya, A.; Kale, P. Use of in silico tools for screening buffers to overcome physical instability of abatacept. *Transpl. Immunol.* **2022**, *71*, 101551. [CrossRef]
37. Pan, X.Y.; Wang, Y.M.; Li, L.; Chi, C.F.; Wang, B. Four antioxidant peptides from protein hydrolysate of red stingray (Dasyatis akajei) cartilages: Isolation, identification, and in vitro activity evaluation. *Mar. Drugs* **2019**, *17*, 263. [CrossRef]
38. Leiva-Portilla, D.; Martínez, R.; Bernal, C. Valorization of shrimp (Heterocarpus reedi) processing waste via enzymatic hydrolysis: Protein extractions, hydrolysates and antioxidant peptide fractions. *Biocatal. Agric. Biotechnol.* **2023**, *48*, 102625. [CrossRef]
39. Abeynayake, R.; Zhang, S.T.; Yang, W.Z.; Chen, L.Y. Development of antioxidant peptides from brewers' spent grain proteins. *Lwt Food Sci. Technol.* **2022**, *158*, 113162. [CrossRef]
40. Zhang, Y.; Shi, Z.Y.; Zhou, Y.J.; Xiao, Q.; Wang, H.Y.; Peng, Y. Emerging substrate proteins of kelch-like ECH associated protein 1 (Keap1) and potential challenges for the development of small-molecule inhibitors of the Keap1-nuclear factor erythroid 2-related factor 2 (Nrf2) protein–protein interaction. *J. Med. Chem.* **2020**, *63*, 7986–8002. [CrossRef]
41. Meng, L.; Yu, D.; Lin, J.; Hu, Y.; Peng, N.; Zhao, S. Structural characterization, HepG2 cell cytoprotective ability, and antioxidant mechanism of novel antioxidant peptides identified from black soldier fly larvae (*Hermetia illucens* L.). *Food Chem.* **2025**, *463*, 141462. [CrossRef] [PubMed]
42. Tong, K.I.; Padmanabhan, B.; Kobayashi, A.; Shang, C.; Hirotsu, Y.; Yokoyama, S.; Yamamoto, M. Different electrostatic Potentials define ETGE and DLG motifs as hinge and latch in oxidative stress response. *Mol. Cell Biol.* **2007**, *27*, 7511–7521. [CrossRef] [PubMed]
43. Mirzaei, M.; Mirdamadi, S.; Ehsani, M.R.; Aminlari, M. Production of antioxidant and ACE-inhibitory peptides from protein hydrolysates: Purification and molecular docking. *J. Food Drug Anal.* **2018**, *26*, 696–705. [CrossRef] [PubMed]
44. McMahon, M.; Lamont, D.J.; Beattie, K.A.; Hayes, J.D. Keap1 perceives stress via three sensors for the endogenous signaling molecules nitric oxide, zinc, and alkenals. *Proc. Natl. Acad. Sci.* **2010**, *107*, 18838–18843. [CrossRef]
45. Pei, X.D.; He, Y.N.; Wu, Q.L.; Zhang, Y.M.; Li, F.; Jiao, D.Q.; Liu, X.L.; Wang, C.H. Novel Antioxidant Peptides Derived from Feather Keratin Alleviate H2O2-Induced Oxidative Damage in HepG2 Cells via Keap1/Nrf2 Pathway. *J. Agr. Food Chem.* **2023**, *71*, 20062–20072. [CrossRef]
46. Liu, H.; Liang, J.; Xiao, G.; Vargas-De-La-Cruz, C.; Simal-Gandara, J.; Xiao, J.; Wang, Q. Active sites of peptides Asp-Asp-Asp-Tyr and Asp-Tyr-Asp-Asp protect against cellular oxidative stress. *Food Chem.* **2022**, *366*, 130626. [CrossRef]
47. Tao, L.; Gu, F.; Liu, Y.; Yang, M.; Wu, X.Z.; Sheng, J.; Tian, Y. Preparation of antioxidant peptides from Moringa oleifera leaves and their protection against oxidative damage in HepG2 cells. *Front. Nutr.* **2022**, *9*, 1062671. [CrossRef]
48. Wen, C.T.; Zhang, J.X.; Zhang, H.H.; Duan, Y.Q.; Ma, H.L. Plant protein-derived antioxidant peptides: Isolation, identification, mechanism of action and application in food systems: A review. *Trends Food Sci. Tech.* **2020**, *105*, 308–322. [CrossRef]

49. Bollati, C.; Cruz-Chamorro, I.; Aiello, G.; Li, J.Q.; Bartolomei, M.; Santos-Sánchez, G.; Ranaldi, G.; Ferruzza, S.; Sambuy, Y.; Arnoldi, A.; et al. Investigation of the intestinal trans-epithelial transport and antioxidant activity of two hempseed peptides WVSPLAGRT (H2) and IGFLIIWV (H3). *Food Res. Int.* **2022**, *152*, 110720. [CrossRef]
50. Harada, A.; Sekido, N.; Akahoshi, T.; Wada, T.; Mukaida, N.; Matsushima, K. Essential involvement of interleukin-8 (IL-8) in acute inflammation. *J. Leukoc. Biol.* **1994**, *56*, 559–564. [CrossRef]
51. Han, X.E.; Liang, Z.Q.; Tian, S.F.; Liu, L.; Wang, S. Epigallocatechin gallate (EGCG) modification of structural and functional properties of whey protein isolate. *Food Res. Int.* **2022**, *158*, 111534. [CrossRef] [PubMed]
52. Li, M.Y.; Fan, W.L.; Xu, Y. Identification of angiotensin converting enzyme (ACE) inhibitory and antioxidant peptides derived from Pixian broad bean paste. *Lwt Food Sci. Technol.* **2021**, *151*, 112221. [CrossRef]
53. Karami, Z.; Peighambardoust, S.H.; Hesari, J.; Akbari-Adergani, B.; Andreu, D. Antioxidant, anticancer and ACE-inhibitory activities of bioactive peptides from wheat germ protein hydrolysates. *Food Biosci.* **2019**, *32*, 100450. [CrossRef]
54. Sharma, S.; Pradhan, R.; Manickavasagan, A.; Thimmanagari, M.; Saha, D.; Singh, S.S.; Dutta, A. Production of antioxidative protein hydrolysates from corn distillers solubles: Process optimization, antioxidant activity evaluation, and peptide analysis. *Ind. Crop Prod.* **2022**, *184*, 115107. [CrossRef]
55. Zhang, T.; Hua, Y.; Zhou, C.Y.; Xiong, Y.Z.; Pan, D.D.; Liu, Z.; Dang, Y.L. Umami peptides screened based on peptidomics and virtual screening from. *Food Chem.* **2022**, *394*, 133504. [CrossRef]
56. Wu, Y.; Guo, S.; Wang, K.; Kang, J. The interaction of peptide inhibitors and Aβ protein: Binding mode analysis, inhibition of the formation of Aβ aggregates, and then exert neuroprotective effects. *Front. Aging Neurosci.* **2023**, *15*, 1139418. [CrossRef]

International Journal of
Molecular Sciences

MDPI

Article

Using Integrated Network Pharmacology and Metabolomics to Reveal the Mechanisms of the Combined Intervention of Ligustrazine and Sinomenine in CCI-Induced Neuropathic Pain Rats

Zhaoyue Yuan [1,†], Xiaoliang Zhao [1,†], Yan Zhang [2], Yue Jiao [1], Yang Liu [1], Chang Gao [1], Jidan Zhang [1], Yanyan Ma [1], Zhiguo Wang [1,*] and Tao Li [1,2,*]

[1] Experimental Research Center, China Academy of Chinese Medical Sciences, Beijing 100700, China; zyueyuan1118@163.com (Z.Y.); zhaoxiaoliang@merc.ac.cn (X.Z.)

[2] Institute of Chinese Materia Medica, China Academy of Chinese Medical Sciences, Beijing 100700, China; 17806297233 6@163.com

* Correspondence: wangzhiguo@merc.ac.cn (Z.W.); litao@merc.ac.cn (T.L.)

† These authors contributed equally to this work.

Abstract: Neuropathic pain (NP) is a type of chronic pain resulting from injury or dysfunction of the nerves or spinal cord. Previous studies have shown that the combination of ligustrazine (LGZ) and sinomenine (SIN) exerts a synergistic antinociceptive effect in peripheral and central NP models. On this basis, a comprehensive analgesic evaluation was performed in a chronic constriction injury (CCI)-induced NP model in rats. Sciatic nerve histopathological changes were observed, and 22 cytokines and chemokines levels were analyzed. We also combined network pharmacology and metabolomics to explore their molecular mechanisms. Results showed that the combination of LGZ and SIN significantly alleviated the pain-like behaviors in CCI rats in a time- and dose-dependent manner, demonstrating superior therapeutic effects compared to LGZ or SIN alone. It also improved pathological damage to sciatic nerves and regulated inflammatory cytokine levels. Network pharmacology identified shared and distinct pain-related targets for LGZ and SIN, while metabolomics revealed 54 differential metabolites in plasma, and 17 differential metabolites in CSF were associated with the combined intervention of LGZ and SIN. Finally, through an integrated analysis of the core targets and differential metabolites, tyrosine metabolism, phenylalanine metabolism, and arginine and proline metabolism were identified as potential key metabolic pathways underlying the therapeutic effects of LGZ and SIN in CCI treatment. In conclusion, our study provides evidence to support the clinical application of LGZ and SIN in the treatment of NP.

Keywords: neuropathic pain; ligustrazine; sinomenine; tyrosine metabolism; phenylalanine metabolism

1. Introduction

Neuropathic pain (NP) is a chronic condition that is characterized by spontaneous and evoked pain, including cold and mechanical allodynia. NP arises from injury or malfunction of the nerves or the spinal cord, and is prevalent in clinical practice. It often occurs secondary to conditions such as trauma, stroke, infection, diabetes, multiple sclerosis, and cancer [1,2]. Globally, the incidence of NP accounts for 6.9% to 10% of the total population [3]. NP leads to significant reductions in both quality of life and behavioral function, placing a heavy physical and psychological burden on patients [4]. The pathophysiology of NP is complex and multifactorial. Despite the availability of various interventions based on different mechanisms, the results of NP treatment are still unsatisfactory [5]. Commonly used medications for NP, such as non-steroidal anti-inflammatory drugs and opioids, often cause adverse effects, including dizziness, drowsiness, arrhythmias, and issues related to tolerance when used alone [6]. Consequently, the development of combination therapies targeting multiple analgesic mechanisms has become a promising strategy in the treatment of NP. Indeed, clinical approaches to chronic pain often evolve from initial monotherapy to combination therapy as a more effective means of addressing the complex nature of NP [7].

Traditional Chinese medicine (TCM) boasts an extensive history in the treatment of NP, with many herbal medicines and their active compounds demonstrating analgesic effects [8]. These therapies represent a rich source for the screening of potential combination drugs for NP treatment. As reported, ligustrazine (LGZ) has exhibited therapeutic effects, such as alleviating migraine [9] and spinal cord injury-induced NP [10], in various pain animal models. Sinomenine (SIN) has demonstrated analgesic effects on inflammatory pain [11], spared nerve injury-induced NP [12], and diabetic peripheral NP [13], exerting both anti-inflammatory and analgesic mechanisms. Our previous studies have shown that the combination of LGZ and SIN effectively alleviates pain in NP models, including sciatic nerve injury, trigeminal neuralgia, and spinal cord injury [14]. Moreover, the combination of LGZ and SIN exhibits a synergistic analgesic effect, allowing for a reduction in the individual dosages of each compound. Both LGZ and SIN have been reported to have numerous potential targets and pathways, but the specific mechanisms underlying their analgesic effects in NP remain unclear.

Network pharmacology is used to analyze the relationships between drugs, targets, and diseases through network-based approaches, making it particularly well-suited for elucidating core targets and pathways in treating complex diseases via TCM, which has "multi-target" and "multi-pathway" characteristics [15]. Through offering a systems-level perspective, network pharmacology uncovers the potential molecular mechanisms of TCM, shedding light on its holistic therapeutic effects [16,17]. Metabolomics, an essential component of systems biology, utilizes high-throughput and highly sensitive instruments to conduct comprehensive profiling of the endogenous components within biological samples. Through integrating multivariate statistical methods, metabolomics can reveal changes in endogenous metabolites under various physiological, pathological, or toxicological conditions [18], enabling the identification of the key metabolic pathways within the organism [19]. Therefore, the integration of metabolomics with network pharmacology can provide a more comprehensive understanding of the metabolic pathways and network regulatory mechanisms of the combined therapy in treating complex diseases.

In this study, we employed network pharmacology to predict the potential targets of LGZ and SIN in treating pain-related diseases, systematically analyzing both their shared and distinct targets. Subsequently, the analgesic effects of LGZ, SIN, and their combination at different time points and dosages were evaluated in a chronic constriction injury (CCI)-induced NP model in rats. Furthermore, the key metabolites in the plasma and cerebrospinal fluid (CSF) were identified and quantified using LC-MS metabolomics technology. Finally, a joint analysis of the potential targets and key metabolites was performed to determine the critical metabolic pathways through which LGZ and SIN exert their therapeutic effects in NP. This study aimed to elucidate the molecular mechanisms underlying the combined use of LGZ and SIN in treating NP. Additionally, this research aimed to offer new experimental evidence to support the clinical combined application of LGZ and SIN in treating NP.

2. Results

2.1. The Combination of LGZ and SIN Pharmacological Effects on CCI Rats

2.1.1. General Animal Condition

Following CCI surgery and treatment, all of the rat groups exhibited an upward trend in body weight (Figure 1A). After administration, the model group had a slower weight gain than the sham group. Conversely, all treated groups exhibited greater body weight gain compared to the model group, with the LGZ+SIN high-dose group showing the most significant increase. These findings suggested that the combined treatment of LGZ and SIN has a restorative effect on the overall status of CCI rats. No significant toxic effects were observed in any of the groups throughout the experiment.

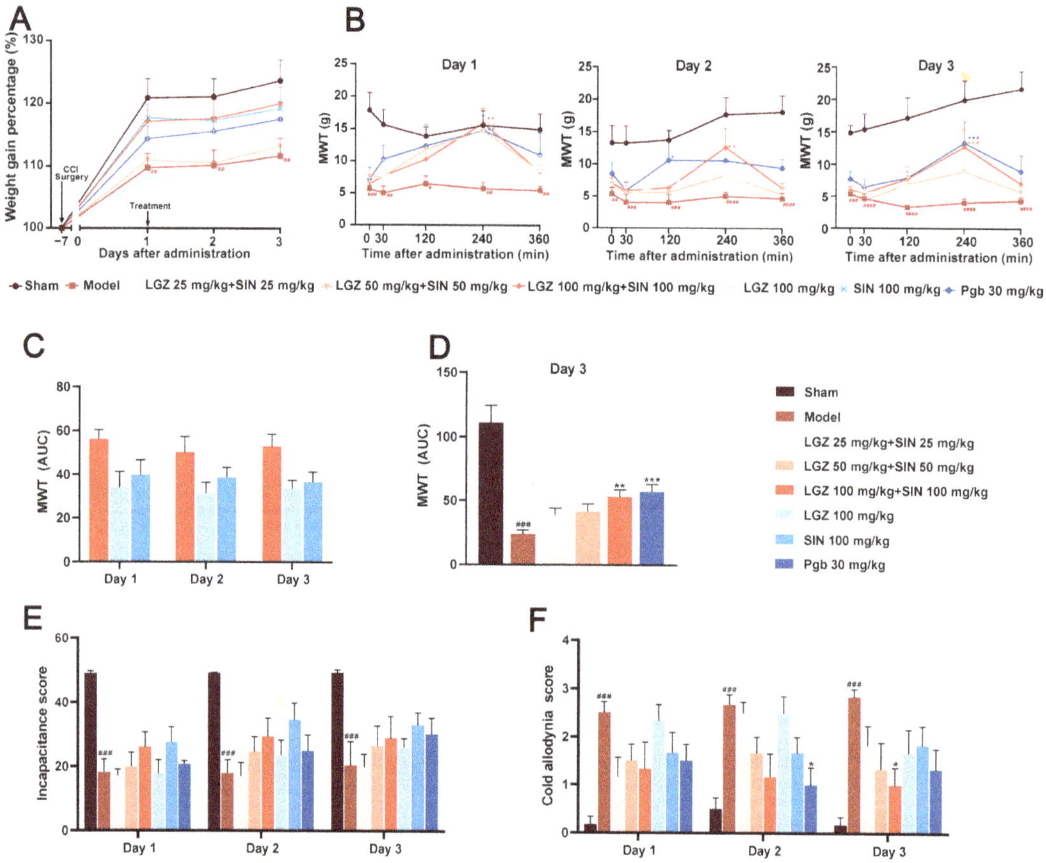

Figure 1. The analgesic effects of different doses of SIN + LGZ (LGZ 25 mg·kg^{-1}·d^{-1} + SIN 25 mg·kg^{-1}·d^{-1}; LGZ 50 mg·kg^{-1}·d^{-1} + SIN 50 mg·kg^{-1}·d^{-1}; and LGZ 100 mg·kg^{-1}·d^{-1} + SIN 100 mg·kg^{-1}·d^{-1}), SIN (100 mg·kg^{-1}·d^{-1}), and LGZ (100 mg·kg^{-1}·d^{-1}) in CCI rats following intragastric administration on days 1, 2, and 3. (**A**) Body weight gain percentages. (**B**) Mechanical withdrawal threshold (MWT) values after administration over 3 days of treatment. (**C**) The AUC of the MWT values on the third day of treatment. (**D**) The AUC of the MWT values per day over 3 days of treatment. (**E**) The cold allodynia score over 3 days of treatment. (**F**) The incapacitance score over 3 days of treatment. $^{\#}$ $p < 0.05$, $^{\#\#}$ $p < 0.01$, $^{\#\#\#}$ $p < 0.001$ and $^{\#\#\#\#}$ $p < 0.001$ (compared to the sham group). * $p < 0.05$, ** $p < 0.01$, and *** $p < 0.001$ (compared to the model group).

2.1.2. The Combination of LGZ and SIN Improve the Pain-Related Behaviors in CCI Rats

The mechanical withdrawal threshold (MWT) results indicated that, as early as 0.5 h post-administration, the LGZ+SIN, LGZ, and SIN groups significantly increased MWT, peaking at 4 h, with statistically significant differences. The pain-relieving trend remained consistent across all treatment groups on days 1, 2, and 3 post-treatment (Figure 1B). To further quantify the effects, we calculated the area under the curve (AUC) of the MWT values at 0, 0.5, 2, 4, and 6 h post-treatment. The MWT values were measured on the first, second, and third days after administration, and they were used to calculate the AUC for the LGZ+SIN high-dose, LGZ, and SIN groups. The results showed that the AUC for the MWT values in the LGZ+SIN groups was consistently superior to that of either the LGZ or SIN groups alone on all days of administration (Figure 1C). Furthermore, the LGZ+SIN high-dose group increased the AUC for MWT more than LGZ+SIN medium- or low-dose group on the third day of treatment (Figure 1D). These results suggest that the combined use of LGZ and SIN effectively alleviated mechanical pain in CCI rats, delivering a superior effect compared to the individual use of LGZ or SIN.

The cold allodynia test results (Figure 1E) revealed that, during the administration period, the cold pain scores of the model group were significantly higher than those of the Sham group ($p < 0.001$), indicating hypersensitivity to cold pain in the CCI rats. Over the 3-day treatment period, the cold pain sensitivity scores in the LGZ+SIN and SIN groups showed a decreasing trend compared to the model group. On the third day of administration, the cold pain score in the LGZ+SIN high-dose group was significantly reduced ($p < 0.05$). These results suggest that the combined use of LGZ and SIN has a significant effect in alleviating the cold pain sensitivity in CCI rats. As shown in Supplementary Figure S1, the cold allodynia score in the LGZ+SIN groups decreased significantly at 4 h post-administration, especially in the LGZ+SIN high-dose group. The results indicate that the combination of LGZ and SIN has the best effect after 4 h of administration.

The results of the incapacitance test (Figure 1F) showed that, compared to the sham group, the model group had a significantly reduced incapacitance score ($p < 0.01$), indicating the presence of spontaneous pain in the CCI rats. Compared to the model group, the incapacitance score in the LGZ+SIN, LGZ, and SIN groups showed an increasing trend following treatment, with more pronounced improvements on the second and third days of administration. These results suggested that the combined use of LGZ and SIN might alleviate spontaneous pain in CCI rats.

2.1.3. The Combination of LGZ and SIN Alleviates the Pathological Structure of the Sciatic Nerves

H&E staining of the sciatic nerves revealed that the sciatic nerve fibers in the sham group rats had a regular, evenly distributed structure, with tightly arranged, intact myelin sheaths. In contrast, the model group rats showed a loose, sparse arrangement of sciatic nerve fibers, with deformed myelin sheaths and a significant infiltration of inflammatory cells. The sciatic nerve structure in the LGZ group was closer to normal, with a more orderly arrangement. Although demyelination was present, it was less severe than in the model group, indicating a degree of pathological recovery. Both the LGZ+SIN and SIN groups exhibited varying degrees of disorganized nerve fiber arrangement, demyelination, and inflammatory cell infiltration, but the extent of these changes was milder than in the model group (Figure 2A). These results suggest that the combined use of LGZ and SIN, LGZ, and SIN has a beneficial effect on the pathological repair of the sciatic nerves in CCI rats.

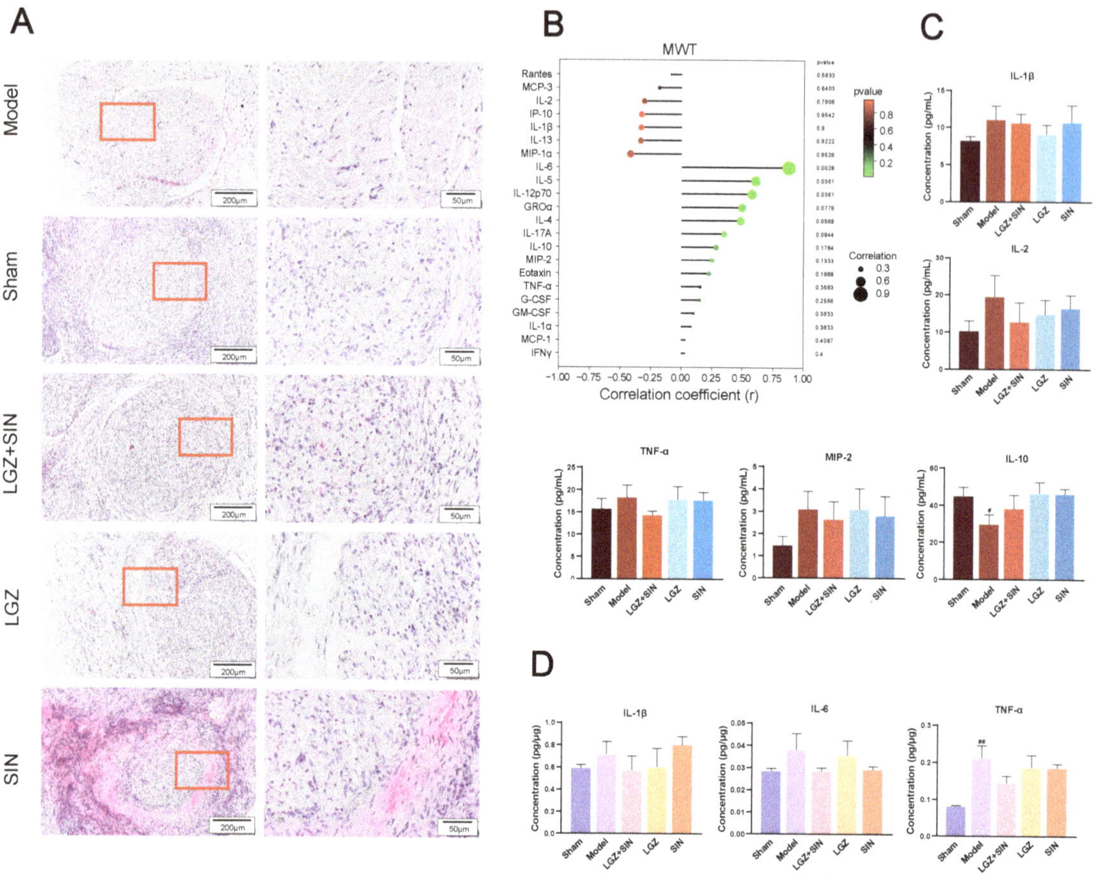

Figure 2. LGZ+SIN-attenuated, CCI-induced sciatic nerve injury. (**A**) Representative images of HE staining of sciatic nerves. The red frame in the left images are enlarged and displayed on the right side. (**B**) Correlation coefficient chart between MWT values and 22 cytokines and chemokines. The size of the dots represents the strength of the association between the MWT values and 22 cytokines and chemokines; a larger dot indicates a stronger correlation. The color represents the p-value, with redder colors indicating larger p-values. (**C**) Inflammatory factor changes in a plasma sample. (**D**) Inflammatory factor changes in a sciatic nerve sample. $^{\#}$ $p < 0.05$, and $^{\#\#}$ $p < 0.01$ (compared to the sham group).

2.1.4. The Combination of LGZ+SIN Alter the Plasma and Sciatic Nerve Inflammatory Cytokine and Chemokine Levels in CCI Rats

As shown in Figure 2B, we performed a correlation analysis of 22 cytokines and chemokines with MWT. The results indicate that MIP-1α, IL-13, IL-1β, IP-10, IL-2, MCP-3, rates are negatively correlated with MWT, while IL-6, IL-5, IL-12p70, GROα, IL-4, IL-17A, IL-10, MIP-2, and others are positively correlated with MWT.

The concentrations of TNF-α, IL-1β, IL-2, and MIP-2 in the plasma of each group (Figure 2C) showed that, compared to the sham group, the model group had higher levels. After three days of treatment, the LGZ+SIN group exhibited a decreased level compared to the model group. Additionally, the concentrations of IL-10 in the model group decreased compared with the sham group. After treatment, the LGZ+SIN group decreased. These results suggest that the combined use of LGZ and SIN has a beneficial effect in alleviating plasma inflammation in CCI rats.

The concentrations of IL-1β, IL-6, and TNF-α in the sciatic nerves of each group (Figure 2D) showed that, compared to the sham group, the model group had higher levels of IL-1β, IL-6, and TNF-α. After three days of treatment, the LGZ+SIN group exhibited a decreased level in the sciatic nerves compared to the model group. These results suggest that the combined use of LGZ and SIN has a beneficial effect in alleviating sciatic nerve inflammation in CCI rats.

2.2. *Results of Network Pharmacology Analysis*

2.2.1. Potential Targets of LGZ and SIN in Pain Treatment

A compound target database search and screening identified 29 targets for LGZ and 100 targets for SIN. The GeneCards database revealed a total of 12,990 pain-related targets, with relevance scores ranging from a maximum of 100.27 to a minimum of 0.08. Targets with scores above the median were screened, resulting in 3303 pain-related targets. After merging the results from multiple databases and removing duplicates, a final set of 3303 disease-related targets was obtained.

The target information of the compounds was intersected with disease-related target genes on the GeneCards database, and a Venn diagram was constructed (Figure 3A). This intersection revealed 22 pain-related targets for LGZ, 58 pain-related targets for SIN, and 8 common targets shared by both compounds. Among these, 16 specific targets were identified for LGZ, and 52 unique targets were identified for SIN. These intersected genes were then entered into the STRING 11.0 platform to construct interaction networks based on protein–protein interactions (Figure 3B).

This analysis revealed that LGZ and SIN exerted their analgesic effects through six key targets: CA2, MPO, HTR6, MAOA, GSK3B, and BDKRB2. Additionally, both compounds exhibited analgesic effects through their respective specific targets.

2.2.2. Pain-Relieving Pathways and Biological Function Differences Between LGZ and SIN

LGZ and SIN exhibited certain differences in their pain-relieving pathways and biological functions. The core targets of both compounds were analyzed using Metascape, and KEGG and GO functional enrichment analyses were performed ($p < 0.01$). The top 30 GO terms were selected for the bar chart, and the top 20 KEGG pathways were visualized in a bubble chart (Supplementary Figure S2A,B).

According to the KEGG analysis, the primary targets of LGZ are related to the nervous system (dopaminergic), signaling pathways (synapse neuroactive ligand–receptor interaction, hedgehog signaling pathway, and ErbB signaling pathway), the immune system (B cell receptor signaling pathway), and the endocrine system (prolactin signaling pathway). It also modulated the amino acid metabolism pathways such as tyrosine metabolism and tryptophan metabolism to exert its analgesic effect. Additionally, SIN exerted its analgesic effects by modulating signaling pathways (PI3K-Akt signaling pathway, TNF signaling pathway, sphingolipid signaling pathway, and ErbB signaling pathway), the nervous system (dopaminergic), the immune system (IL-17 signaling pathway and T cell receptor signaling pathway), and the endocrine system (prolactin signaling pathway, insulin signaling pathway, growth hormone synthesis, secretion, and action).

Venn diagram analysis revealed that LGZ and SIN share six common analgesic targets. By integrating these targets with the KEGG analysis, their intersecting genes primarily exerted analgesic effects by influencing the nervous system (serotonergic synapse and dopaminergic synapse), signaling pathways (calcium signaling pathway, neuroactive ligand–receptor interaction, and hedgehog signaling pathway), and the endocrine system (prolactin signaling pathway). These pathways, along with energy metabolism and amino acid metabolism pathways, were key to their analgesic effects.

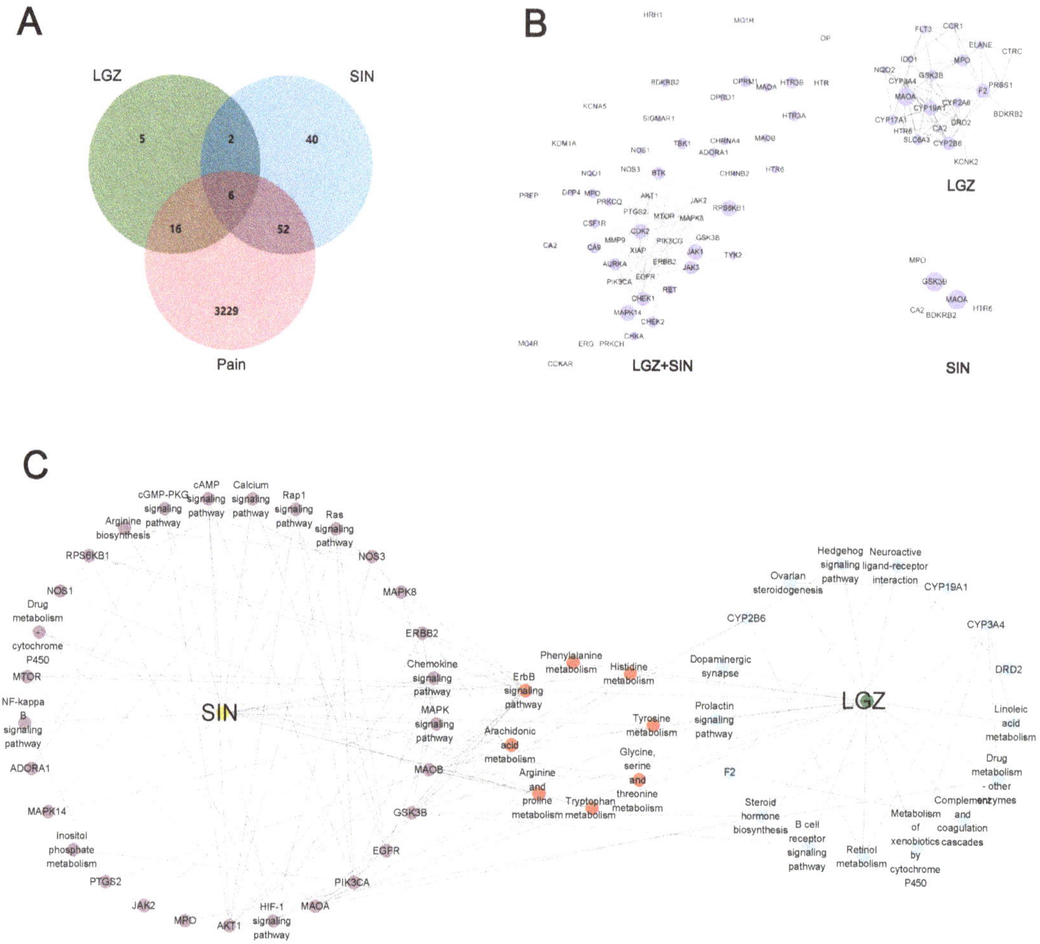

Figure 3. Exploring the effects of LGZ and SIN on pain based on network pharmacology. (**A**) Venn diagram. (**B**) Protein–protein network diagram. (**C**) Target–pathway networks of LGZ and SIN through KEGG enrichment analysis of the targets.

To further clarify the combination pathways of LGZ and SIN for treating NP, all of the targets were mapped onto KEGG pathways to identify the pathways with $p < 0.05$. The pathways were selected, resulting in a target–pathway network (Figure 3C). The results showed that LGZ and SIN shared eight signaling pathways. These results demonstrated that the combined treatment of LGZ and SIN for NP was a form of combination therapy.

In summary, LGZ and SIN share six common pain-relieving target genes: CA2, HTR6, MAOA, GSK3B, BDKRB2, and MPO. An analysis of the core targets revealed that both compounds primarily affect the nervous system, immune system, and signaling pathways to exert their analgesic effects. However, a closer examination of the intersecting genes reveals that HTR6 and BDKRB2 are highly expressed in signaling pathways, while CA2 shows strong expression in the neuroactive ligand–receptor interaction pathway. Furthermore, the biological processes of these common genes suggest that they mainly influence neurotransmitter breakdown and ion transport processes. This indicates that the combined use of LGZ and SIN might, to some extent, enhance the neuroprotective and signaling function of the system.

2.3. Metabolomic Analysis

A widely targeted metabolomics approach was employed to investigate the therapeutic mechanism of the combined use of LGZ and SIN, as well as their individual applications on CCI rats. To gain more detailed insights, plasma and CSF samples were analyzed separately, in both positive and negative ion modes. The results from the positive and negative ion detection were subsequently merged and further analyzed to explore the underlying therapeutic mechanisms of treatment.

2.3.1. Differential Metabolites Analysis

Principal component analysis (PCA) was performed on the data from each group. The PCA score plot revealed that the QC samples exhibited excellent stability, indicating that the analytical system had strong reliability, and that the experimental data were of high quality. This suggests that the data met the criteria for metabolomics analysis. The details can be found in Supplementary Figure S3.

Orthogonal partial least squares discriminant analysis (OPLS-DA) was used to further identify the potential biomarkers between the groups in the plasma and CSF samples, respectively (Figure 4A,B). The OPLS-DA score plots clearly distinguished between the groups, indicating significant differences in metabolic profiles and suggesting that endogenous metabolites underwent noticeable changes among the groups.

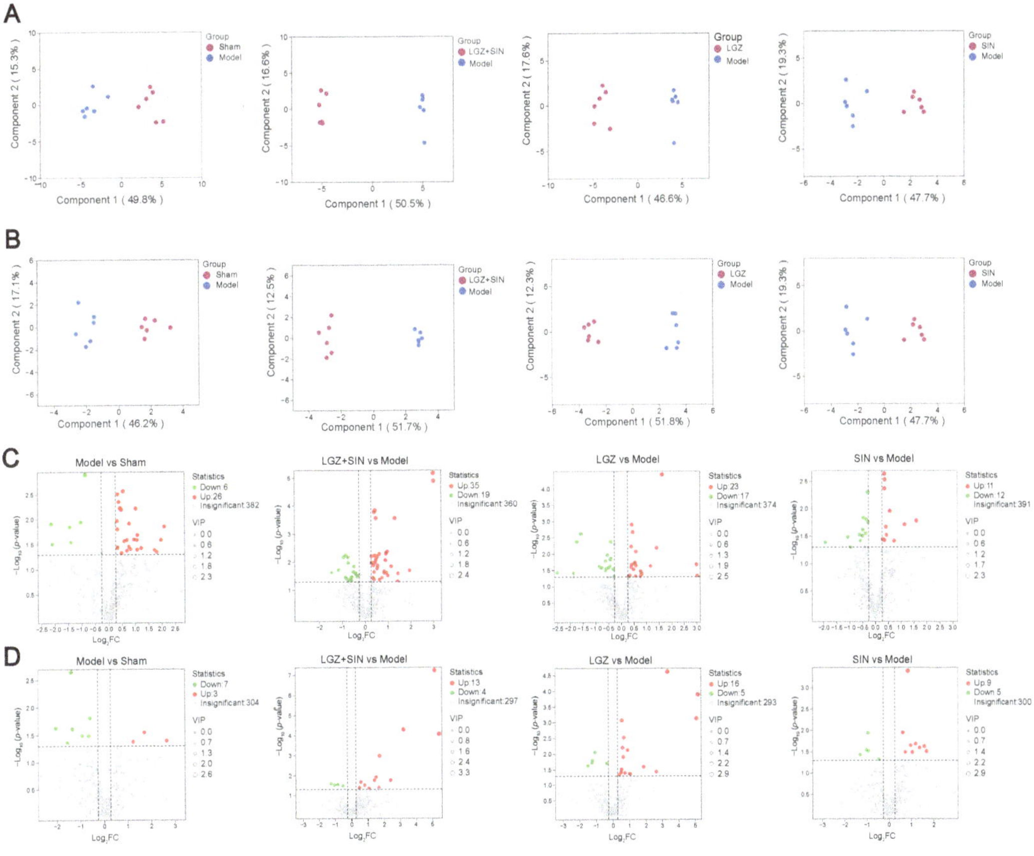

Figure 4. *Cont.*

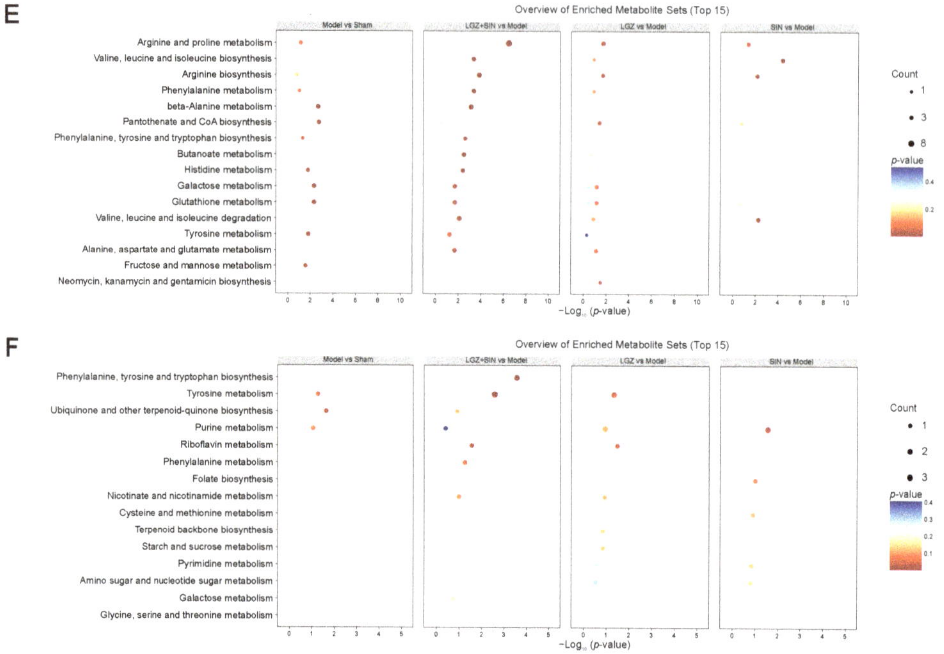

Figure 4. The metabolite changes in CCI rats following LGZ+SIN intervention. (**A**) Results of the OPLS-DA scores' scatter plot of the plasma samples. (**B**) Results of the OPLS-DA scores' scatter plot of the CSF samples. (**C**) Volcano plot of the metabolites in the plasma samples. (**D**) Volcano plot of the metabolites in the CSF samples. (**E**) Top 15 enriched pathways for differential metabolites upon plasma samples. (**F**) Top 15 enriched pathways for differential metabolites upon CSF samples. Data were analyzed by MetaboAnalyst 6.0 using KEGG metabolite sets library.

To assess the robustness and validity of the OPLS-DA models, a random permutation test ($n = 200$) was performed (Supplementary Figure S4). The results indicate that none of the models were overfitted, confirming that the modeling outcomes were reliable. A model is considered valid when both the R^2Y and Q^2 values are greater than 0.5, which demonstrates good predictive power and confirms that the data are reliable.

To further validate the results, a *t*-test was conducted to examine whether significant differences existed between the groups. Differential metabolites between the sham and model groups, as well as between the treatment groups and the model group, were identified. As shown in Figure 4C,D, the metabolites that increased significantly ($p < 0.05$) with a FC > 1.2 are marked in red, and metabolites that decreased significantly ($p < 0.05$) and exhibited a FC < 0.83 are marked in green.

In the plasma samples, a total of 32 differential metabolites were identified between the model and sham groups, with 26 metabolites upregulated and 6 downregulated. Compared to the model group, 54 differential metabolites were identified in the LGZ+SIN group, including 35 upregulated and 19 downregulated. In the LGZ group, 40 differential metabolites were detected compared to the model group, with 23 upregulated and 17 downregulated. In the SIN group, 23 differential metabolites were identified compared to the model group, with 11 upregulated and 12 downregulated. From the perspective of the number of differential metabolites, the LGZ+SIN group exhibited a total of 54 differential metabolites, which was 14 more than those in the LGZ group and 31 more than those in the SIN group. In the CSF samples, 10 differential metabolites were identified between the model and sham groups, with 3 metabolites upregulated and 7 downregulated. In the LGZ+SIN group, compared to the model group, 17 differential metabolites were identified, with 13 upregulated and 4 downregulated. In the LGZ group compared to the model group, 21 differential metabolites were identified, with 16 upregulated and 5 downregulated. In the SIN group, 14 differential metabolites were identified, with 9 upregulated and 5 downregulated. From the perspective of the number of differential metabolites, the LGZ+SIN group had a greater effect on the changes in the metabolic profile in CSF.

We performed correlation analysis between differential metabolites and MWT in plasma and CSF of the LGZ+SIN group, respectively (Supplementary Figure S5A,B). As shown in the figure, the differential metabolites in plasma, including phenylacetaldehyde, L-Ornithine, and 5-hydroxytryptophan, were significantly correlated with MWT (Mantel's $r \geq 0.4$, $p < 0.01$). Similarly, the differential metabolites in CSF, including N, N-dimethyl-1,4-phenylenediamine, L-gulose, dethiobiotin, and phenylpyruvate, also showed significant correlations with MWT (Mantel's $r \geq 0.4$, $p < 0.01$). Additionally, we performed correlation analysis between the differential metabolites and 22 cytokines and chemokines in the plasma and CSF of the LGZ+SIN group (Supplementary Figure S5C,D). Differential metabolites in plasma, including 1-aminocyclopropanecarboxylate, succinate semialdehyde, pentanoate and cortexolone, were correlated with cytokines and chemokines (Mantel's $r \geq 0.4$, $p < 0.05$). And riboflavin in CSF was correlated with cytokines and chemokines (Mantel's $r \geq 0.4$, $p < 0.01$). These results suggest that changes in differential metabolites are closely associated with pain-like behavior and inflammatory factors in plasma.

2.3.2. Metabolomic Pathway Analysis

To identify the metabolic pathways most relevant to the pain-relieving effect of the combined use of LGZ and SIN in CCI rats, MetaboAnalyst 6.0 software was used to analyze the metabolic pathways. Pathways with an impact value (impact > 0) were selected as the screening criterion.

In the plasma samples, the model group primarily interfered with 28 metabolic pathways, including tyrosine metabolism; phenylalanine, tyrosine and tryptophan biosynthesis; and beta-alanine metabolism, which contribute to the development of neuropathic pain and related symptoms. The LGZ+SIN group primarily modulated 29 metabolic pathways, including arginine and proline metabolism, phenylalanine metabolism, phenylalanine, tyrosine and tryptophan biosynthesis, arginine biosynthesis, alanine, aspartate and glutamate metabolism, beta-alanine metabolism, and tyrosine metabolism. The LGZ group primarily modulated 28 pathways, including arginine biosynthesis, arginine and proline metabolism, and phenylalanine metabolism. The SIN group primarily regulated 14 pathways, including cysteine and methionine metabolism, phenylalanine, tyrosine and tryptophan biosynthesis, phenylalanine metabolism, and arginine biosynthesis. From the perspective of the number of metabolic pathways regulated, the LGZ+SIN group influenced one more pathway than LGZ and fourteen more than SIN. Moreover, LGZ+SIN regulated several metabolic pathways that involved four or more differential metabolites, facilitating the identification of key metabolic pathways (Figure 4E).

In the CSF samples, the model group primarily modified three pathways, including tyrosine metabolism and purine metabolism, contributing to the occurrence of neuropathic pain-related symptoms. The LGZ+SIN group primarily modulated 12 pathways, including tyrosine metabolism, riboflavin metabolism, and phenylalanine metabolism. The LGZ group regulated 11 pathways, including tyrosine metabolism, riboflavin metabolism, and tryptophan metabolism. The SIN group primarily affected five pathways, including purine metabolism (Figure 4F).

Based on these results and in conjunction with the KEGG database, the key metabolic pathways involved in LGZ and SIN intervention in NP include the tyrosine metabolism pathway and the phenylalanine metabolism pathway (Figure 5).

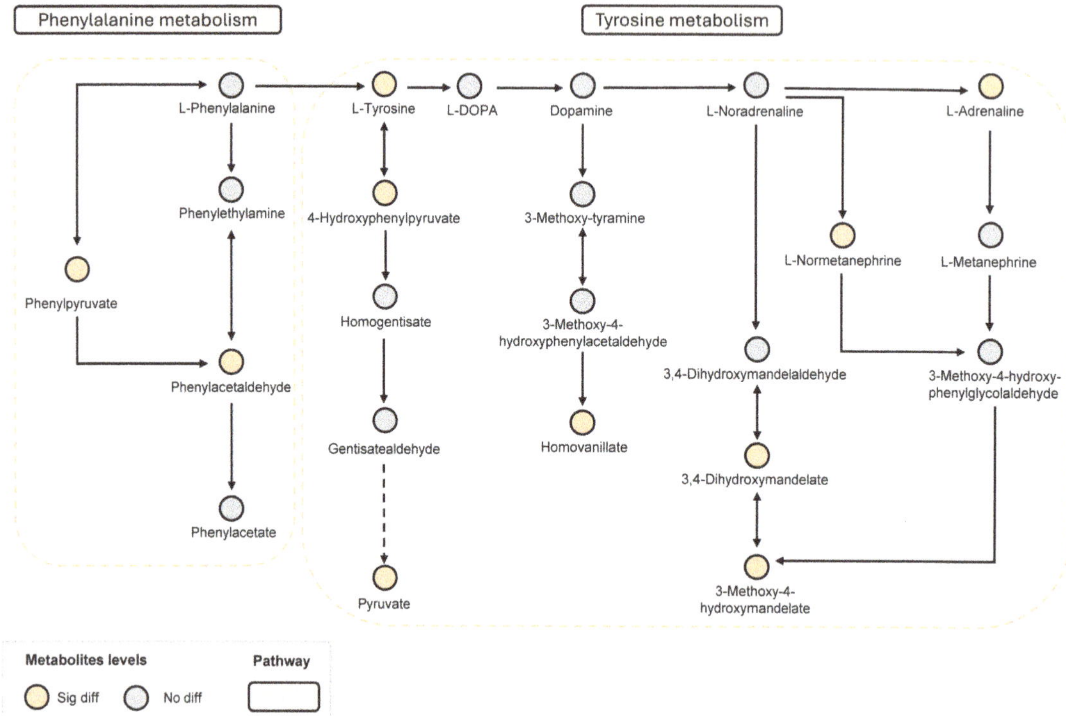

Figure 5. LGZ and SIN intervened in the key metabolic pathways of the CCI rats. Orange circles indicate metabolites that have significant difference. Gray circles indicate metabolites that are not different.

2.3.3. The Combination of LGZ and SIN Altered the Metabolic Profile of CCI Rats

In the plasma samples, LGZ+SIN significantly increased the levels of 5-hydroxytryptophan, normetanephrine, anserine, and carnosine, while significantly reducing the levels of phenylpyruvate and N-methylaspartate. Additionally, LGZ+SIN significantly increased the levels of epinephrine, tryptophan, and spermine, while decreasing the levels of 3-methoxy-4-hydroxymandelate and N-acetylphenylalanine. LGZ significantly increased the levels of 5-hydroxytryptophan and notably decreased the levels of 5-hydroxytryptophol and 3-methoxy-4-hydroxymandelate. LGZ also increased the levels of epinephrine, normetanephrine, tryptophan, spermine, anserine, and carnosine, while decreasing the levels of phenylpyruvate, N-acetylphenylalanine, and N-methylaspartate. SIN significantly elevated the levels of 5-hydroxytryptophan, epinephrine, tryptophan, spermine, anserine, and carnosine, while notably decreasing the levels of 3-methoxy-4-hydroxymandelate, N-acetylphenylalanine, and N-methylaspartate. In the targeted analysis of neurotransmitter-related metabolites in plasma samples, LGZ+SIN significantly decreased the levels of 4-aminobutyric acid and serotonin, while notably improving the levels of 3-methoxytyramine, 5-hydroxytryptophan, choline, dopamine, glutamine, picolinic acid, and quinolinic acid. LGZ and SIN both significantly decreased the levels of 4-aminobutyric acid. LGZ also significantly reduced the level of dopamine, while improving the levels of 3-methoxytyramine, aspartate, choline, glutamine, norepinephrine, picolinic acid, quinolinic acid, serine, and serotonin. SIN notably improved the levels of 3-methoxytyramine, aspartate, dopamine, glutamate, glutamine, kynurenine, norepinephrine, phenylalanine, quinolinic acid, and serotonin (Figure 6A).

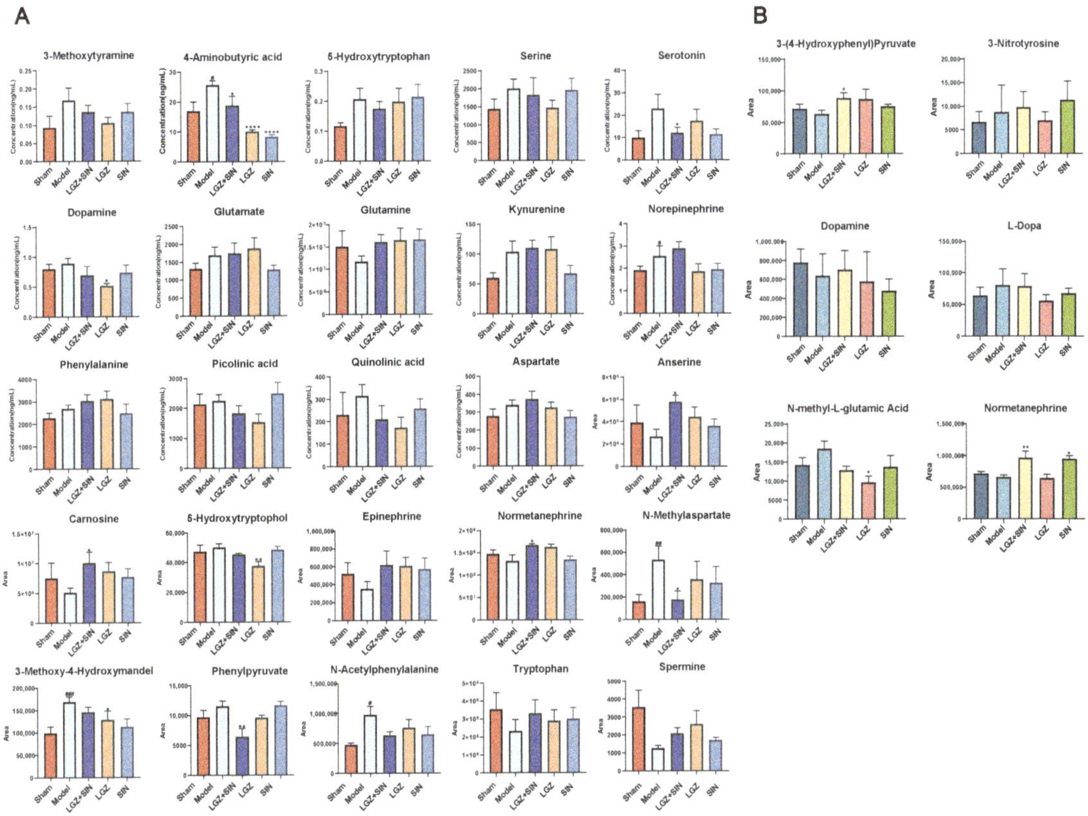

Figure 6. LGZ+SIN altered the metabolites of the biological samples in the CCI rats. (**A**) The differential metabolites in plasma. (**B**) The differential metabolites in CSF. # $p < 0.05$, ## $p < 0.01$, and ### $p < 0.001$ (compared to the sham group). * $p < 0.05$, ** $p < 0.01$ and **** $p < 0.0001$ (compared to the model group).

In the CSF samples, the combination of LGZ and SIN significantly rebalanced the levels of 3-(4-hydroxyphenyl) pyruvate and normetanephrine, and notably adjusted the levels of dopamine and L-dopa, while reducing the level of N-methyl-L-glutamic acid. LGZ significantly reduced the level of N-methyl-L-glutamic acid, while also rebalancing 3-(4-hydroxyphenyl) pyruvate and reducing 3-nitrotyrosine, dopamine, and L-dopa levels. SIN significantly rebalanced normetanephrine levels, and notably adjusted the levels of L-dopa and N-methyl-L-glutamic acid (Figure 6B).

2.4. Joint Pathway Analysis

Using the "Joint-Pathway Analysis" module in MetaboAnalyst 6.0, we conducted an association analysis between the 34 core targets of LGZ and SIN, as predicted by network pharmacology, and the differentially regulated metabolites identified in plasma metabolomics following their combined administration (Figure 7A). The analysis identified 11 pathways with a $p < 0.05$, among which five pathways were enriched with both network pharmacology targets and differential metabolites from metabolomics. These pathways included arginine and proline metabolism, phenylalanine metabolism, arginine biosynthesis, histidine metabolism, and tyrosine metabolism. Further analysis of nine core targets of LGZ with the plasma differential metabolites revealed five pathways with a $p < 0.05$, of which two pathways—phenylalanine metabolism, and arginine and proline metabolism—were enriched with both targets and metabolites. For SIN, the analysis of its 25 core targets with plasma differential metabolites resulted in eight pathways with a $p < 0.05$, with three pathways simultaneously enriched with network pharmacology targets and metabolomics with differential metabolites. These included arginine and proline metabolism, arginine biosynthesis, and tyrosine metabolism.

These combined findings suggest that the phenylalanine metabolism, tyrosine metabolism, and arginine and proline metabolism pathways may be key metabolic pathways in the pain-relieving effects of LGZ and SIN. LGZ may primarily influence the tyrosine metabolism in the cerebrospinal fluid and phenylalanine metabolism in the plasma, while SIN appears to mainly regulate the tyrosine metabolism in the plasma.

The analysis of the number of pathways with a $p < 0.05$, enriched by both network pharmacology and plasma metabolomics, showed that LGZ+SIN regulated six more pathways than ligustrazine and three more than SIN. Furthermore, LGZ+SIN enriched three more pathways, with both network pharmacology targets and differential metabolites compared to LGZ and two more than SIN.

An association analysis of the 34 core targets of LGZ and SIN, as predicted by network pharmacology, with the differentially regulated metabolites in CSF metabolomics following LGZ+SIN treatment was conducted (Figure 7B). The results revealed eight pathways with a $p < 0.05$, of which three pathways—tyrosine metabolism; phenylalanine metabolism; and glycine, serine, and threonine metabolism—were significantly enriched with both network pharmacology targets and differential metabolites. These three pathways were considered the primary metabolic pathways through which LGZ+SIN regulates pain. Among these, tyrosine metabolism contained the highest number of targets and differential metabolites, indicating its pivotal role. For LGZ, the association analysis with the CSF metabolites revealed three pathways with a $p < 0.05$, with only one pathway (tyrosine metabolism) being enriched with both network pharmacology targets and differential metabolites. For SIN, the analysis showed five pathways with a $p < 0.05$, but none of these pathways were enriched with both network pharmacology targets and CSF differential metabolites.

Overall, pathway enrichment analysis demonstrated that LGZ+SIN regulated more pathways with a $p < 0.05$ than either LGZ (by five pathways) or SIN (by three pathways). Additionally, pathways enriched with both network pharmacology targets and differential metabolites were more abundant in LGZ+SIN, with three more pathways than LGZ and two more pathways than SIN.

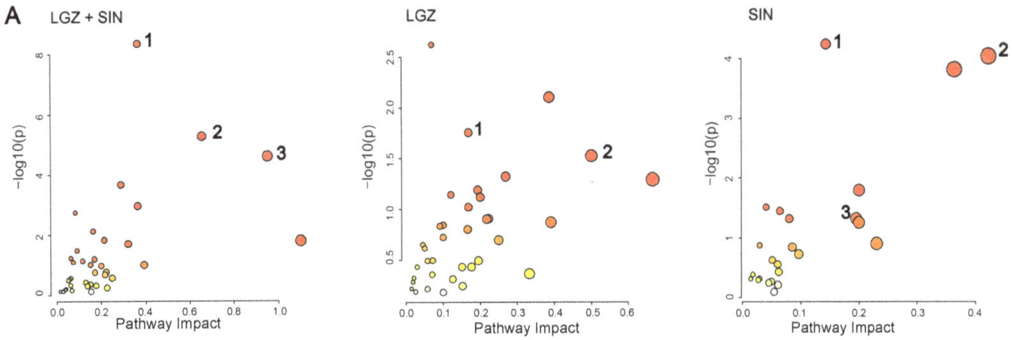

Group	Pathways name	Matched targets	Matched metabolites
LGZ+SIN	1.Arginine and proline metabolism	NOS1, NOS2, MAOA, MAOB	HMDB0000517, HMDB0000128, HMDB0001256, HMDB0001257, HMDB0001366, HMDB0000214, HMDB0003464, HMDB0000243
	2.Arginine biosynthesis	NOS1, NOS2	HMDB0000517, HMDB0000904, HMDB0000214, HMDB0000208
	3.Phenylalanine metabolism	MAOA, MAOB	HMDB0006236, HMDB0000205, HMDB0000158
	4.Histidine metabolism	MAOA, MAOB	HMDB0000033, HMDB0000194, HMDB0002024
	5.Tyrosine metabolism	MAOA, MAOB	HMDB0000208, HMDB0000118, HMDB0000243
LGZ	1.Arginine and proline metabolism	MAOA	HMDB0000064, HMDB0000162, HMDB0000214
	2. Phenylalanine metabolism	MAOA	HMDB0006236
SIN	1. Arginine and proline metabolism	NOS1, NOS2, MAOA, MAOB	HMDB0000214, HMDB0003464
	2. Arginine biosynthesis	MAOA, MAOB	HMDB0000214, HMDB0000904
	3.Tyrosine metabolism	MAOA, MAOB	HMDB0000291

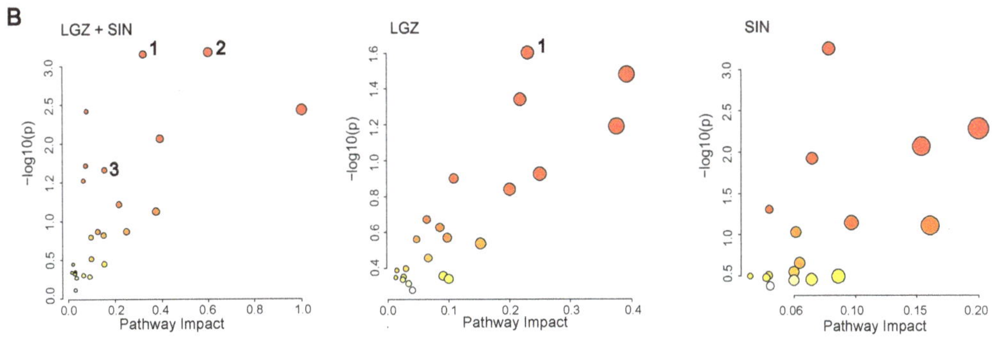

Group	Pathways name	Matched targets	Matched metabolites
LGZ+SIN	1. Tyrosine metabolism	MAOA, MAOB	HMDB0001866, HMDB0000819, HMDB0000707
	2. Phenylalanine metabolism	MAOA, MAOB	HMDB0000205
	3. Glycine, serine and threonine metabolism	MAOA, MAOB	HMDB0001721
LGZ	1. Tyrosine metabolism	MAOA	HMDB0001866, HMDB0000819
SIN	-	-	-

Figure 7. Joint pathway analysis: matched targets and metabolites. (**A**) Joint pathway analysis of the plasma samples. (**B**) Joint pathway analysis of the CSF samples. The size of the bubbles represents the number of enriched metabolites—the larger the bubble, the greater the number of enriched metabolites. The color of the bubbles indicates enrichment significance, where a darker color represents a smaller p-value and higher significance.

3. Discussion

NP is a secondary condition associated with various clinical disorders which significantly impacts the quality of life of patients. However, the precise pathogenesis of NP remains poorly understood and involves complex interactions between multiple signaling pathways. Currently, several widely accepted mechanisms are believed to contribute to the development of NP, including inflammation, peripheral sensitization, central sensitization, dysfunction of the descending inhibitory system, and changes in ion channels. Clinical metabolomics studies have recently reported significant alterations in the plasma metabolite profiles of patients with NP, particularly in the levels of amino acid analogs such as histidine, asparagine, glutamine, tyrosine, phenylalanine, proline, and choline [20]. Furthermore, phenylalanine and tyrosine levels in the CSF of patients with localized pain syndromes have been shown to be markedly elevated [21]. Some important metabolites, such as tyrosine, purine, asparagine, histidine, serine, and glutamate, have been shown to be involved in the onset and development of neuropathic pain [22]. Therefore, exploring pathological mechanisms from the perspective of metabolomics, identifying clinical biomarkers, and further developing therapeutic drugs are highly necessary.

Multi-drug combination therapies, encompassing both combined pharmaceutical agents and multi-target TCM, can partially address clinical treatment needs. Chuanxiong Rhizoma, derived from the rhizome of *Ligusticum chuanxiong* Hort., has been widely used in traditional medicine since the Han dynasty (~1800 years ago), though it is typically employed as an adjunctive or supporting medicine according to TCM theory. Its formulations, such as ligustrazine injection and salvia miltiorrhiza ligustrazine injection, are primarily used in China for the treatment of occlusive cerebrovascular diseases [23,24]. Sinomenii Caulis, sourced from the stems of *Sinomenium acutum* (Thunb.) Rehd. et Wils., is used clinically for the treatment of rheumatism, rheumatoid arthritis, and related pain symptoms [25,26]. Chuanxiong Rhizoma and Sinomenii Caulis, widely used in clinical practice as TCM, have long been recognized for their substantial efficacy in treating various pain-related conditions. Based on these findings, we selected their active components, LGZ and SIN, for combined application, to explore their potential therapeutic effects on NP. These natural products, or TCM, frequently exhibit multi-target properties, complicating the precise identification of their therapeutic effects [27]. However, for drugs with unclear mechanisms of action, identifying their key therapeutic targets is essential.

In previous studies, we investigated the analgesic effects of the combined use of LGZ and SIN in models of inflammatory pain, sciatic nerve injury, and spinal cord injury NP [14]. Given that the CCI model simulated both neuropathic and inflammatory pain characteristics, we examined the analgesic effects of LGZ and SIN, both in combination and individually, in the CCI model rats to comprehensively assess the benefits of their combined use. In previous research, LGZ and SIN have been administered via intraperitoneal injection [12], even though both have established oral administration protocols [28,29]. Therefore, in this study, we evaluated the analgesic effects of the oral administration of LGZ and SIN in combination. Additionally, the experimental design evaluated the analgesic effects of different time points (0, 0.5, 2, 4, and 6 h), dosages (LGZ 25 $mg{\cdot}kg^{-1}{\cdot}d^{-1}$ + SIN 25 $mg{\cdot}kg^{-1}{\cdot}d^{-1}$; LGZ 50 $mg{\cdot}kg^{-1}{\cdot}d^{-1}$ + SIN 50 $mg{\cdot}kg^{-1}{\cdot}d^{-1}$; and LGZ 100 $mg{\cdot}kg^{-1}{\cdot}d^{-1}$ + SIN 100 $mg{\cdot}kg^{-1}{\cdot}d^{-1}$) and days (1, 2, and 3 days) for both combined and single-drug treatments to comprehensively characterize the analgesic properties of the LGZ-SIN combination. Results from the MWT test, cold allodynia test, and incapacitance test demonstrated that both the combined and individual treatments of LGZ and SIN effectively alleviated mechanical allodynia, cold pain sensitivity, and spontaneous pain in CCI-induced NP. Furthermore, the combination of LGZ and SIN exhibited significant greater analgesic effects than single-drug treatments, reinforcing the rationale for their combined use. We also found a dose dependence of the combination of LGZ and SIN in the MWT test. Moreover, the results suggested that the combined use of LGZ and SIN also had beneficial effects on plasma inflammation, sciatic nerve inflammation and repair in CCI rats. Clinical studies have shown that IL-6 levels are elevated in the plasma of NP patients, which is consistent with the trend observed in our CCI rat model, where IL-6 levels were detected in both the sciatic nerve and plasma. However, we did not observe a significant change in IL-6 levels, which may be attributed to the relatively short treatment duration, which may not have allowed sufficient time for the drug's regulatory effects to fully manifest [30].

Following the evaluation of LGZ, SIN, and their combination, we conducted network pharmacology and metabolomics studies to explore their potential mechanisms in treating NP. The network pharmacology approach elucidated the analgesic mechanism of LGZ and SIN by examining their individual contributions. First, both LGZ and SIN demonstrated multi-target properties. Second, pathway analysis confirmed that both LGZ and SIN could regulate multiple signaling pathways to exert their synergistic effects. Based on network pharmacology results, modulation of the tyrosine metabolism and phenylalanine metabolism pathways may be the key mechanisms through which the combined use of LGZ and SIN exerted its analgesic effects. In addition, the combination of LGZ and SIN regulate arginine and proline metabolism, as well as histidine metabolism.

As NP affects both the peripheral and central systems, this study analyzed plasma and CSF samples to investigate the metabolic regulatory effects of LGZ and SIN, both in combination and individually, on CCI rats. The results of metabolic pathway analysis showed that the combined treatment of LGZ and SIN regulated more metabolic pathways in both CSF and plasma samples compared to either LGZ or SIN used alone, exhibiting a synergistic effect. Finally, joint pathway analysis revealed that tyrosine metabolism and phenylalanine metabolism were the key pathways enriched in both CSF and plasma samples. These pathways were considered the most critical. Among these, LGZ had a greater impact on tyrosine metabolism in CSF, while SIN exhibited a stronger effect on the tyrosine metabolism in plasma. The arginine and proline metabolism pathways contained the most targets and differential metabolites enriched by the combined treatment of LGZ and SIN in plasma samples. Therefore, the combined treatment of LGZ and SIN may alleviate pain in CCI model rats by co-regulating tyrosine metabolism and phenylalanine metabolism in both the CSF and plasma, as well as by modulating the arginine and proline metabolism in the plasma. Moreover, the number of differential metabolites in the metabolic pathways regulated by the combination of LGZ and SIN was much higher than that of LGZ and SIN alone, and interestingly, some of the differential metabolites were not present in LGZ or SIN alone, which were new differential metabolites generated by the combination. The enhanced effect of combining the two also suggests that we may be able to achieve the same effect of LGZ and SIN alone at a lower dose when they are combined. Thus, the combination of LGZ and SIN may produce a synergistic effect.

Tyrosine is an essential amino acid, and phenylalanine serves as its precursor. Both tyrosine and phenylalanine serve as precursors for monoamine neurotransmitters, including dopamine, norepinephrine, and epinephrine. The descending monoaminergic pathways, particularly those involving norepinephrine and serotonin transmission, play a crucial role in the endogenous pain modulation system, a mechanism that is well documented in NP [31]. Studies have shown that CCI leads to a reduction in the neurotransmitters crucial for descending pain regulation pathways, such as serotonin and norepinephrine [32]. Serotonin and dopamine potentiate noradrenergic effects to inhibit neuropathic pain. Moreover, antidepressants that inhibit the reuptake of norepinephrine and serotonin have been shown to be effective in chronic neuropathic pain [5]. The metabolomics results showed that serotonin and its precursor, tryptophan, increased in the plasma of the model group, but were restored after treatment. The LGZ+SIN group exhibited a more pronounced recovery compared to the individual treatments. Arginine, a non-essential amino acid, serves as a precursor for nitric oxide, proline, and glutamate. Studies have demonstrated that arginine could increase pain sensitivity in animal models [33]. Small-scale patient studies have suggested that L-arginine might have an analgesic effect on chronic pain [34]. The central glutamatergic system plays a critical role in the onset and persistence of persistent pain, including both neuropathic and inflammatory pain [35]. Following nerve injury, the downregulation of GABA and opioid receptors in the spinal cord leads to increased glutamate release, which may contribute to the development of neuropathic pain [36]. Studies have shown that CCI-induced NP reduces the GABA levels and neuronal activity in the dorsal horn [37]. Furthermore, the glutamatergic system could exacerbate chronic neuropathic pain by activating N-methyl-D-aspartate receptors (NMDARs) [38]. Studies have demonstrated that NMDARs play a crucial role in modulating both peripheral and central sensitization in NP [39]. The metabolomics analysis revealed a decrease in glutamine levels in the model group, which was subsequently restored following treatment, potentially contributing to this effect. In addition, reduced arginine levels may lead to neurotransmitter depletion, contributing to inflammatory pain [40]. Meanwhile, histidine plays a crucial role in the inflammatory process by regulating the synthesis of histamine neurotransmitters [41]. Therefore, arginine and histidine metabolism may be closely linked to the anti-inflammatory effects of the LGZ and SIN combination. In our experiment, the improvement of inflammatory factors in the sciatic nerve and plasma of CCI rats after treatment might be related to this.

To the best of our knowledge, this is the first report on the therapeutic effects and potential mechanisms of combining LGZ and SIN for the treatment of neuropathic pain induced by CCI using metabolomics and network pharmacology approaches. The combination of two drugs, LGZ and SIN, also offers a new combination of clinical treatment options for neuropathic pain. On one hand, special attention should be given to their specific targets and related metabolic signaling pathways to uncover the molecular-level regulatory mechanisms. On the other hand, extending the treatment duration and integrating in-depth studies at the cytokine level could provide a more comprehensive assessment of their anti-inflammatory, analgesic, and other potential effects. This multi-layered research approach will contribute to a more thorough understanding of the pharmacological mechanisms of LGZ and SIN combination therapy, thereby providing a stronger scientific foundation for its clinical application.

4. Materials and Methods

4.1. Chemicals and Materials

The Easyflow independent ventilation cage was purchased from Tecniplast, Italy. The Von Frey filaments were obtained from Ugo Basile Biological Apparatus Company. The Incapacitance Meter (BIO-SWB-TOUCH-R) was purchased from Bioseb, French. The pain testing frame was made in our laboratory.

AB Sciex HPLC-MS/MS system (Framingham, MA, USA) comprised an ExionLC-20AC high-performance liquid chromatography (HPLC) system, Ion Drive^{TM} Turbo V ion source, Sciex 6500^{+} triple quadrupole mass spectrometer, Analyst 1.7 data acquisition software, and MultiQuant 3.0.3 data processing software. The Targin VX-III multi-tube vortexer was purchased from Beijing Targin Technology Co., Ltd. (Beijing, China). The Forma 88,000 Series −86 °C ultra-low temperature freezer was obtained from Thermo Scientific (Waltham, MA, USA). The Rotanta 460R high-speed refrigerated centrifuge was acquired from Hettich (Kirchlengern, Germany). The MC-8 integrated cryogenic centrifuge concentrator was obtained from Beijing JM Technology Co., Ltd. (Beijing, China). The Synergy2 multifunctional microplate reader was purchased from Bio Tek (Winooski, VT, USA). The desktop anesthetic machine was supplied by Harvard Apparatus (Cambridge, MA, USA). The ThermoStar body temperature maintenance device was purchased from RWD Life Science Co., Ltd, (Shenzhen, China). The optical microscope (Olympus BX50) was purchased from Olympus Optical Co. (Tokyo, Japan).

Ligustrazine (ligustrazine hydrochloride, lot number: DT201803-19) and sinomenine (sinomenine hydrochloride, lot number: DT201806-22), both with a purity of ≥98%, were provided by Shanxi Datian Biotechnology Ltd. (Xi'an, China). Pregabalin (lot number: 295422) was provided by Beijing J&K Scientific Ltd. (Beijing, China). The isoflurane (lot number: 217180801) was purchased from RWD Life Science Co., Ltd.

IL-6, IL-1β, and TNF-α ELISA kits were purchased from Raybiotech (Peachtree Corners, GA, USA). The tissue lysis buffer (EL-lysis) was obtained from Raybiotech. The BCA protein assay kit was purchased from Thermo Fisher (Waltham, MA, USA) and used to calibrate the content of inflammatory factors. The ProcartaPlex^{TM} Rat Cytokine and Chemokine Panel was purchased from Thermo Fisher. The 4-0 chromic gut sutures were obtained from Shandong Boda Medical Products Co., Ltd. (Heze, China).

Mass spectrometry library kits and reference standards for glucose metabolism, amino acids, bile acids, and others were purchased from Sigma for the establishment of a widely targeted metabolomics analysis platform in our laboratory. Internal standards, including d-3 norepinephrine, d-4 dopamine, d-5 serotonin, and MSK-A2, were obtained from Cambridge Isotope Laboratories. Reference standards for metabolic pathways, including tyrosine, sodium borate, benzoyl chloride, and d-5 benzoyl chloride were purchased from Sigma. All reference and internal standards had a purity greater than 99%. LC/MS-grade acetonitrile, methanol, formic acid, and ammonium formate were obtained from Beijing Dikma Technologies Inc. (Beijing, China).

4.2. Animals and Treatment

Adult male Sprague Dawley rats (180–200 g, 6–7 weeks) were obtained from Beijing HFK Bioscience Co., Ltd. (Beijing, China). A total of 3 rats were housed per cage in an SPF-grade lab at a constantly maintained temperature (22 ± 2 °C) with a 12 h light/dark cycle and free access to food and water.

Following the successful establishment of the model, 42 rats were randomly assigned into 7 groups, with 6 animals per group. The groups were as follows: the model group (model, 10 $mL \cdot kg^{-1} \cdot d^{-1}$ saline); LGZ+SIN low-dose group (LGZ 25 $mg \cdot kg^{-1} \cdot d^{-1}$ + SIN 25 $mg \cdot kg^{-1} \cdot d^{-1}$); LGZ+SIN medium-dose group (LGZ 50 $mg \cdot kg^{-1} \cdot d^{-1}$ + SIN 50 $mg \cdot kg^{-1} \cdot d^{-1}$); LGZ+SIN high-dose group (LGZ 100 $mg \cdot kg^{-1} \cdot d^{-1}$ + SIN 100 $mg \cdot kg^{-1} \cdot d^{-1}$); Ligustrazine group (LGZ, 100 $mg \cdot kg^{-1} \cdot d^{-1}$); Sinomenine group (SIN, 100 $mg \cdot kg^{-1} \cdot d^{-1}$); and Pregabalin-positive control group (Pgb, 30 $mg \cdot kg^{-1} \cdot d^{-1}$). In addition, 6 healthy rats were set as the sham operation group (sham, 20 $mL \cdot kg^{-1} \cdot d^{-1}$ saline). All animals were orally administered their respective treatments twice daily (morning and evening) for a period of three consecutive days.

4.3. CCI Model Establishment

The CCI model in rats was established following the method described by Bennett [42]. After anesthetizing the rats with isoflurane, they were placed on a heating pad to maintain a body temperature of approximately 37 °C. The skin below the left femur was incised and the left sciatic nerve was exposed following blunt dissection of the surrounding tissue. The nerve was then ligated with 4–0 chromic gut sutures tied in four knots, each approximately 1 mm apart. The degree of ligation was adjusted to induce slight twitching of the calf muscles without compromising the blood supply to the nerve epineurium. In the sham group, the sciatic nerve was exposed but left unligated. The MWT test was conducted both prior to surgery and on day 7 post-surgery to evaluate the success of the model.

4.4. Pharmacodynamic Research

The body weight of the rats was recorded daily. Behavioral tests were conducted on days 1, 2, and 3 following drug administration. The behavioral tests included the MWT test, cold allodynia test, and incapacitance test. The MWT and cold allodynia tests were conducted at 0, 0.5, 2, 4, and 6 h after drug administration each day. The incapacitance test was conducted 4 h after drug administration each day. After the final behavioral test, samples of the affected sciatic nerve were collected for hematoxylin and eosin (H&E) staining and enzyme-linked immunosorbent assay (ELISA) analysis. Plasma and CSF samples were also collected for subsequent metabolomic analysis.

The MWT test was assessed using Von Frey filaments [43]. The rats were placed in a plastic chamber (20 cm × 20 cm × 15 cm) with a transparent acrylic lid, and they were allowed to acclimate for 30 min. Von Frey filaments, ranging from 4 g to 26 g, were used during the test. The "up-and-down" method was employed to determine the MWT value of each rat [44].

The cold allodynia test was performed by spraying 0.1 mL of acetone on the affected hind paw of the rat. The responses of the rats, including paw withdrawal and licking behavior, were observed, and these were then scored based on the degree of reaction: 0 points for no response, 1 point for mild reaction or rapid withdrawal of the hind paw, 2 points for repeated paw shaking, and 3 points for sustained or repeated lifting and licking of the hind paw [45].

The incapacitance test was conducted by placing the rats in a transparent box with an inclined platform, on which the rats stood on their hind feet. The left and right hind feet were positioned on separate sensor panels. Care was taken to ensure that the rats maintained an exploratory posture without leaning against the sides of the box. A capacimeter was used to measure the weight (in grams) on each panel over a 3 s period. Each rat underwent three measurements, with a minimum of 1 min between readings. The average of the readings for each hind foot was used to calculate the weight distribution. The bipedal balance bearing value was recorded as the percentage of total body weight supported by each hind foot. In normal rats, the weight distribution is nearly symmetrical (50:50%), whereas pain resulting from injury leads to a reduction in the load-bearing capacity of the injured hind foot. The incapacitance test result was calculated using the following formula: result = weight on the affected hind foot/(weight on the left hind foot + weight on the right hind foot) × 100% [46].

The sciatic nerve tissue was fixed in 4% paraformaldehyde and subsequently embedded in paraffin to prepare 5 μm thick sections. The sections were stained with hematoxylin for 5 min, followed by eosin for 3 min. Changes in the sciatic nerve were observed under an optical microscope.

The concentrations of IL-1β, IL-6, and TNF-α in the sciatic nerve were measured by ELISA. The experimental procedure was strictly followed according to the instructions provided with the kits. The assay of 22 cytokines and chemokines in plasma was performed and analyzed independently by Laizee Biotech (Shanghai, China) via a Luminex200 instrument and ProcartaPlex Analyst 1.0 software.

4.5. Network Pharmacology Analysis

First, the potential targets of LGZ and SIN were identified using the SWISS Target Prediction database (http://swisstargetprediction.ch/, accessed on 11 April 2024). These targets were then verified and refined using the UniProt database to obtain accurate target names for each compound. Subsequently, pain-related target information was retrieved from the Genecards database (https://www.genecards.org/, accessed on 11 April 2024) and the OMIM database (https://omim.org/, accessed on 15 April 2024). After removing duplicates, the remaining targets were considered pain-related targets for further analysis. The intersection of LGZ and SIN alkaloid targets with those associated with pain was identified using the Bioinformatics platform (http://www.bioinformatics.com.cn, accessed on 15 April 2024), yielding common genes across the different compounds. This gene set was then analyzed based on the uniqueness of the drug–target interactions.

Next, the intersecting target genes were entered into the STRING database to construct a protein–protein interaction (PPI) network. The network was visualized using Cytoscape 3.8.2, and the CytoHubba plugin was used to identify the core targets for further differential analysis.

Finally, the core target genes underwent Gene Ontology (GO) and Kyoto Encyclopedia of Genes and Genomes (KEGG) enrichment analysis using the Metascape database. A significance threshold of $p < 0.01$ was established for all analyses. The GO analysis covered three subcategories: biological process (BP), molecular function (MF), and cellular component (CC). Furthermore, based on the relationships between protein targets and signaling pathways, a compound–target–pathway association network was built.

4.6. Plasma and CSF Metabolomics Analysis

4.6.1. Plasma and CSF Sample Preparation

For sample preparation, 50 μL of the test sample was mixed with 450 μL of ice-cold extraction solvent containing internal standards (methanol: acetonitrile: water = 2:2:1). The mixture was vortexed for 3 min and then placed at −20 °C for 2 h. The samples were then centrifuged at 20,000× *g* for 15 min at 4 °C. The supernatant was carefully transferred to a 1.5 mL Eppendorf tube and subjected to vacuum concentration at 35 °C and 1500 rpm for 2 h. The residue was reconstituted with 100 μL of extraction solvent that was devoid of internal standards. The sample was centrifuged again at 20,000× *g* for 15 min at 4 °C, and 80 μL of the supernatant was collected for analysis. Additionally, 10 μL of each sample was pipetted to pool a quality control (QC) sample.

4.6.2. Widely Targeted Metabolomics Analysis

The metabolites were identified using an in-house reference database. An ACQUITY UPLC BEH Amide column (2.1 × 50 mm, 1.7 μm, Waters, Milford, MA, USA) and a pre-column (2.1 mm × 5 mm, 1.7 μm, Waters, USA) were used for sample separation. The mobile phase consisted of Solvent A (95% water: 5% acetonitrile, with 5 mM of ammonium formate and 0.01% formic acid) and Solvent B (95% acetonitrile: 5% water, containing 5 mM of ammonium formate and 0.01% formic acid). The gradient elution program was as follows: 0–2 min, 95–95% B; 2–4 min, 95–90% B; 4–6 min, 90–90% B; 6–9 min, 90–85% B; 9–12 min, 85–85% B; 12–15 min, 85–75% B; 15–16 min, 75–75% B; 16–18 min, 75–50% B; 18–20 min, 50–50% B; 20–22 min, 50–25% B; 22–24 min, 25–25% B; 24–25 min, 25–95% B; and 25–30 min, 95–95% B. Flow rate: 0.3 mL/min; column temperature: 35 °C; temperature: 4 °C; and injection volume: 5 μL.

Electrospray ionization (ESI) was used as the ionization source. The curtain gas (N_2) was set to 40 psi, the collision gas (N_2) to 9 psi, and the spray voltage was set at +5500 V and −4500 V for positive and negative ion modes, respectively. The nebulizer temperature was set to 550 °C, with ion source gas (Ion Source Gas1, N_2) and auxiliary gas (Ion Source Gas2, N_2) both maintained at 55 psi. Scanning was performed in both positive and negative ion modes. Optimized ion pairs and mass spectrometry parameters were applied for each metabolite.

4.6.3. Targeted Metabolomics Analysis

The method for measuring the tyrosine pathway was adapted from previously published protocols [47], with the necessary modifications outlined below.

Sample preparation: a 50 μL aliquot of the sample was mixed with 150 μL of acetonitrile (1:3, *v*/*v*). The mixture was vortexed at 8000 rpm for 5 min, followed by centrifugation at 20,000× *g* for 10 min. Subsequently, 10 μL of the supernatant was transferred and added to 10 μL of 100 mM sodium borate and 10 μL of 1% benzoyl chloride. The mixture was vortexed for 5 min, incubated at 25 °C for 5 min, and then centrifuged at 20,000× *g* for 10 min. The resulting supernatant (24 μL) was mixed with 6 μL of an internal standard solution (a mixture of tyrosine pathway standards and d-5 benzoyl chloride for derivatization). The mixture was then vortexed and prepared for injection.

A PFP C18 column (2.1 × 50 mm, 1.8 μm, Waters, Milford, MA, USA) was used to separate the samples. Water with 0.1% formic acid and 5 mM of ammonium formate served as Mobile Phase A and acetonitrile served as Mobile Phase B. The gradient programs were as follows: 0–1 min, 20–20% B; 1–2 min, 20–50% B; 2–6 min, 50–70% B; 6–6.5 min, 70–95% B; 8–8.1 min, 95–20% B; and 8.1–10 min, 20–20% B. Flow rate: 0.3 mL/min; column temperature: 35 °C; sample temperature: 4 °C; and injection volume: 2 μL.

Electrospray ionization (ESI) was used as the ionization source. The curtain gas (N_2) was set to 35 psi, collision gas (N_2) to 9 psi, and the spray voltage was set at 5500. The nebulizer temperature was set to 550 °C, with ion source gas (Ion Source Gas1, N_2) and auxiliary gas (Ion Source Gas2, N_2) both at 55 psi. The analysis was performed in multiple reaction monitoring (MRM) mode with positive ion scanning. The specific ion pair parameters used for the analysis are provided in Supplementary Table S1.

4.6.4. Metabolomics Analysis

To ensure QC for the metabolomics analysis, a QC sample was injected after every ten experimental sample during the chromatography run. All LC-MS data were meticulous preprocessing using MultiQuant 3.0.3 software, including key steps such as peak detection, peak identification, peak area calculation, and calibration.

PCA was initially performed to identify the major variability patterns within the dataset. OPLS-DA was then applied to identify metabolites that might differentiate between groups. The quality of the OPLS-DA model was evaluated using the parameters R^2Y and Q^2. Additionally, permutation testing was conducted to assess the risk of false positives in the OPLS-DA model. Potential biomarkers with significant statistical and biological relevance were selected based on the following criteria: VIP > 1, *t*-test (p) < 0.05, and fold change (FC) > 1.2 or <0.83. Finally, metabolic pathways associated with the differentially expressed metabolites were determined using a significance threshold of $p < 0.05$. Metabolomics data were analyzed using the Metware Metabolomics Cloud Platform (https://cloud.metware.cn/, accessed on 20 December 2024) and MetaboAnalyst 6.0 (https://www.metaboanalyst.ca/, accessed on 27 December 2024).

4.7. Joint Pathway Analysis

A joint pathway analysis was performed using the "Joint-Pathway Analysis" module in MetaboAnalyst 6.0, in order to correlate the key targets predicted by network pharmacology with the differential metabolites identified in the metabolomics analysis. Pathways exhibiting the highest enrichment of both targets and metabolites were considered key pathways.

4.8. Statistical Analysis

Statistical analyses were performed using SPSS 20.0 and GraphPad Prism 8.0. All data are presented as the mean ± standard error of the mean (SEM). The significance analysis of differences between two groups was assessed using a *t*-test, while multiple comparison was conducted using one-way or two-way analysis of variance (ANOVA). A p value < 0.05 indicated statistical significance, and $p < 0.01$ indicated highly statistical significance.

Supplementary Materials: The following supporting information can be downloaded at: https://www.mdpi.com/article/10.3390/ijms26062604/s1.

Author Contributions: Conceptualization, T.L., Z.W. and Z.Y.; methodology, T.L., Z.Y., X.Z., J.Z. and Y.M.; software, Z.Y.; validation, X.Z.; formal analysis, X.Z., Y.J. and Y.L.; writing—original draft preparation, Z.Y. and Y.Z.; investigation, Y.Z.; resources, C.G.; data curation, Z.Y.; supervision, C.G. and Z.W.; writing—review and editing, T.L., Z.Y. and Y.Z. All authors have read and agreed to the published version of the manuscript.

Funding: This research was supported by the Fundamental Research Funds for the Central Public Welfare Research Institutes of China (grant numbers ZZ14-YQ-041, JJPY2022003, JJPY2022025, JBGS2023004, and ZZ2019004); the Scientific and Technological Innovation Project of China Academy of Chinese Medical Sciences (grant number CI2021B015); the National Natural Science Foundation of China (grant numbers 81503278 and T2341017).

Institutional Review Board Statement: The animal study protocol was approved by the Institutional Animal Care and Use Committee at the Experimental Research Center, China Academy of Chinese Medical Sciences (No. ERCCACMS21-2111-01).

Informed Consent Statement: Not applicable.

Data Availability Statement: The original contributions presented in this study are included in the article/Supplementary Material. Further inquiries can be directed to the corresponding author.

Conflicts of Interest: The authors declare no conflicts of interest.

References

1. Finnerup, N.B.; Kuner, R.; Jensen, T.S. Neuropathic Pain: From Mechanisms to Treatment. *Physiol. Rev.* **2021**, *101*, 259–301. [CrossRef] [PubMed]
2. Alles, S.R.A.; Smith, P.A. Etiology and Pharmacology of Neuropathic Pain. *Pharmacol. Rev.* **2018**, *70*, 315–347. [CrossRef]
3. Orhurhu, M.S.; Chu, R.; Claus, L.; Roberts, J.; Salisu, B.; Urits, I.; Orhurhu, E.; Viswanath, O.; Kaye, A.D.; Kaye, A.J.; et al. Neuropathic Pain and Sickle Cell Disease: A Review of Pharmacologic Management. *Curr. Pain Headache Rep.* **2020**, *24*, 52. [CrossRef] [PubMed]
4. Hao, S.; Lin, S.; Tao, W.; Zhuo, M. Cortical Potentiation in Chronic Neuropathic Pain and the Future Treatment. *Pharmaceuticals* **2025**, *18*, 363. [CrossRef]
5. Baron, R.; Binder, A.; Wasner, G. Neuropathic pain: Diagnosis, pathophysiological mechanisms, and treatment. *Lancet Neurol.* **2010**, *9*, 807–819. [CrossRef] [PubMed]
6. Jiang, B.-C.; Liu, T.; Gao, Y.-J. Chemokines in chronic pain: Cellular and molecular mechanisms and therapeutic potential. *Pharmacol. Ther.* **2020**, *212*, 107581. [CrossRef]
7. Yoshimoto, Y.; Okai, H.; Namba, H.; Taguchi, K.; Yamauchi, Y.; Wakita, J.; Okazaki, R. Combined antiallodynic effects of Neurotropin®–tramadol and Neurotropin®–mirogabalin in rats with L5-spinal nerve ligation. *J. Pharmacol. Sci.* **2024**, *156*, 30–37. [CrossRef]
8. Hung, Y.C.; Kuthati, Y.; Zhang, X.; Gao, T. Analgesic Alkaloids Derived From Traditional Chinese Medicine in Pain Management. *Front. Pharmacol.* **2022**, *13*, 851508. [CrossRef]
9. Xing, Z.; Chen, Y.; Chen, J.; Peng, C.; Peng, F.; Li, D. Metabolomics integrated with mass spectrometry imaging reveals novel action of tetramethylpyrazine in migraine. *Food Chem.* **2024**, *460*, 140614. [CrossRef]
10. Rao, Y. Tetramethylpyrazine and Astragaloside IV have synergistic effects against spinal cord injury-induced neuropathic pain via the OIP5-AS1/miR-34a/Sirt1/NF-κB axis. *Int. Immunopharmacol.* **2023**, *115*, 109546. [CrossRef]
11. Gao, T.; Hao, J.; Wiesenfeld-Hallin, Z.; Wang, D.-Q.; Xu, X.-J. Analgesic effect of sinomenine in rodents after inflammation and nerve injury. *Eur. J. Pharmacol.* **2013**, *721*, 5–11. [CrossRef]

12. Gao, T.; Li, T.; Jiang, W.; Fan, W.; Xu, X.-J.; Zhao, X.; Yin, Z.; Guo, H.; Wang, L.; Gao, J.; et al. Antinociceptive Effects of Sinomenine Combined With Ligustrazine or Paracetamol in Animal Models of Incisional and Inflammatory Pain. *Front. Physiol.* **2021**, *11*, 523769. [CrossRef]
13. Chen, J.; Guo, P.; Liu, X.; Liao, H.; Chen, K.; Wang, Y.; Qin, J.; Yang, F. Sinomenine alleviates diabetic peripheral neuropathic pain through inhibition of the inositol-requiring enzyme 1 alpha–X-box binding protein 1 pathway by downregulating prostaglandin-endoperoxide synthase 2. *J. Diabetes Investig.* **2023**, *14*, 364–375. [CrossRef] [PubMed]
14. Gao, T.; Shi, T.; Wiesenfeld-Hallin, Z.; Li, T.; Jiang, J.-D.; Xu, X.-J. Sinomenine facilitates the efficacy of gabapentin or ligustrazine hydrochloride in animal models of neuropathic pain. *Eur. J. Pharmacol.* **2019**, *854*, 101–108. [CrossRef]
15. Meng, X.; Ma, J.; Kang, S.Y.; Jung, H.W.; Park, Y.-K. Jowiseungki decoction affects diabetic nephropathy in mice through renal injury inhibition as evidenced by network pharmacology and gut microbiota analyses. *Chin. Med.* **2020**, *15*, 24. [CrossRef] [PubMed]
16. Hu, S.; Zuo, H.; Qi, J.; Hu, Y.; Yu, B. Analysis of Effect of Schisandra in the Treatment of Myocardial Infarction Based on Three-Mode Gene Ontology Network. *Front. Pharmacol.* **2019**, *10*, 232. [CrossRef] [PubMed]
17. Zhang, P.; Zhang, D.; Zhou, W.; Wang, L.; Wang, B.; Zhang, T.; Li, S. Network pharmacology: Towards the artificial intelligence-based precision traditional Chinese medicine. *Brief. Bioinform.* **2024**, *25*, bbad518. [CrossRef]
18. Yan, X.-Y.; Xiang, P.; Yu, Z.-G.; Yan, H. Application of Metabonomics in Substance Abuse Toxicology Research. *J. Forensic. Sci.* **2022**, *38*, 400–407. [CrossRef]
19. Lei, C.; Chen, Z.; Fan, L.; Xue, Z.; Chen, J.; Wang, X.; Huang, Z.; Men, Y.; Yu, M.; Liu, Y.; et al. Integrating Metabolomics and Network Analysis for Exploring the Mechanism Underlying the Antidepressant Activity of Paeoniflorin in Rats With CUMS-Induced Depression. *Front. Pharmacol.* **2022**, *13*, 904190. [CrossRef]
20. Chen, P.; Wang, C.; Ren, Y.; Ye, Z.; Jiang, C.; Wu, Z. Alterations in the gut microbiota and metabolite profiles in the context of neuropathic pain. *Mol. Brain* **2021**, *14*, 50. [CrossRef]
21. Meissner, A.; Van Der Plas, A.A.; Van Dasselaar, N.T.; Deelder, A.M.; Van Hilten, J.J.; Mayboroda, O.A. 1H-NMR metabolic profiling of cerebrospinal fluid in patients with complex regional pain syndrome–related dystonia. *Pain* **2014**, *155*, 190–196. [CrossRef]
22. Ghafouri, B.; Thordeman, K.; Hadjikani, R.; Bay Nord, A.; Gerdle, B.; Bäckryd, E. An investigation of metabolome in blood in patients with chronic peripheral, posttraumatic/postsurgical neuropathic pain. *Sci. Rep.* **2022**, *12*, 21714. [CrossRef] [PubMed]
23. Shao, H.; He, X.; Zhang, L.; Du, S.; Yi, X.; Cui, X.; Liu, X.; Huang, S.; Tong, R. Efficacy of Ligustrazine Injection as Adjunctive Therapy in Treating Acute Cerebral Infarction: A Systematic Review and Meta-Analysis. *Front. Pharmacol.* **2021**, *12*, 761722. [CrossRef] [PubMed]
24. Ma, Z.; Zhang, H.; Zhao, F.; Li, K.; Dong, N.; Sang, W. Safety and effectiveness of Salvia miltiorrhiza and ligustrazine injection for acute cerebral infarction in Chinese population: A PRISMA-compliant meta-analysis. *Front. Pharmacol.* **2024**, *15*, 1425053. [CrossRef] [PubMed]
25. Huang, Z.; Mao, X.; Chen, J.; He, J.; Shi, S.; Gui, M.; Gao, H.; Hong, Z. Sinomenine hydrochloride injection for knee osteoarthritis: A protocol for systematic review and meta-analysis. *Medicine* **2022**, *101*, e28503. [CrossRef]
26. Li, J.-M.; Yao, Y.-D.; Luo, J.-F.; Liu, J.-X.; Lu, L.-L.; Liu, Z.-Q.; Dong, Y.; Xie, Y.; Zhou, H. Pharmacological mechanisms of sinomenine in anti-inflammatory immunity and osteoprotection in rheumatoid arthritis: A systematic review. *Phytomedicine* **2023**, *121*, 155114. [CrossRef]
27. Gan, X.; Shu, Z.; Wang, X.; Yan, D.; Li, J.; Ofaim, S.; Albert, R.; Li, X.; Liu, B.; Zhou, X.; et al. Network medicine framework reveals generic herb-symptom effectiveness of traditional Chinese medicine. *Sci. Adv.* **2023**, *9*, eadh0215. [CrossRef]
28. Zhang, S.; Zheng, Y.; Du, H.; Zhang, W.; Li, H.; Ou, Y.; Xu, F.; Lin, J.; Fu, H.; Ni, X.; et al. The Pathophysiological Changes and Clinical Effects of Tetramethylpyrazine in ICR Mice with Fluoride-Induced Hepatopathy. *Molecules* **2023**, *28*, 4849. [CrossRef]
29. Gao, T.; Shi, T.; Wang, D.-Q.; Wiesenfeld-Hallin, Z.; Xu, X.-J. Repeated sinomenine administration alleviates chronic neuropathic pain-like behaviours in rodents without producing tolerance. *Scand. J. Pain* **2014**, *5*, 249–255. [CrossRef]
30. Jönsson, M.; Gerdle, B.; Ghafouri, B.; Bäckryd, E. The inflammatory profile of cerebrospinal fluid, plasma, and saliva from patients with severe neuropathic pain and healthy controls-a pilot study. *BMC Neurosci.* **2021**, *22*, 6. [CrossRef]
31. Zhao, X.; Wang, C.; Zhang, J.-F.; Liu, L.; Liu, A.-M.; Ma, Q.; Zhou, W.-H.; Xu, Y. Chronic curcumin treatment normalizes depression-like behaviors in mice with mononeuropathy: Involvement of supraspinal serotonergic system and GABAA receptor. *Psychopharmacology* **2014**, *231*, 2171–2187. [CrossRef]
32. Chaplan, S.R.; Bach, F.W.; Pogrel, J.W.; Chung, J.M.; Yaksh, T.L. Quantitative assessment of tactile allodynia in the rat paw. *J. Neurosci. Methods* **1994**, *53*, 55–63. [CrossRef] [PubMed]
33. Severyanova, L.A.; Bobyntsev, I.I.; Kir'yanova, N.A.; Dolgintsev, M.E. Effects of L-arginine on various types of pain sensitivity. *Bull. Exp. Biol. Med.* **2006**, *141*, 567–570. [CrossRef] [PubMed]
34. Harima, A.; Shimizu, H.; Takagi, H. Analgesic effect of L-arginine in patients with persistent pain. *Eur. Neuropsychopharmacol.* **1991**, *1*, 529–533. [CrossRef]

35. Osaka, H.; Mukherjee, P.; Aisen, P.S.; Pasinetti, G.M. Complement-derived anaphylatoxin C5a protects against glutamate-mediated neurotoxicity. *J. Cell. Biochem.* **1999**, *73*, 303–311. [CrossRef]
36. Kohno, T.; Ji, R.-R.; Ito, N.; Allchorne, A.J.; Befort, K.; Karchewski, L.A.; Woolf, C.J. Peripheral axonal injury results in reduced μ opioid receptor pre- and post-synaptic action in the spinal cord. *Pain* **2005**, *117*, 77–87. [CrossRef]
37. Moon, H.C.; Park, Y.S. Reduced GABAergic neuronal activity in zona incerta causes neuropathic pain in a rat sciatic nerve chronic constriction injury model. *J. Pain Res.* **2017**, *10*, 1125–1134. [CrossRef]
38. Medeiros, P.; Negrini-Ferrari, S.E.; Palazzo, E.; Maione, S.; Ferreira, S.H.; De Freitas, R.L.; Coimbra, N.C. N-methyl-d-aspartate Receptors in the Prelimbic Cortex are Critical for the Maintenance of Neuropathic Pain. *Neurochem. Res.* **2019**, *44*, 2068–2080. [CrossRef]
39. Liu, Y.J.; Li, Y.L.; Fang, Z.H.; Liao, H.L.; Zhang, Y.Y.; Lin, J.; Liu, F.; Shen, J.F. NMDARs mediate peripheral and central sensitization contributing to chronic orofacial pain. *Front. Cell. Neurosci.* **2022**, *16*, 999509. [CrossRef]
40. Fung, T.S.; Ryu, K.W.; Thompson, C.B. Arginine: At the crossroads of nitrogen metabolism. *EMBO J.* **2025**, *44*, 1275–1293. [CrossRef]
41. Brosnan, M.E.; Brosnan, J.T. Histidine Metabolism and Function. *J. Nutr.* **2020**, *150*, 2570S–2575S. [CrossRef] [PubMed]
42. Bennett, G.J.; Xie, Y.-K. A peripheral mononeuropathy in rat that produces disorders of pain sensation like those seen in man. *Pain* **1988**, *33*, 87–107. [CrossRef]
43. Shen, Y.; Ding, Z.; Ma, S.; Ding, Z.; Zhang, Y.; Zou, Y.; Xu, F.; Yang, X.; Schäfer, M.K.E.; Guo, Q.; et al. SETD7 mediates spinal microgliosis and neuropathic pain in a rat model of peripheral nerve injury. *Brain. Behav. Immun.* **2019**, *82*, 382–395. [CrossRef] [PubMed]
44. Deuis, J.R.; Dvorakova, L.S.; Vetter, I. Methods Used to Evaluate Pain Behaviors in Rodents. *Front. Mol. Neurosci.* **2017**, *10*, 284. [CrossRef]
45. Chen, J.; Joshi, S.K.; DiDomenico, S.; Perner, R.J.; Mikusa, J.P.; Gauvin, D.M.; Segreti, J.A.; Han, P.; Zhang, X.-F.; Niforatos, W.; et al. Selective blockade of TRPA1 channel attenuates pathological pain without altering noxious cold sensation or body temperature regulation. *Pain* **2011**, *152*, 1165–1172. [CrossRef] [PubMed]
46. Buys, M.J.; Alphonso, C. Novel Use of Perineural Pregabalin Infusion for Analgesia in a Rat Neuropathic Pain Model. *Anesth. Analg.* **2014**, *119*, 481–488. [CrossRef]
47. Wong, J.-M.T.; Malec, P.A.; Mabrouk, O.S.; Ro, J.; Dus, M.; Kennedy, R.T. Benzoyl chloride derivatization with liquid chromatography–mass spectrometry for targeted metabolomics of neurochemicals in biological samples. *J. Chromatogr. A* **2016**, *1446*, 78–90. [CrossRef]

International Journal of
Molecular Sciences

MDPI

Article

Enhancing the Therapeutic Efficacy of Berberine and Quercetin Through Salt Formulation for Liver Fibrosis Treatment

Yangyang Cheng, Haoyang Yu, Sitong Yang †, Xiaolian Tian †, Mengyu Zhao, Ling Ren, Xiuping Guo, Chujuan Hu, Jiandong Jiang * and Lulu Wang *

Institute of Medicinal Biotechnology, Chinese Academy of Medical Science & Peking Union Medical College, Beijing 100050, China; yy550258560@outlook.com (Y.C.); yuhy1014@163.com (H.Y.); yangst0323@163.com (S.Y.); 18852072531@163.com (X.T.); zhaomengyu85@163.com (M.Z.); renling0225@163.com (L.R.); gxp2000429@163.com (X.G.); hcj20000616@163.com (C.H.)

* Correspondence: jiangjiandong@imb.cams.cn (J.J.); wanglulu@imb.cams.cn (L.W.)

† These authors have contributed equally to this work.

Abstract: Liver fibrosis, caused by chronic hepatic injury, is a major threat to human health worldwide, as there are no specific drugs available for its treatment. Natural compounds, such as berberine (BBR) and quercetin (QR), have shown the ability to regulate energy metabolism and protect the liver without significant adverse effects. Additionally, combination therapy (the cocktail therapy approach), using multiple drugs, has shown promise in treating complicated conditions, including liver injury. In this study, we prepared a salt formulation of BBR and QR (BQS) to enhance their combined effect on liver fibrosis. The formation of BQS was confirmed using various analytical techniques, including nuclear magnetic resonance spectroscopy (NMR), differential scanning calorimetry (DSC), Fourier-transform infrared spectroscopy (FTIR), powder X-ray diffractometry (PXRD), and scanning electron microscopy (SEM). The results demonstrated that the dissolution efficiency and bioavailability of QR significantly increased in the BQS form, aligning with that of BBR, compared to the physically mixed (BQP) form. Moreover, BQS exhibited a superior inhibitory effect on fibrosis compared to BQP in the human hepatic stellate cell line LX-2 by modulating lipid accumulation, inflammation, apoptosis, and the cell cycle. Furthermore, in a mouse model of hepatic fibrosis induced by methionine and choline-deficient (MCD) diets, BQS demonstrated enhanced anti-fibrotic activities compared to BQP. These findings suggest that BQS holds promise as a potential alternative treatment for liver fibrosis. Importantly, this study provides novel insights into achieving a cocktail effect through the salt formation of two or more drugs. The results highlight the potential of salt formulations in enhancing the therapeutic efficacy and consistent biological processes of drug combinations.

Keywords: liver fibrosis; berberine; quercetin; cocktail therapy

1. Introduction

Liver diseases, which result from a multitude of factors, including unhealthy lifestyles, viral infections, genetic susceptibility, and drug-induced injury, pose a significant global health threat. These diseases commonly progress through various stages, starting with the accumulation of lipids, known as nonalcoholic fatty liver disease (NAFLD). Chronic NAFLD can then lead to nonalcoholic steatohepatitis (NASH), cirrhosis, hepatocellular carcinoma (HCC), and ultimately, mortality [1,2]. Liver fibrosis, a reversible progression of cirrhosis, represents a promising target for therapeutic intervention in cirrhosis. However, currently, there is no specific therapy available for liver fibrosis [3]; hence, the search for safe and effective drugs to treat it is of utmost importance.

In recent decades, there has been a growing focus on natural compounds due to their potential for liver protection and low toxicity. Furthermore, the synergistic effects of combining herbs with similar pharmacological properties, often referred to as the cocktail effect, have shown promising outcomes. Historically, cocktail treatments have demonstrated improved efficacy, not only in infectious diseases such as Sudan virus infection [4], human immunodeficiency virus infection [5], Staphylococcus Aureus infection [6] and COVID-19 infection [7], but also in chronic diseases like chronic asthma [8] and Alzheimer's disease [9]. By reducing side effects and drug resistance, cocktail treatment offers a favorable approach to enhance therapeutic outcomes through the combined effects of multiple drugs.

Berberine (BBR) is an isoquinoline alkaloid extracted from *Coptidis rhizoma*, which is used in traditional Chinese medicine. It possesses diverse biological activities, including anticancer [10], metabolic improvement [11], and antibacterial effects [12]. Recent studies have highlighted the potential of BBR in treating liver fibrosis through various direct and indirect mechanisms. From the direct perspective, BBR can modulate the apoptosis, proliferation, and activation of hepatic stellate cells (HSCs). It induces ferroptosis through ROS-mediated activation of HSCs [13] and promotes HSCs apoptosis by reducing mitochondrial membrane potential [14]. BBR also arrests HSCs in the G1 phase [15] and reduces the expression of matrix metalloproteinases (MMPs) and tissue inhibitors of MMPs (TIMPs), thereby facilitating collagen degradation [16]. Indirectly, BBR improves lipid metabolism [17], reduces endoplasmic reticulum stress [18], modulates gut microbiota [19], inhibits oxidative stress [20] and suppresses inflammation [21], all of which contribute to its anti-fibrotic effect. However, the low intestinal permeability and aqueous solubility of BBR limit its oral bioavailability.

Quercetin (QR), a flavonoid compound widely found in plants and fruits, has limited bioavailability due to its poor solubility, permeability, and stability. Despite these challenges, QR exhibits a wide range of pharmacological actions, including anticancer [22], neuroprotection [23] and liver protection [24]. In the context of liver fibrosis, QR regulates MMP-9 and TIMP-1 to inhibit extracellular matrix formation. It also inhibits autophagy through modulation of the HMGB1-TLRs-NF-kB signaling pathways [25], activates the PI3K/Akt signaling pathway, and suppresses the TGF-β1/Smads signaling pathway [26]. Additionally, QR exhibits liver-protective functions, such as oxidation resistance and anti-inflammatory action, which contribute to its effectiveness in improving liver fibrosis [27].

BBR and QR, derived from natural sources, have demonstrated the potential to improve liver fibrosis through various mechanisms. Moreover, they exhibit a favorable safety profile with minimal adverse effects at standard doses. Interestingly, clinical studies have shown that combining flavonoids with BBR can enhance their pharmaceutical effects [28,29]. A meta-analysis revealed that the combination of BBR and QR improved lipid and glucose levels in patients. Traditional prescriptions incorporating both BBR and QR further support the feasibility and effectiveness of a cocktail therapy approach [30–32].

However, the efficacy of BBR and QR is hindered by their disparate dissolution and absorption characteristics. To address this issue and enhance their combined effects, we synthesized a BBR-QR salt (BQS). This formulation aims to improve the bioavailability of both compounds, thereby facilitating their simultaneous absorption. We evaluated its efficacy in treating liver fibrosis through both in vitro and in vivo experiments, anticipating enhanced therapeutic outcomes.

2. Results

2.1. Synthesis and Characterization of BBR-QR Salt

As shown in Figure 1A, BQS was successfully synthesized using BBR and QR. The BBR acts as the cation, and the QR as the anion, forming the BQS salt through acid-base interactions. The obtained BQS was 7.4 g, and the yield was 87%. The ^{1}H NMR, FTIR, PXRD, DSC and SEM were used to prove the formation of BQS.

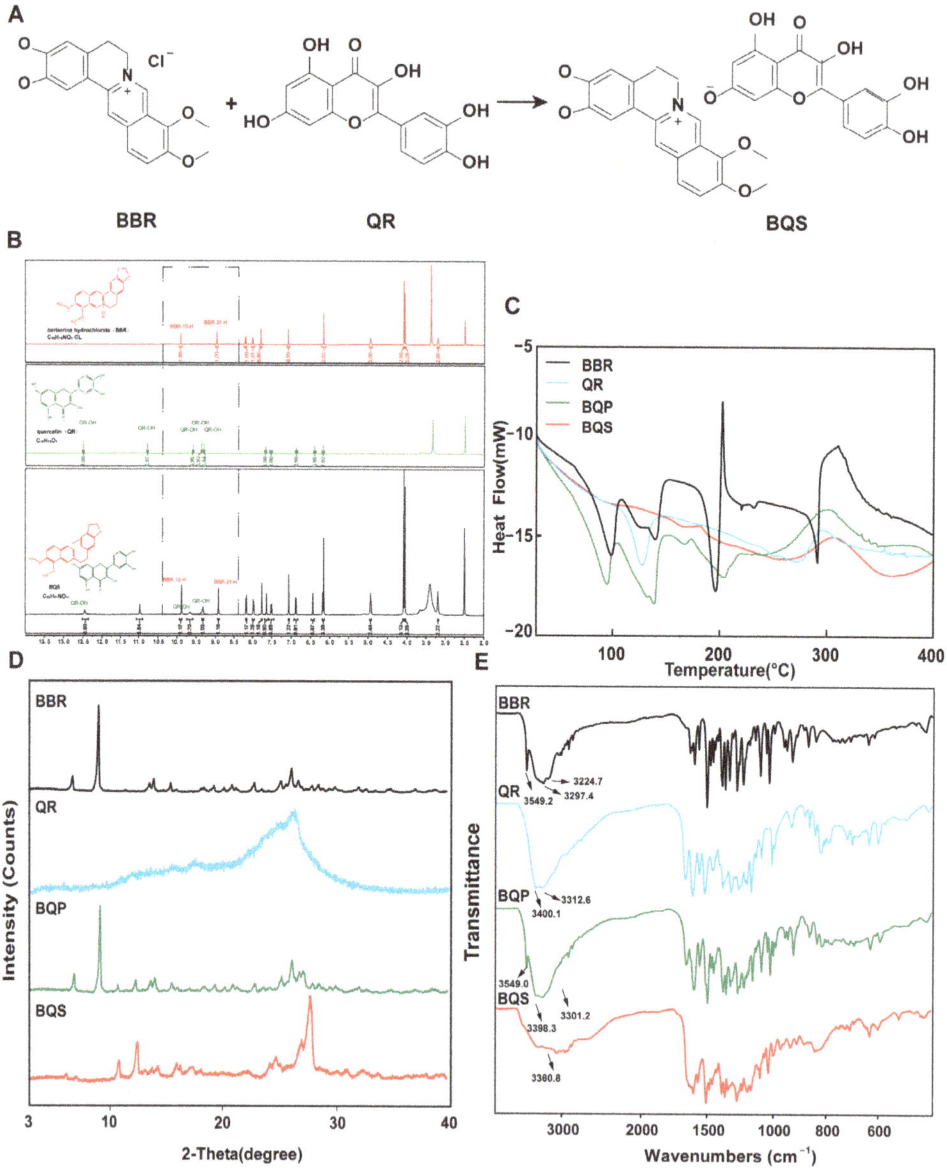

Figure 1. Synthesis and characterization of BQS (Berberine-quercetin salt). (**A**) The synthetic route of BQS. (**B**) The ^{1}H NMR spectra of the BBR (Berberine), QR (Quercetin) and BQS. (**C**) The DSC spectra of BBR, QR, BQP (Berberine-quercetin physically mixed), and BQS. (**D**) The PXRD of BBR, QR, BQP, and BQS. (**E**) The FTIR of BBR, QR, BQP, and BQS.

2.1.1. Hydrogen-1 Nuclear Magnetic Resonance (^{1}H NMR) Results

As seen in Figure 1B, the synthesized compound BQS was characterized as a mono-substituted product of BBR and QR based on ^{1}H NMR analysis. The molecular formula of BQS is $C_{35}H_{27}NO_{14}$, and its ^{1}H NMR spectrum indicates the presence of 27 hydrogen atoms, consistent with this structure. When comparing the ^{1}H NMR spectra of BQS, BBR, and QR, it is observed that the chemical shifts in BQS include signals corresponding to both BBR and QR, indicating the combination of these two compounds. Importantly, the ^{1}H NMR spectrum of BQS lacks one specific proton signal near δ 9.33, which corresponds to one of the hydroxyl groups of QR. This absence is indicative of the formation of salt between the protonated nitrogen of BBR and a single hydroxyl group of QR, which conclusively supports the formation of a 1:1 mono-substituted quercetin-berberine salt (Supplementary Figure S1).

2.1.2. Differential Scanning Calorimetry (DSC) Results

We can conclude, from the DSC results shown in Figure 1C, that the curve of BQS is different from BQP, BBR, and QR, respectively. To be specific, BBR had three peaks at different temperatures: 98.37 °C (free water), 139.48 °C (crystal water), and 196.20 °C (the melting point peak). There was an endothermic peak at 278.42 °C in the QR curve, which was similar to its melting point. As for the curve of BQP, it appeared to be a superposition of two respective curves. Three exothermic peaks were observed at 94.91 °C, 139.09 °C, and 150.08 °C. However, there were no characteristic peaks in the curve of BQS, which was another piece of evidence for the formation of the amorphous substance.

2.1.3. Powder X-Ray Diffraction (PXRD) Results

Figure 1D represents the PXRD results. The diffractogram of BBR and QR showed sharp peaks of crystals. The spectrum of BQP consisted of the superposition of BBR and QR. The diffraction spectra of BQS had no sharp peaks, displaying a typical diffuse diffraction halo of the crystalline state, indicating the formation of an amorphous substance, which may have contributed to the improvement of the solubility and bioavailability.

2.1.4. Fourier-Transform Infrared Spectroscopy (FTIR) Evaluation

Figure 1E represents the FTIR spectra of BBR, QR, BQP, and BQS. As illustrated in the picture, the infrared spectrum of BQS had no characteristic absorption peaks of BBR and QR at 3200–3600 cm^{-1}, indicating that the O-H stretching vibration disappeared. The spectrum of BQP had the infrared signatures of the two pure drugs. This difference between the spectra of BQS and BQP supports the hypothesis that BQS is a novel compound resulting from the synthesis of the two drugs. A deviation of ± 2 cm^{-1} was allowed in the infrared spectra.

2.1.5. Scanning Electron Microscopy (SEM) Results

The appearance of BQS is deep yellow, which is obviously different from the appearance of BQP, BBR, and QR (Figure 2). As shown in Figure 2, both BBR and QR are presented as elongated rod-shaped crystals. BQP is also presented as elongated rod-shaped crystals. In comparison, BQS was distinctive in form, showing no features identical to BBR and QR at all.

To sum up, these characterization means have proved that we synthesized BQS successfully.

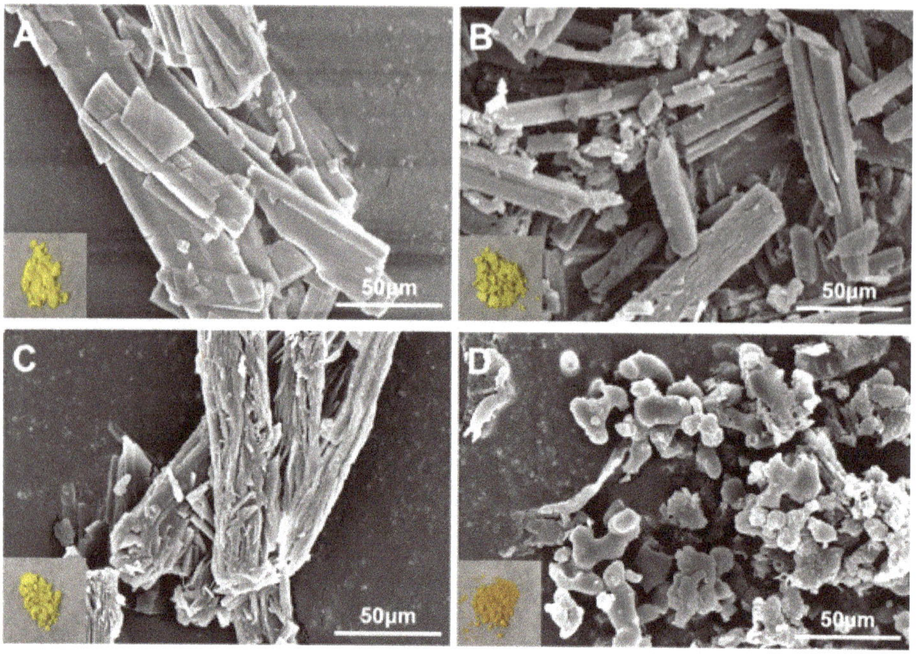

Figure 2. SEM image and appearance of BBR, QR, BQP, and BQS. (**A**) BBR, (**B**) QR, (**C**) BQP, (**D**) BQS.

2.2. BQS Improved the Dissolution and Bioavailability of QR

The dissolution studies showed that (Figure 3A) BBR in BQP dissolved well with rapid release within 2 h, while a sustained release of BBR was found in BQS. The dissolution of QR was markedly promoted after the formation of salt by combining it with BBR. It was noteworthy that the salt formation realized synchronous release of BBR and QR.

LC-MS/MS was used to determine the concentration of BBR and QR in the livers of mice who were given the drugs via intragastric administration. Figure 3B represents the concentrations of BBR and QR, respectively. As shown in the Figure 3B, the concentration of BBR remains higher in the BQS group than in the BQP group, which may lead to a long-lasting therapeutic effect. The area under the curve for the BQS group was 17% higher than that for the BQP group. Furthermore, the concentration of QR was markedly promoted after salt formation with BBR (approximately 2.2-fold higher than that observed in the BQP group).

The dissolution and pharmacokinetic properties of QR were improved by salt formation.

2.3. BQS Ameliorated Liver Fibrosis In Vitro

The activation of HSCs plays a vital role in liver fibrosis. After liver injury, HSCs were activated and produced extracellular matrix (ECM) and pro-inflammatory mediators, triggering progressive liver fibrosis [33]. Both BBR and QR can suppress inflammation [34] in parenchymal cells. Furthermore, numerous studies have proven that BBR is able to induce apoptosis and inhibit proliferation in HSCs [14,35]. It is possible to combine BBR and QR for a better therapeutic effect [28,29].

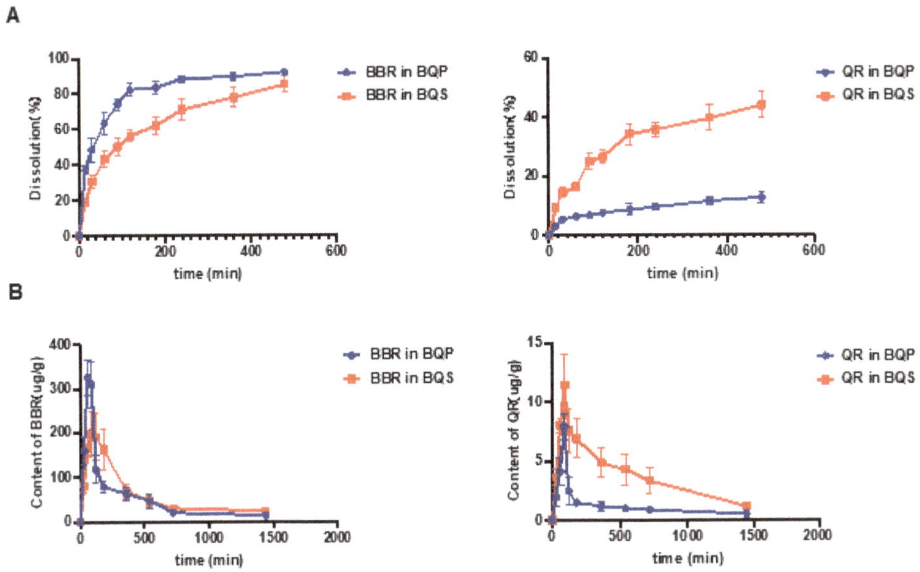

Figure 3. The dissolution and bioavailability of BBR and QR. (**A**) The dissolution of BBR and the dissolution of QR. (**B**) BBR content in the liver, QR content in the liver, Mean ± SEM, n = 5.

2.3.1. BQS Inhibited Lipid Accumulation and Inflammation

In the process of liver fibrosis, the undue accumulation of lipids and inflammation are two risk factors. Thus, we established a HepG2 hepatic steatosis model by treating the cells with 0.5 mmol/L sodium oleate to investigate the effects of BQS on lipid accumulation and inflammation.

As shown in Figure 4A, there are a few lipid droplets in the NC group; however, the lipid droplets in the MC (model control) group are much more frequent than those of the NC group. This phenomenon was attenuated not by BQP (10 μg/mL of BBR and 9.0 μg/mL of QR) but by BQS (19 μg/mL). Excessive lipid accumulation may trigger the development of NAFLD, which may eventually progress to liver fibrosis [36]. Thus, inhibiting lipid accumulation is a promising method for the prevention and treatment of liver fibrosis. BQS exhibited a remarkable effect in this area.

Inflammation plays an imperative role in the progress of liver fibrosis [37]. Modulating inflammation is a promising method for the treatment of liver fibrosis. TNF-α, IL-1β, and IL-6 are common inflammation factors, and they were therefore used to study the anti-inflammatory effects of the drugs (Figure 4B). TNF-α was dyed red, IL-1β was dyed green, and IL-6 was dyed pink. Compared to the NC group, the intensity of fluorescence of the MC group was much higher, indicating the success of our model. BQP attenuated the inflammation to some extent. BQS improved inflammation by a large margin to almost the same level as the NC group. The qPCR and ELISA results further confirmed the consistent trend observed in the treatment outcomes (Figure 4C,D).

Given the favorable effects of BQS on inhibiting lipid accumulation and inflammation, it may become a curative strategy for liver fibrosis.

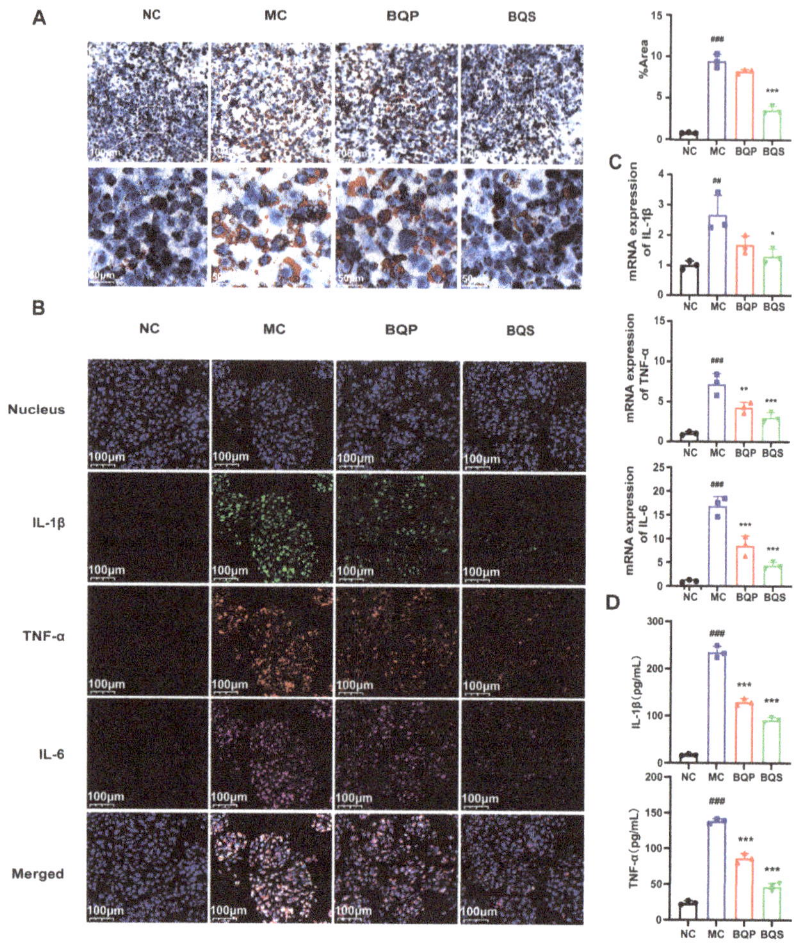

Figure 4. Lipid accumulation and inflammatory expression in HepG2 cells. (**A**) The results of Oil Red O staining. And the quantification of Oil Red O staining. Mean ± SEM, n = 3. (**B**) Immunofluorescence analysis showing DAPI (blue), IL-1β (green), TNF-α (red), and IL-6 (pink) in HepG2 cells of each group. (**C**) The relative mRNA level of IL-1β, TNF-α, IL-6. (**D**) The productions of IL-1β and TNF-α in supernatants of LX-2 cells were measured by ELISA. Mean ± standard error of the mean (SEM), n = 3. *: vs. MC group; #: vs. NC group. Significant differences are indicated as * $p < 0.05$, ** $p < 0.01$, *** $p < 0.001$; ## $p < 0.01$, ### $p < 0.001$.

2.3.2. BQS Alleviated Extra Collagen Formation

As downstream proteins of the TGF-β signaling pathway [38,39], α-SMA and collagen I are markers of the occurrence of fibrosis. In this study, we measured the expression of α-SMA and collagen I in LX-2 cells (Figure 5A). α-SMA was dyed red, and collagen I was dyed green. As the picture shows, little α-SMA and collagen I were expressed in the NC group. While inducing the TGF-β pathway, the expression of α-SMA and collagen I evidently increased in the MC group, indicating that our model was successful. When we came to BQP and BQS, the intensity of fluorescence in BQP and BQS was lower than that in the MC group. The difference was that the intensity of fluorescence in BQP only attenuated a little, while the intensity of fluorescence in BQS reduced a lot. This phenomenon enlightens us that BQS had a much better anti-fibrosis effect than BQP. The PCR and Western blot analyses showed similar results (Figure 5B,C).

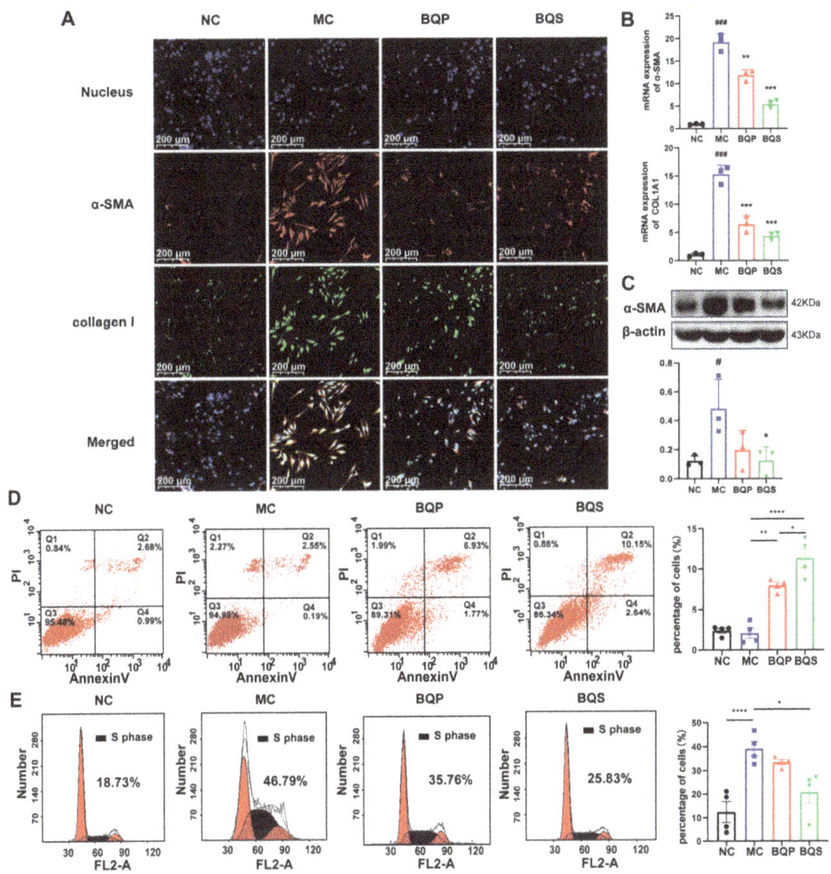

Figure 5. The in vitro effects of BQS on collagen formation and the cell cycle evaluated in LX-2 cells. (**A**) Immunofluorescence analysis showing DAPI (blue), α-SMA (red), collagen I (green) in LX-2 cells of each group.(**B**) The relative mRNA level of α-SMA, COL1A1 in LX-2 cell, Mean ± standard error of the mean (SEM), n = 3. (**C**) Protein levels of α-SMA were detected in LX-2 cells by western blot. (**D**) The results of apoptosis of LX-2 cells, Mean ± standard error of the mean (SEM), n = 4. (**E**) Cell cycle of LX-2 cells. Mean ± standard error of the mean (SEM), n = 4. *: vs. MC group; #: vs. NC group. Significant differences are indicated as * $p < 0.05$, ** $p < 0.01$, *** $p < 0.001$; **** $p < 0.0001$; # $p < 0.05$, ### $p < 0.001$.

2.3.3. BQS Modulated Proliferation and Apoptosis

As an important cell in the development of liver fibrosis, activated LX-2 is an important target for treatment. Managing the number of activated HSCs is an ideal treatment strategy. In this experiment, we investigated the pro-apoptotic and anti-proliferative effects of BQS.

Previous studies have shown that inducing the apoptosis of activated HSCs is beneficial to the amelioration of liver fibrosis [40,41]. Flow cytometry was used to explore the apoptosis and cell cycle of the LX-2 cells incubating with BQP and BQS. The percentage of apoptotic cells in each group is 2.37 ± 0.61, 2.09 ± 1.27, 7.98 ± 0.76 and 11.34 ± 2.4 (Figure 5C). The pro-apoptotic effect of BQS was obviously stronger than BQP, which may lead to better therapeutic effects for liver fibrosis.

Moreover, besides promoting apoptosis, the BQS was also capable of inhibiting the proliferation of LX-2 cells. Cells in the synthesis (S) phase of cell division have the strongest division potential. The more cells in this phase, the more likely they are to divide. As displayed in Figure 5D, the cells in the S phase accounted for 18.73%, 46.79%, 35.76%, and 25.83% of the NC group, MC group, BQP group, and BQS group, respectively. The BQP demonstrated a small inhibition effect on the proliferation of LX-2 cells, while BQS showed a significant effect. Overall, BQS was capable of inducing the apoptosis of LX-2 cells and inhibiting their proliferation. BQP, however, did not have these capabilities. In the process of liver fibrosis, HSCs are activated by TGF-β secreted by Kuffer cells, and then they secrete large amounts of TGF-β themselves and transform into myofibroblasts to secrete collagen and other proteins to form the ECM [42]. By inducing the apoptosis of HSCs and inhibiting their proliferation, the progression of liver fibrosis can be alleviated. Thus, we speculate that our BQS exhibits an excellent anti-fibrosis effect in vivo, which lays the foundation for our following research.

2.4. *BQS Improved Liver Fibrosis In Vivo*

Encouraged by the results of in vitro studies, the potential therapeutic effect of BQS on liver fibrosis was next investigated in a mice model. Methionine and Choline Deficit (MCD) diets are recognized methods for modeling liver fibrosis. We took photographs of livers from typical samples from four groups of animals (Figure 6A). The appearance of the BQP group was somewhere between the NC and MC groups. The appearance of the BQS group was similar to the NC group, indicating a better effect of BQS than BQP, which was also proven by other results displayed in the following sections.

2.4.1. BQS Protected Liver Function

H and E staining (Figure 6B) showed that the histological structure of the liver was severely damaged in the MC group, significantly improved in the BQS group, and slightly improved in the BQP group. The indicators ALP and TBA were improved for the BQS group only (Figure 6C,D), meaning that BQS alleviated liver damage. Moreover, BQS also improved the levels of ALT, AST, and GGT, which means that BQS can contribute to liver function, indicating that BQS has an improvement effect on liver fibrosis, while BQP showed influence only on AST and GGT (Figure 6E,F). Lipid accumulation is another risk factor for liver fibrosis [43]. We compared the lipid accumulation in the liver of different groups (Figure 6G,H). The results showed that BQP and BQS reduced the TG levels, while the CHO level was only reduced by BQS. These changes remind us that BQS can inhibit lipid accumulation, which is beneficial for anti-inflammation and anti-fibrosis purposes.

2.4.2. BQS Ameliorated Liver Fibrosis

The result of Masson staining is shown in Figure 7A. After staining, the collagenous fiber became blue, enabling its visualization. Sirius Red staining was also used to visualize collagenous fiber and evaluate the degree of fibrosis. It is obvious that BQS reduced the accumulation of collagen in the liver; however, BQP showed nearly no effect. Meanwhile, we can draw the same conclusion from Figure 7B, where the TGF-β was dyed red, collagen I was dyed green, and α-SMA was dyed gray. These three markers, which are closely related to fibrosis, were significantly inhibited by BQS. The PCR analysis showed a similar result (Figure 7C). Western blot further confirmed these findings, and BQS exhibits a more potent anti-fibrotic effect (Figure 7D). These results are highly similar to the in vitro results.

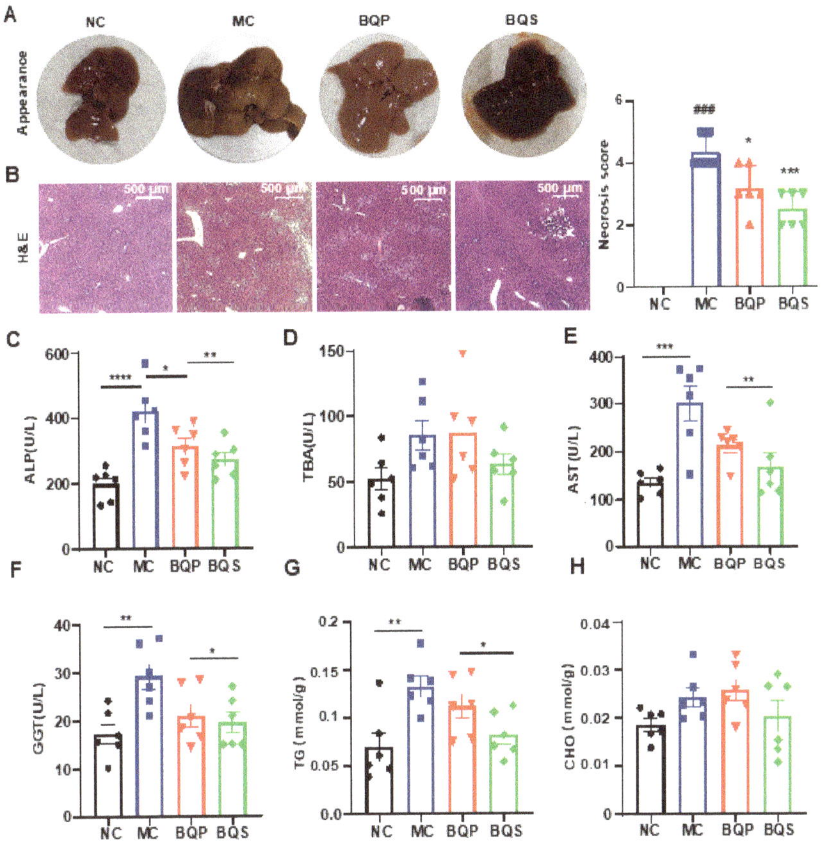

Figure 6. The hepatoprotective effects of BQS in mice model of liver fibrosis. (**A**) The appearances of mice liver tissue in different groups. (**B**) H and E staining of mouse liver in each group. (**C–F**) Blood biochemical results (ALP, TBA, AST and GGT). (**G**,**H**) The TG and CHO content in mice liver. Mean ± standard error of the mean (SEM), n = 6, *: vs. MC group; #: vs. NC group. * $p < 0.05$; ** $p < 0.01$; *** $p < 0.001$; **** $p < 0.0001$. ### $p < 0.001$.

2.4.3. BQS Alleviated Inflammation in Liver Tissue

The results of the immunohistochemistry analysis (Figure 8A) showed that inflammation was suppressed by BQS. TNF-α was dyed red, and IL-1β was dyed green. After modeling, there were significant differences between the MC group and the NC group. Both BQP and BQS improved the inflammation, but BQS had a more dramatic effect. The PCR and ELISA analyses showed similar results (Figure 8B,C).

Taken together, the in vivo experiments showed the same results as the in vitro experiments, implying that our BQS could exert anti-fibrotic pharmacological effects in vivo.

In every aspect we investigated, BQS had better effects than BQP.

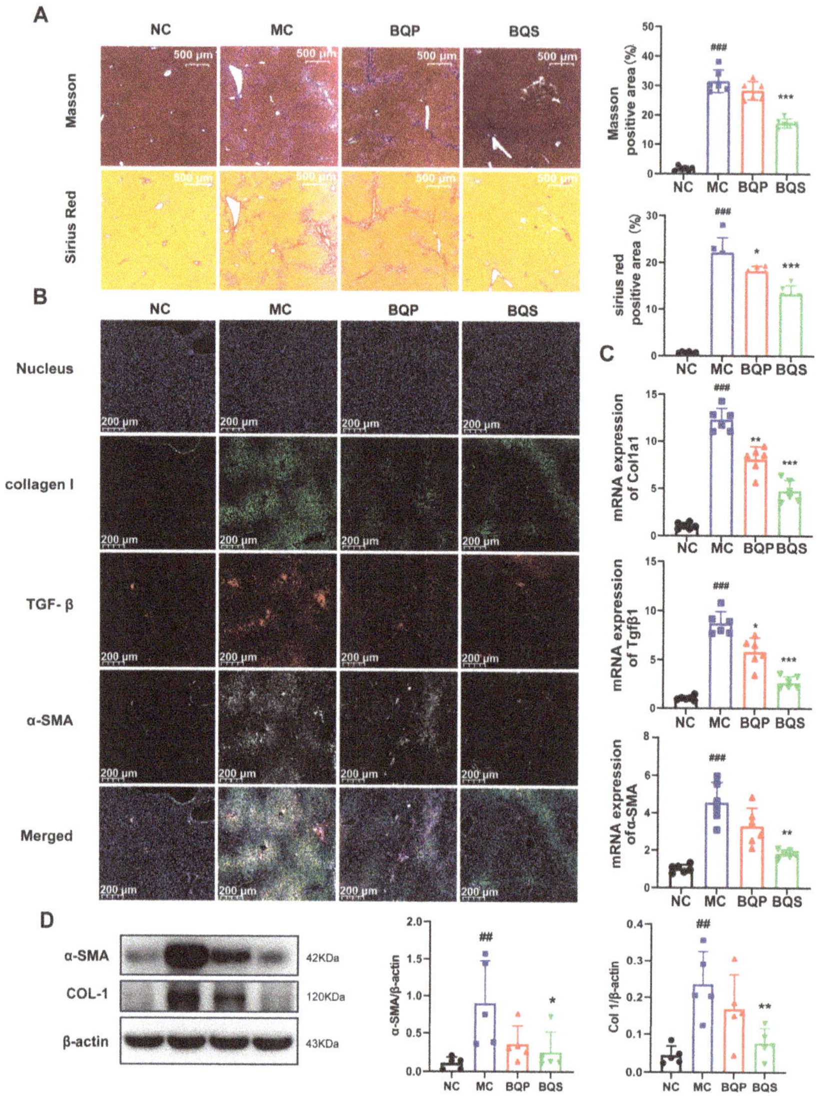

Figure 7. The antifibrotic effects of BQS in mice model of liver fibrosis. (**A**) Masson and Sirius Red staining of mice liver in each group. (**B**) Immunofluorescence analysis showing DAPI (blue), collagen I (green), TGF-β (red), and α-SMA (gray) of mouse liver in each group.(**C**) The relative mRNA level of Col1a1, Tgfβ, α-SMA in mice liver. (**D**) Protein levels of α-SMA COL-1 were detected in mice liver by western blot. Mean ± standard error of the mean (SEM), n = 5 or n = 6, *: vs. MC group; #: vs. NC group. * $p < 0.05$, ** $p < 0.01$; *** $p < 0.001$; ## $p < 0.01$, ### $p < 0.001$.

2.5. Network Pharmacology Analysis and Mechanism Validation

In the current investigation, a network pharmacological approach was employed to delineate the molecular mechanisms underpinning the therapeutic effects of BQS on liver fibrosis. A comprehensive intersectional target analysis identified a set of 96 potential targets associated with both BQS components and liver fibrosis (Figure 9A). The subsequent construction of a protein–protein interaction (PPI) network using the STRING database, and analysis with Cytoscape 3.10.1, identified 87 nodes and 472 edges, representing a complex interaction landscape (Figure 9B). Core targets were extracted using Centiscape 2.2, with visual analytics further refining these to 18 pivotal nodes within the network (Figure 9C). Gene ontology (GO) functional annotations indicated the significant modulation of molecular functions following BQS treatment, particularly noting enhancements in protein kinase activity, receptor binding, and oxidoreductase activity, which are crucial for cellular signaling and metabolic regulation (Figure 9D, top panel). Moreover, the Kyoto Encyclopedia of Genes and Genomes (KEGG) pathway analysis highlighted the importance of the 'Pathways in Cancer' and 'PI3K-Akt signaling pathway' in the mechanistic action of BQS, suggesting its central role in the amelioration of hepatic fibrosis (Figure 9E).

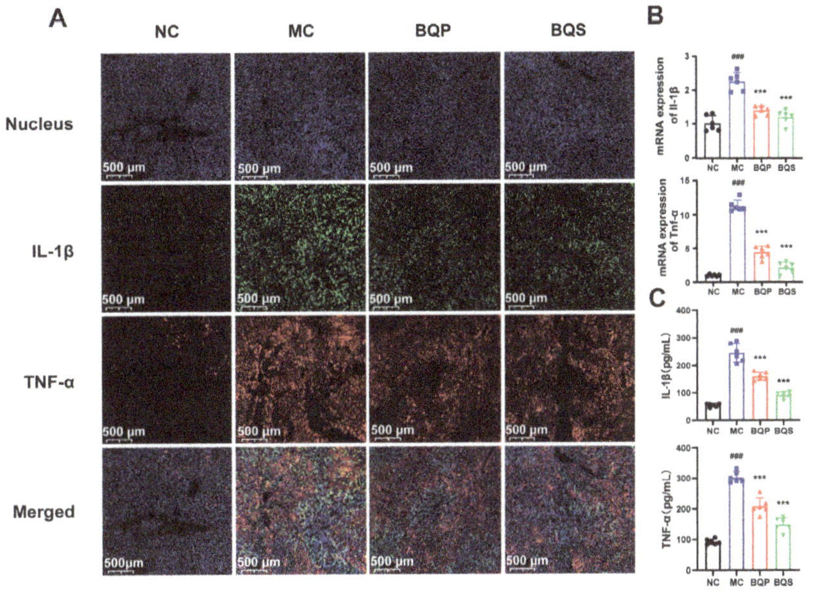

Figure 8. The image of two inflammatory factors of animals. (**A**) Immunofluorescence analysis showing DAPI (blue), IL-1β (green), TNF-α (red) of mouse liver in each group. (**B**) The relative mRNA level of the inflammatory marker IL-1β, TNF-α in mice liver tissue. (**C**) The productions of IL-1β and TNF-α in supernatants of indicated cells were measured by ELISA. Mean ± standard error of the mean (SEM), n = 6. *: vs. MC group; #: vs. NC group. Significant differences are indicated as *** $p < 0.001$; ### $p < 0.001$.

Protein expression levels of phospho-Akt (p-Akt) and phospho-FoxO1 (p-FoxO1) in hepatic tissues were quantified using Western blotting assays. As shown in Figure 9F, comparative analysis revealed that in contrast to the NC group, the phosphorylation of both Akt and FoxO1 was significantly elevated in the MC group, indicating enhanced activity within this signaling axis. Notably, treatment with BQS markedly attenuated these phosphorylation levels. These findings suggest that the therapeutic effect of BQS on liver fibrosis can be attributed to the inhibition of the Akt/FoxO1 signaling pathway. Furthermore, BQS demonstrated superior efficacy in modulating these pathways compared to BQP.

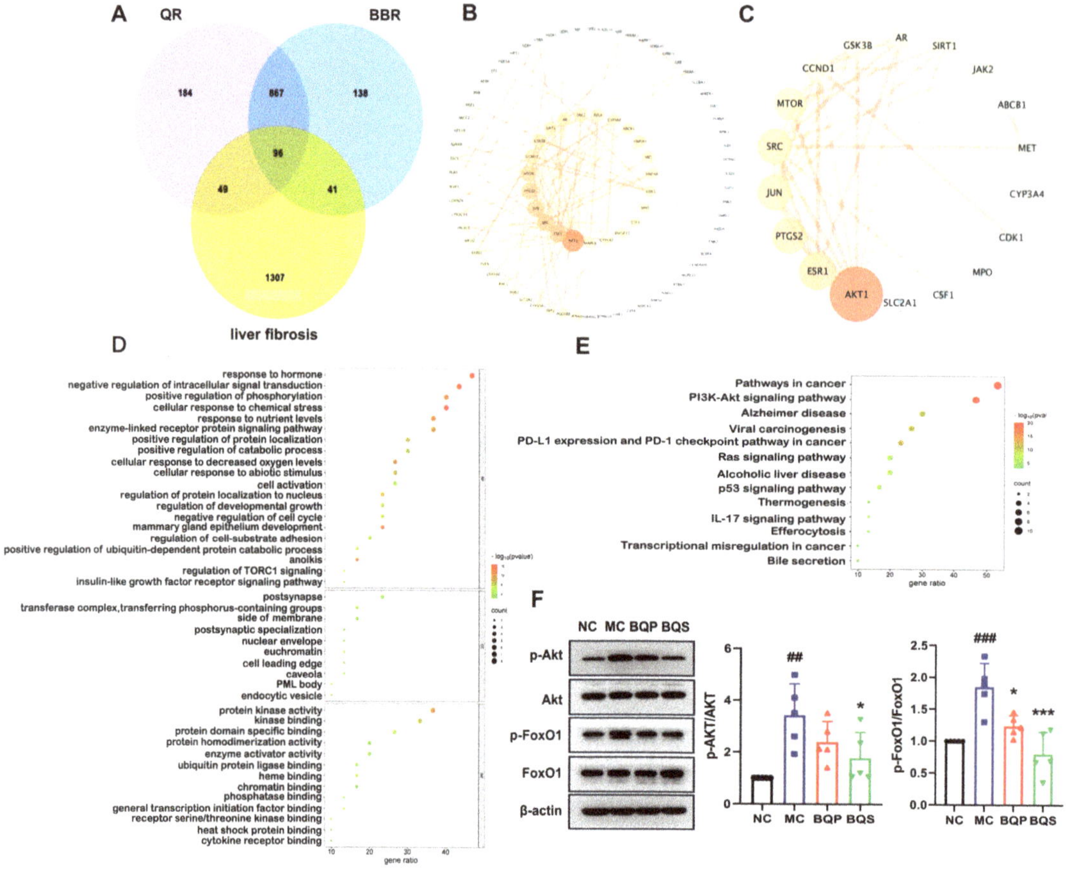

Figure 9. Network pharmacological analysis and protein pathway validation of BQS in fibrosis. (**A**) Acquisition of drug–disease intersection targets. (**B**) Construction of protein interaction network (PPI). (**C**) Core target screening and visualization analysis. (**D**) Enrichment analysis of gene ontology (GO). (**E**) Enrichment analysis of Kyoto Encyclopedia of Genes and Genomes (KEGG). (**F**) The protein expression of p-Akt/Akt and p-FoxO1/FoxO1 in mouse liver tissue was evaluated by Western blotting, with β-actin used as the control. Mean ± SEM, n = 5. *: vs. MC group; #: vs. NC group. Significant differences are indicated as * $p < 0.05$, *** $p < 0.001$; ## $p < 0.01$, ### $p < 0.001$.

3. Discussion

Liver fibrosis is a progressive medical condition characterized by the excessive accumulation of ECM proteins, notably collagen [44]. This pathological buildup disrupts liver architecture and impairs its function. The condition commonly arises from chronic liver damage due to factors such as hepatitis infections, alcohol abuse, NAFLD, and other metabolic disorders. As fibrosis advances, it may progress to cirrhosis—a severe form of liver scarring that significantly impairs liver function and escalates the risk of liver failure and liver cancer. Globally, the incidence of liver fibrosis is increasing, primarily due to escalating rates of obesity, diabetes, and metabolic syndrome, which are key contributors to NAFLD, now recognized as one of the most prevalent causes of liver disease worldwide [45]. Consequently, the rising prevalence and severity of liver fibrosis underscore the urgent need for the development of more effective treatments.

BBR, an isoquinoline alkaloid, has been increasingly recognized for its hepatoprotective effects, including its efficacy against liver fibrosis. Research has shown that BBR ameliorates liver fibrosis through multiple mechanisms. It possesses anti-inflammatory properties and counteracts oxidative stress, which are crucial steps in halting the progression of the disease. Additionally, BBR inhibits the activation of HSCs, which is central to the deposition of fibrotic tissue [14,46]. Furthermore, it regulates metabolic pathways, notably by improving insulin resistance and modulating lipid metabolism, thus offering a protective effect on the liver [19]. This multifaceted strategy highlights the potential of BBR as a therapeutic agent in the management of liver fibrosis.

QR has been demonstrated to possess potent anti-inflammatory properties. Studies suggest that in liver fibrosis, QR significantly reduces the expression of pro-inflammatory cytokines, including TNF-α and IL-6. Moreover, it inhibits the activation of HSCs and promotes their apoptosis. QR exerts its therapeutic effects by modulating several critical signaling pathways associated with liver fibrosis, notably the TGF-β pathway [47]. Such modulation results in the downregulation of α-SMA and collagen expression, effectively impeding the progression of liver fibrosis. Recognizing the established hepatoprotective effects of BBR and QR, which operate via distinct mechanisms and are inspired by the principles of principal-assistant synergy and cocktail therapy from traditional Chinese medicine, we proposed a combination of these two compounds to enhance therapeutic outcomes. This strategic combination aims to harness the complementary mechanisms of BBR and QR, potentially offering a more effective therapeutic profile for liver protection.

The formation of salts from two different compounds, termed pharmaceutical salt preparation, offers numerous advantages, particularly within the pharmaceutical sector. This technique is widely used to enhance the physicochemical properties and biological efficacy of drugs. In this study, we initially focused on the synthesis and characterization of BQS. The synthesis yielded 7.4 g of BQS at an efficiency of 87%, as confirmed by various analytical techniques, including ^{1}H NMR, FTIR, PXRD, DSC, and SEM. The ^{1}H NMR data confirmed the successful synthesis, evidenced by a reduction in the number of hydrogen peaks from 28 in BQP to 27 in BQS, with peak shifts and shape alterations indicating interactions between BBR and QR in the salt form. DSC analysis revealed distinct thermal behavior of BQS, differing from BQP and individual components by showing no characteristic peaks, indicative of the formation of an amorphous substance. This finding was corroborated by PXRD results, which displayed diffuse diffraction patterns typical of amorphous materials, likely enhancing solubility and bioavailability. The FTIR spectra further validated the synthesis, with the absence of characteristic O-H stretch vibrations in BQS, suggesting interactions that alter typical absorption peaks of the individual components. The distinct deep yellow color of BQS, different from either component or their physical mixture, visually confirms these findings.

QR is renowned for its potent antioxidant and anti-inflammatory properties. It may reduce inflammatory responses, thereby minimizing damage to the intestinal barrier and enhancing the absorption of BQS. BBR, known for its synergistic pharmacological effects with QR, is a substrate for various efflux transporters, such as P-glycoprotein (P-gp), which expels BBR from cells, reducing its absorption and bioavailability. QR has been shown to inhibit the activity of these transporters, thereby increasing the retention and intestinal absorption of BBR [48,49]. Additionally, QR is known to inhibit certain enzymes in the CYP450 family, particularly CYP3A4 and CYP2D6, which are essential for metabolizing many drugs, and BBR is also a substrate for these enzymes [50]. By inhibiting these enzymes, QR can slow the metabolism of BBR, resulting in elevated plasma concentrations. LC-MS/MS analysis has demonstrated that BQS achieves higher and more sustained concentrations in liver tissue, suggesting improved pharmacokinetic properties compared to BQP. This improvement is reflected in both accelerated peak times and higher AUC values for BQS, indicating a more prolonged therapeutic effect. Overall, the successful synthesis and enhanced characteristics of BQS underscore its potential as a more effective therapeutic formulation, combining the benefits of BBR and QR through advanced pharmaceutical techniques.

In the in vitro experiment assessing the anti-hepatic fibrosis effects of BQS, pharmacological evaluations demonstrated that BQS modulates lipid accumulation, inflammatory responses, collagen formation, and cell proliferation. Experimental results showed a significant reduction in lipid droplets and the expression of inflammatory cytokines (TNF-α, IL-1β, and IL-6) in HepG2 cells, as confirmed by fluorescence staining and qPCR analysis. These findings are particularly significant considering the pivotal roles of lipid accumulation and inflammation in the progression of NAFLD and fibrosis. Additionally, BQS was effective in reducing collagen and α-SMA expression in LX-2 cells, thereby attenuating ECM formation, which is crucial in fibrogenesis. BQS not only demonstrated superior anti-inflammatory effects compared to BQP but also significantly inhibited LX-2 cell proliferation and induced apoptosis. Cell cycle analysis indicated a notable reduction in the proportion of cells in the S phase after BQS treatment, underscoring its potent effect on inhibiting cell division. Furthermore, BQS significantly increased the proportion of apoptotic cells compared to either individual compounds or BQP. By targeting multiple biological pathways and modulating the activation of HSCs and fibrosis, BQS shows promise as a potent anti-fibrotic compound. Therefore, further in vivo studies are warranted to validate its efficacy and elucidate its mechanism of action in the treatment of liver fibrosis.

In vivo experiments utilizing a MCD diets-induced mouse model of liver fibrosis demonstrated that BQS exhibits potent therapeutic effects. Results, including liver photographs and histological stains (Masson and Sirius Red), indicated that BQS significantly reduced collagen accumulation and improved the histological architecture of the liver compared to the BQP group and the MC group. Furthermore, BQS notably decreased key fibrotic markers, such as TGF-β, collagen I, and α-SMA, underscoring its robust anti-fibrotic properties. Additional analyses confirmed that BQS preserved liver function, evidenced by improved biochemical markers, including ALP, TBA, ALT, AST, and GGT. Remarkably, BQS outperformed BQP, which exhibited limited efficacy. Moreover, BQS effectively inhibited lipid accumulation—a critical risk factor for liver fibrosis—and demonstrated superior efficacy in reducing TG and CHO levels in liver tissue compared to BQP. The immunohistochemical findings and qPCR analysis further validated that BQS significantly reduced inflammation in liver tissues, evidenced by the marked suppression of inflammatory cytokines such as TNF-α and IL-1β. The in vivo results closely mirrored those observed in vitro, suggesting consistent anti-fibrotic effects of BQS across both experimental settings.

The compounding of Chinese medicines follows the complex traditional principles of 'four qi and five flavors' alongside the roles of 'ruler, minister, auxiliary, and envoy.' These principles guide the alteration of medicinal properties and toxicity in herbal combinations. Considering the crucial role of herb pairing in formulations, this article introduces a novel method that refines herb preparation and decoction to potentially enhance the efficacy of the resulting compounds. Specifically, we propose a salt form of BBR and QR, termed BBR–QR salt. A bioinformatics approach was employed to identify disease-related targets and pathways to substantiate the rationality of this drug pairing. Liver fibrosis, characterized by chronic inflammatory damage, currently lacks specific pharmacological treatments, with general strategies focusing on addressing the underlying cause. Previous studies have indicated that reducing lipid droplet aggregation, alleviating inflammation, and promoting apoptosis could positively affect the progression of liver fibrosis. Our bioinformatics analysis underscores the therapeutic potential of BQS, identifying 96 potential targets associated with the components of BQS and liver fibrosis. Subsequent PPI network construction revealed a complex interaction landscape with 18 pivotal nodes. Functional annotations and pathway analysis, especially targeting the 'Pathways in cancer' and 'PI3K-Akt signaling pathway', suggest that BQS primarily mitigates hepatic fibrosis through these pathways.

The Akt/FoxO1 signaling pathway critically influences liver fibrosis by modulating cell survival and proliferation [51]. Phosphorylation of FoxO1 by Akt promotes fibrogenesis, allowing fibrogenic cells, such as HSCs, to survive and proliferate, thereby contributing to the accumulation of scar tissue. This progression can lead to severe hepatic conditions, including cirrhosis and liver failure. Under normal conditions, FoxO1 regulates the cell cycle and inflammation; however, its inactivation through phosphorylation triggers uncontrolled cell growth and heightened inflammation, further exacerbating liver damage [52]. Moreover, FoxO1 is involved in metabolic regulation, and its dysfunction is linked to metabolic disorders associated with liver diseases, such as steatosis and NAFLD, thereby increasing the risk of fibrosis [53,54]. Therapeutically targeting this pathway with inhibitors that block FoxO1 phosphorylation or restore its function may effectively reduce liver fibrosis by diminishing HSC activation, enhancing apoptosis, and alleviating inflammation. Such an approach holds potential for the development of new treatments for liver fibrosis. Western blot analysis has confirmed that BQS modulates the Akt/FoxO1 signaling pathway, demonstrating enhanced efficacy in regulating this pathway compared to BQP.

4. Materials and Methods

4.1. Materials

Berberine hydrochloride (BBR, MW: 371.81) was purchased from Nanjing Zelang Biological Technology Co., Ltd. (Nanjing, China); quercetin (QR, MW: 302.24) was purchased from Aladdin (Shanghai, China); the HepG2 cell line was obtained from the Cell Resource Center, Peking Union Medical College (Beijing, China). The human HSC line LX-2 was obtained from the Shanghai Institutes for Biological Sciences (Shanghai, China). Trypsin EDTA (0.25%), cell culture media, penicillin/streptomycin, and FBS were obtained from Thermo Fisher Scientific (Waltham, MA, USA). All other reagents were of analytical grade.

Male 6-week-old specific-pathogen-free (SPF) C57BL/6N mice were obtained from Beijing Vital River Laboratory Animal Technology Co., Ltd. (Beijing, China). All experimental procedures were approved by the ethics committee of the Institute of Medicinal Biotechnology, Chinese Academy of Medical Sciences and Peking Union Medical College (Beijing, China; No. IMB-20231016D1).

4.2. Preparation and Characterization of BBR–QR Samples

First, we dissolved 5 g BBR hydrochloride in 300 mL of water at a temperature of 80 °C until we obtained a clear and transparent light yellow solution. Second, 4.06 g QR was added to a mixed solution of water and ethanol (150 mL: 450 mL); 15 mL of sodium hydroxide aqueous solution containing 0.54 g of sodium hydroxide was also added to this system. Finally, the BBR solution was added, drop by drop, to the system prepared in the previous step while stirring at room temperature. This resulted in the formation of deep yellow precipitation. We continued stirring the mixture at room temperature for one hour after the dropwise addition was complete. Next, we let the sample stand for two hours, then filtered it to obtain the deep yellow precipitation. We left the precipitation at room temperature for two days and then dried it at 50 °C for five hours. This yielded the BBR–QR sample.

Additionally, a physical mixture of BBR and QR was prepared by grinding 9.2 g of BBR hydrochloride with 7.5 g of QR in a mortar until they were thoroughly mixed. The resulting solid, weighing 16.4 g, represented the physical mixture of BBR and QR.

4.2.1. NMR Analysis

A Bruker ASCEND 300 spectrometer was used to record the ^{1}HNMR of BBR, QR, BQP, and BQS. The instrument was set to standard mode, 500 MHz, and the sample was dissolved in DMSO.

4.2.2. DSC Analysis

A differential scanning calorimeter (DSC1, Mettler Toledo Zurich, Switzerland) was used to measure the thermodynamic characteristics of the obtained samples. The samples were accurately weighted. The instrument was heated from 30 °C to 800 °C at 10 °C/min in a nitrogen atmosphere of 50 mL/min.

4.2.3. FTIR Analysis

We accurately weighed 10 mg each of BBR, QR, BQP, and BQS for Fourier-transform infrared spectra (Nicolet 5700, Waltham, MA, USA) measurement in the spectral range of 4000–400 cm^{-1}. All samples were mixed with KBr at a 1/100 ratio (w/w) and formed in the KBr disks.

4.2.4. PXRD Analysis

We accurately weighed BBR, QR, BQP, and BQS, which were placed on a vitreous sample holder in a D8 advance X-ray diffractometer with Cu/Ka radiation (λ 0.154 nm) operating at 40 kV, 30 mA while monitoring the reflection angle 2θ from 3° to 40° at a scan speed of 10°/min.

4.2.5. SEM Analysis

Material microscopic morphology characterization was obtained by using a scanning electron microscope (AKASHI SX-40, Akashi, Japan). BBR, QR, BQP, and BQS were, respectively, placed on the platinum tape, and then images were obtained under an excitation voltage of 20 kV under vacuum.

4.2.6. Dissolution Experiment

The powder of QR (18 mg), BBR (20 mg), BQP (20 mg BBR + 18 mg QR) and BQS (38 mg) were respectively placed in 900 mL water and stirred at 50 rpm at 37 ± 0.5 °C through paddle-rotating method. A volume of 3 mL of the medium was withdrawn through a 0.45 μm membrane at time points of 15, 30, 60, 90, 120, 180, 240, 360 and 480 min and immediately added with 3 mL of fresh dissolution medium. There were 6 repeats under each point. The amount of dissolved BBR or QR was determined using the method described in Section 4.3.

4.3. Bioavailability Analysis

All animals were allowed to acclimatize for a week. Ninety healthy male C57BL/6N mice were stratified into eighteen groups (n = 5 per group) based on body weight to ensure a similar average weight across groups. Then, the animals fasted for 12 h. Mice in each group were intragastrically administered BQP or BQS (equivalent to 100 mg/kg BBR and 90 mg/kg QR) dissolved in water. Approximately 50 mg of liver tissue was obtained from each mouse at different time points (0.5, 1, 1.5, 2, 3, 6, 9, 12, and 24 h). Liver tissues collected were stored at −80 °C until analysis. The concentration of BBR and QR in the liver tissues was determined using the UHPLC–MS/MS method, as described below.

In this study, we established a mathematical model-assisted UHPLC-MS/MS method for the targeted quantification of BBR or QR in liver samples of normal C57BL/6N mice. The internal standard solution, containing Palmatine, was added to 50 mg liver tissue for homogenization. After centrifugation at 12,000 rpm for 10 min, 300 μL of the supernatant was transferred into 675 μL of methanol. After vortexing them for 30 s, the samples were centrifuged at 12,000 rpm for 10 min. The supernatant was taken as the final sample.

The LC-MS analysis was performed using a TripleTOF 4600 analyzer (AB SCIEX, Framingham, MA, USA). A C18 column was used for separation. The mobile phase comprised water (mobile phase A) and acetonitrile (mobile phase B). The gradient program was 0.00−0.50 min (10% B), 0.50−4.00 min (10−95% B), 4.00−6.00 min (95% B), 6.00−6.10 min (95–10% B), and 6.10−9.00 min (10% B). The experiment was performed in a negative mode of ionization. The quantitative ion pair of BBR, QR, and the internal standard (Palmatine) quantitative were m/z = 336.0/292.1, m/z = 303.1/229.1, and m/z = 352.1/308.1, respectively. The qualitative ion pair of BBR, QR, and the internal standard qualitative were m/z = 336.0/321.1, m/z = 303.1/153.1, and m/z = 352.1/322.1, respectively.

4.4. In Vitro Anti-Fibrosis Activity in LX-2 Cells

Lx-2 cells were cultured in DMEM at 37 °C in an atmosphere of 5% CO_2. Once the cells reached 90–95% confluence, they were starved by incubating them in DMEM containing 2% FBS for 12 h. To create a liver fibrosis model, 2 ng/mL of TGF-β1 (R&D Systems, Minneapolis, MN, USA) was added to the cell culture. Additionally, the BQP and BQS (equivalent to a concentration of 10 μg/mL of BBR and 9.0 μg/mL of QR) were also added to the culture.

4.4.1. Apoptosis

As described by Crowley et al. [55], the harvested cells were co-stained with fluorescein isothiocyanate (FITC) labeled Annexin V and propidium iodide (PI) (Beyotime Biotechnology, Shanghai, China) for 15 min at room temperature in the dark. Then, flow cytometric analysis was carried out on a flow cytometer (BD Biosciences, Franklin Lake, New Jersey, USA) and the results were analyzed using FlowJo_V10 software.

4.4.2. Cell Cycle

The harvested cells were co-stained with propidium iodide (PI) (Beyotime Biotechnology, Shanghai, China) for 30 min at room temperature in the dark. The results were analyzed using FlowJo_V10 software.

4.4.3. Immunofluorescence (IF) Analysis

The tyramide signal amplification (TSA) method was used for α-SMA (Affinity Biosciences, Liyang, Jiangsu, China) and collagen I (Affinity Biosciences, Liyang, Jiangsu, China) double staining. In brief, the samples were treated successively with rabbit primary antibodies and HRP-conjugated goat anti-rabbit secondary antibodies. A fluorescent reagent was incubated with the samples after washing to develop the color. The antibodies used are shown in Table S1.

4.5. In Vitro Analysis of HepG2 Cells

HepG2 (ATCC) cells were cultured in DMEM at 37 °C in an atmosphere of 5% CO_2. The cells were seeded into six-well plants. We did not stimulate the cells until they reached 90–95% confluence. HepG2 cells were incubated with 0.5 mmol/L sodium oleate (SO, Sigma-Aldrich, St. Louis, MO, USA), together with the BQP and BQS (equivalent to 10 μg/mL of BBR and 9.0 μg/mL of QR). We harvested the cells after 12 h.

4.5.1. Oil Red O Staining

After being fixed in paraformaldehyde, HepG2 cells were stained with 0.1% Oil Red O to detect lipid droplets. The results were analyzed using Fiji (ImageJ 1.53c) software.

4.5.2. IF Analysis

The tyramide signal amplification (TSA) method was used for TNFα (Novus Biologicals, Littleton, CO, USA). IL-1β (ProteinTech Group, Chicago, IL, USA) and IL-6 (ProteinTech Group, Chicago, IL, USA) triple staining. The operation refers to the method mentioned in Section 4.4.3.

4.6. *In Vivo Study*

Six-week-old male C57BL/6N mice were purchased from the Vital River Laboratory Animal Technology (Beijing, China). All experimental procedures mentioned below were approved by the ethics committee of the Institute of Medicinal Biotechnology, Academy of Medical Sciences, and Peking Union Medical College (Beijing, China). The mice were allowed to acclimate in the specific-pathogen-free (SPF) grade animal facility (20 ± 1 °C on a 12 h light/dark cycle) for 7 days, with ad libitum access to water and food. The mice were randomly categorized into four groups (n = 10): a model group (MC, fed with Methionine and Choline-Deficient Diets, diet research, for six weeks) [56], a normal control group (NC, fed with a standard chow diet, for six weeks), a physical mixture group (BQP, fed with Methionine and Choline-Deficient Diets, diet research, for six weeks), and a salt group (BQS, fed with Methionine and Choline-Deficient Diets, diet research, for six weeks). Additionally, during the induction process, mice in the two drug administration groups were administered BQP (100 mg/kg/day of BBR and 90 mg/kg/day of QR) and BQS (190 mg/kg/day) by gavage, respectively. Both groups received a dosage equivalent to 100 mg/kg/day of BBR [57]; the mice in the MC and NC groups were administered a commensurable control vehicle intragastrically. The trial was conducted for two weeks.

At the end of the experiment, the mice were fasted for 12 h. Plasma was collected for biochemical analyses. The livers were harvested and weighed. We divided the liver into two parts; one was fixed with polyethylene, and the other was frozen at −80 °C.

4.6.1. Biochemical Analysis

Plasma aspartate aminotransferase (AST), alanine aminotransferase (ALT), total bile acid (TBA), alkaline phosphatase (ALP), and gamma-glutamyltransferase (GGT) were measured using a TOSHIBA automatic biochemical analyzer (Toshiba, Tokyo, Japan), according to the manufacturer's protocol and using commercially available kits (Biosino Biotechnology, Beijing, China).

4.6.2. Tissue Staining

Liver samples were embedded in paraffin and then were sectioned into 4 μm thick sections. Hematoxylin–eosin, Sirius Red and Masson were used to stain the samples for microscopic observation.

4.6.3. Immunohistochemistry (IHC) Analysis

The tyramide signal amplification (TSA) method was used for TNF-α, IL-1β double staining and for TGF-β, α-SMA, and collagen I triple staining. The operation refers to the method mentioned in Section 4.4.3.

4.6.4. Enzyme-Linked Immunosorbent Assay

TG and CHO levels in mouse liver tissues were quantified using the TG assay kit and the CHO Assay Kit, respectively, both from Nanjing Jiancheng Bioengineering Institute

(A110-1-1, A111-1-1, Nanjing, China). The liver tissue was accurately weighed and treated with the T-PER™ tissue protein extraction reagent at a ratio of 1:20 (g:mL). The samples were then centrifuged at 2500 rpm at 4 °C for 10 min. The supernatant obtained was used for subsequent analysis according to the manufacturer's instructions.

4.7. ELISA Measurement

Levels of IL-1β and TNF-α from mouse liver tissue extracts or cultured supernatants of LX-2 cells were measured using ELISA kits for mouse or human IL-1β and TNF-α, respectively (MeilunBio, Dalian, China) according to the manufacturer's protocols.

4.8. Quantitative Real-Time PCR (qPCR) Analysis

Total RNA was extracted from tissues or cells utilizing the RaPure Total RNA Kit (R4011, Magen, Guangzhou, China), strictly adhering to the manufacturer's instructions. The quantitative polymerase chain reaction (qPCR) assays were conducted on the 7500 Fast Real-time PCR System (Thermo Fisher Scientific, Waltham, MA, USA) employing the HiScript® II One Step gRT-PCR SYBR® Green Kit (Q22101, Vazyme, Nanjing, China). The assays were performed under thermal cycling conditions as specified in the kit's protocol. The sequences of primers used are listed in Table 1. Glyceraldehyde-3-phosphate dehydrogenase served as the internal reference gene. The qPCR data were analyzed using the comparative cycle threshold (Ct) method, and details regarding the qRT-PCR primers are provided in the accompanying table.

Table 1. Primer sequence information used for qRT–PCR.

Origin	Name	Forward (5′-3′)	Reverse (5′-3′)
Human	*IL-1β*	TATCATCTTTCAACACGCAGGACAG	TATCATCTTTCAACACGCAGGACAG
	TNF-α	GCCGTCTCCTACCAGACCAAG	ATGGGCTCATACCAGGGCTTG
	IL-6	ACAGACAGCCACTCACCTCTTC	AGTGCCTCTTTGCTGCTTTCAC
	COL1A1	AGGGCGACAGAGGCATAAAGG	AGGACCAGAGGCTCCAGAGG
	α-SMA	CCGGGAGAAAATGACTCAAA	GCAAGGCATAGCCCTCATAG
	GAPDH	GAACATCATCCCTGCCTCTACTGG	CCTCCGACGCCTGCTTCAC
Murine	*Tgfβ1*	CCGCTTCTGCTCCCACTCC	CATGTCGATGGTCTTGCAGGTG
	Col1a1	GGTCCTGCTGGTCCTGCTG	GAGAAGCCACGATGACCCTTTATG
	Il-1β	CAAACCTTTGACCTGGGCTGTC	GCCTGCCTGAAGCTCTTGTTG
	Tnf-α	GCCTCTTCTCATTCCTGCTTGTG	GTGTGAGGGTCTGGGCCATAG
	α-SMA	CTTCGTGACTACTGCCGAGC	AGGTGGTTTCGTGGATGCC
	Gapdh	CTCCCACTCTTCCACCTTCG	TAGGGCCTCTCTTGCTCAGT

4.9. Network Pharmacology-Based Analysis

Initially, drug targets for BBR and QR were collected using keywords "berberine" and "quercetin" from databases like Drugbank, Swiss TargetPrediction, Targetnet, and batman-TCM. After removing duplicates, a substantial number of potential targets for both drugs were established. Targets related to hepatic fibrosis were then collected from the DisGenet and Gene Cards databases, utilizing criteria based on protein-coding relevance and scoring to refine the target list. Venn diagrams were employed to ascertain the intersection of drug targets with those associated with hepatic fibrosis, identifying crucial targets for further analysis. The intersecting genes were analyzed using the String database to construct a protein interaction network, which was then examined in Cytoscape to determine network dynamics. Core targets were identified using CentiScape, based on specific criteria for closeness, degree, and betweenness, and subsequently visualized. Enrichment analyses of these core targets were performed in the Metascape database for gene ontology (GO) and

Kyoto Encyclopedia of Genes and Genomes (KEGG), elucidating their biological functions and pathways, with a focus on processes related to hepatic fibrosis.

4.10. Western Blot Analysis

Total liver proteins were extracted using RIPA lysis buffer, supplemented with protease and phosphatase inhibitors. Protein concentrations were determined using the BCA Protein Assay Kit (Thermo Fisher Scientific, Waltham, MA, USA). Twenty micrograms of total protein were subjected to electrophoresis on a 12% SDS-polyacrylamide gel and subsequently transferred to a polyvinylidene fluoride (PVDF) membrane. Following the transfer, the membrane was blocked with 5% non-fat milk and then incubated overnight at 4 °C with primary antibodies against phospho-Akt (1:1000, CST, Danvers, MA, USA), Akt (1:1000, Cell Signaling Technology Inc., Danvers, MA, USA), phospho-FoxO1 (1:1000, Cell Signaling Technology Inc., Danvers, MA, USA), Foxo-1 (1:1000, Cell Signaling Technology Inc., Danvers, MA, USA), Collagen I (1:3000, ProteinTech Group, Chicago, IL, USA), α-SMA (1:1000, Cell Signaling Technology Inc., Danvers, MA, USA) and β-actin (1:5000). After washing with TBST, the membrane was incubated with horseradish peroxidase-conjugated secondary antibodies (1:5000) at room temperature for one hour. The protein bands were visualized using a chemiluminescent detection reagent and an imaging system.

4.11. Statistical Analysis

All data are presented as mean ± standard error of mean with respect to the number of samples in each group. One-way ANOVA was used for comparisons among multiple groups, followed by Student's t-test for independent group comparisons. Statistical significance was determined as follows: * $p < 0.05$; ** $p < 0.01$; *** $p < 0.001$; **** $p < 0.0001$; # $p < 0.05$; ## $p < 0.01$; ### $p < 0.001$. A p-value of <0.05 was considered statistically significant.

5. Conclusions

Liver fibrosis, a prevalent pathological condition associated with chronic liver diseases, currently lacks effective therapeutic options. Without timely and effective intervention, this condition may progress to several irreversible complications that significantly endanger human life. In this study, we successfully synthesized a novel compound, BBR–QR salt, which demonstrates outstanding biological and chemical properties. Comprehensive in vitro and in vivo studies have demonstrated that BQS possesses enhanced anti-inflammatory, anti-lipid accumulation, and anti-fibrotic capabilities compared to both the individual drugs and their physical mixture. This increased efficacy is likely mediated through complex molecular interactions and the modulation of critical biochemical pathways, particularly the Akt/FoxO1 pathway. Our results suggest that BQS interferes with the signaling cascade that promotes fibrogenesis, thus attenuating the pathological process at multiple levels. These findings underscore the potential of BQS as a potent therapeutic agent for liver fibrosis, highlighting its ability to act on key mechanistic pathways involved in the disease process. Given the promising results observed, further clinical trials are warranted to validate BQS's efficacy and safety in human subjects. This could potentially lead to the development of a new, more effective treatment strategy for managing liver fibrosis, offering hope to millions affected by this debilitating condition.

Supplementary Materials: The following supporting information can be downloaded at: https://www.mdpi.com/article/10.3390/ijms26052193/s1.

Author Contributions: Conceptualization, J.J. and L.W.; investigation, M.Z., L.R., X.G. and X.T.; methodology, validation, visualization, formal analysis, Y.C., H.Y., S.Y., X.T., M.Z. and C.H.; writing—review and editing, Y.C., H.Y. and S.Y.; writing—review & editing, supervision, funding acquisition, project administration, J.J. and L.W.; resources, X.T. and M.Z.; All authors have read and agreed to the published version of the manuscript.

Funding: This work was supported by the CAMS Innovation Fund for Medical Sciences (No. 2023-I2M-2-006; 2022-I2M-2-002; 2021-I2M-1-070).

Institutional Review Board Statement: The animal study was reviewed and approved by the Ethics Committee of the Institute of Medicinal Biotechnology, Chinese Academy of Medical Sciences and Peking Union Medical College (IMB-20231016D1).

Informed Consent Statement: Not applicable.

Data Availability Statement: The original contributions presented in the study are included in the article/Supplementary Materials.

Conflicts of Interest: The authors declare no conflict of interest.

References

1. Younossi, Z.; Tacke, F.; Arrese, M.; Chander Sharma, B.; Mostafa, I.; Bugianesi, E.; Wai-Sun Wong, V.; Yilmaz, Y.; George, J.; Fan, J.; et al. Global Perspectives on Nonalcoholic Fatty Liver Disease and Nonalcoholic Steatohepatitis. *Hepatology* **2019**, *69*, 2672–2682. [CrossRef] [PubMed]
2. Friedman, S.L.; Neuschwander-Tetri, B.A.; Rinella, M.; Sanyal, A.J. Mechanisms of NAFLD development and therapeutic strategies. *Nat. Med.* **2018**, *24*, 908–922. [CrossRef] [PubMed]
3. Damiris, K.; Tafesh, Z.H.; Pyrsopoulos, N. Efficacy and safety of anti-hepatic fibrosis drugs. *World J. Gastroenterol.* **2020**, *26*, 6304–6321. [CrossRef]
4. Herbert, A.S.; Froude, J.W.; Ortiz, R.A.; Kuehne, A.I.; Dorosky, D.E.; Bakken, R.R.; Zak, S.E.; Josleyn, N.M.; Musiychuk, K.; Jones, R.M.; et al. Development of an antibody cocktail for treatment of Sudan virus infection. *Proc. Natl. Acad. Sci. USA* **2020**, *117*, 3768–3778. [CrossRef] [PubMed]
5. Li, W.; Yu, F.; Wang, Q.; Qi, Q.; Su, S.; Xie, L.; Lu, L.; Jiang, S. Co-delivery of HIV-1 entry inhibitor and nonnucleoside reverse transcriptase inhibitor shuttled by nanoparticles: Cocktail therapeutic strategy for antiviral therapy. *AIDS* **2016**, *30*, 827–838. [CrossRef]
6. Plumet, L.; Ahmad-Mansour, N.; Dunyach-Remy, C.; Kissa, K.; Sotto, A.; Lavigne, J.P.; Costechareyre, D.; Molle, V. Bacteriophage Therapy for *Staphylococcus Aureus* Infections: A Review of Animal Models, Treatments, and Clinical Trials. *Front. Cell. Infect. Microbiol.* **2022**, *12*, 907314. [CrossRef]
7. Weinreich, D.M.; Sivapalasingam, S.; Norton, T.; Ali, S.; Gao, H.; Bhore, R.; Musser, B.J.; Soo, Y.; Rofail, D.; Im, J.; et al. REGN-COV2, a Neutralizing Antibody Cocktail, in Outpatients with COVID-19. *New Engl. J. Med.* **2021**, *384*, 238–251. [CrossRef]
8. Lin, L.J.; Huang, H.Y. DFSG, a novel herbal cocktail with anti-asthma activity, suppressed MUC5AC in A549 cells and alleviated allergic airway hypersensitivity and inflammatory cell infiltration in a chronic asthma mouse model. *Biomed. Pharmacother. = Biomed. Pharmacother.* **2020**, *121*, 109584. [CrossRef]
9. McLarnon, J.G. Consideration of a Pharmacological Combinatorial Approach to Inhibit Chronic Inflammation in Alzheimer's Disease. *Curr. Alzheimer Res.* **2019**, *16*, 1007–1017. [CrossRef]
10. Zhu, Y.; Xie, N.; Chai, Y.; Nie, Y.; Liu, K.; Liu, Y.; Yang, Y.; Su, J.; Zhang, C. Apoptosis Induction, a Sharp Edge of Berberine to Exert Anti-Cancer Effects, Focus on Breast, Lung, and Liver Cancer. *Front. Pharmacol.* **2022**, *13*, 803717. [CrossRef]
11. Xu, X.; Yi, H.; Wu, J.; Kuang, T.; Zhang, J.; Li, Q.; Du, H.; Xu, T.; Jiang, G.; Fan, G. Therapeutic effect of berberine on metabolic diseases: Both pharmacological data and clinical evidence. *Biomed. Pharmacother. = Biomed. Pharmacother.* **2021**, *133*, 110984. [CrossRef] [PubMed]
12. Li, T.; Wang, P.; Guo, W.; Huang, X.; Tian, X.; Wu, G.; Xu, B.; Li, F.; Yan, C.; Liang, X.J.; et al. Natural Berberine-Based Chinese Herb Medicine Assembled Nanostructures with Modified Antibacterial Application. *ACS Nano* **2019**, *13*, 6770–6781. [CrossRef]
13. Yi, J.; Wu, S.; Tan, S.; Qin, Y.; Wang, X.; Jiang, J.; Liu, H.; Wu, B. Berberine alleviates liver fibrosis through inducing ferrous redox to activate ROS-mediated hepatic stellate cells ferroptosis. *Cell Death Discov.* **2021**, *7*, 374. [CrossRef]
14. Wang, N.; Xu, Q.; Tan, H.Y.; Hong, M.; Li, S.; Yuen, M.F.; Feng, Y. Berberine Inhibition of Fibrogenesis in a Rat Model of Liver Fibrosis and in Hepatic Stellate Cells. *Evid. Based Complement. Altern. Med.* **2016**, *2016*, 8762345. [CrossRef] [PubMed]
15. Zhou, M.; Deng, Y.; Liu, M.; Liao, L.; Dai, X.; Guo, C.; Zhao, X.; He, L.; Peng, C.; Li, Y. The pharmacological activity of berberine, a review for liver protection. *Eur. J. Pharmacol.* **2021**, *890*, 173655. [CrossRef] [PubMed]
16. Wang, Y.; Zhao, Z.; Yan, Y.; Qiang, X.; Zhou, C.; Li, R.; Chen, H.; Zhang, Y. Demethyleneberberine Protects against Hepatic Fibrosis in Mice by Modulating NF-κB Signaling. *Int. J. Mol. Sci.* **2016**, *17*, 1036. [CrossRef]

17. Li, D.; Yang, C.; Zhu, J.Z.; Lopez, E.; Zhang, T.; Tong, Q.; Peng, C.; Lin, L.G. Berberine remodels adipose tissue to attenuate metabolic disorders by activating sirtuin 3. *Acta Pharmacol. Sin.* **2022**, *43*, 1285–1298. [CrossRef]
18. Nasiri-Ansari, N.; Nikolopoulou, C.; Papoutsi, K.; Kyrou, I.; Mantzoros, C.S.; Kyriakopoulos, G.; Chatzigeorgiou, A.; Kalotychou, V.; Randeva, M.S.; Chatha, K.; et al. Empagliflozin Attenuates Non-Alcoholic Fatty Liver Disease (NAFLD) in High Fat Diet Fed ApoE$^{(-/-)}$ Mice by Activating Autophagy and Reducing ER Stress and Apoptosis. *Int. J. Mol. Sci.* **2021**, *22*, 818. [CrossRef]
19. Liu, X.; Wang, L.; Tan, S.; Chen, Z.; Wu, B.; Wu, X. Therapeutic Effects of Berberine on Liver Fibrosis are associated with Lipid Metabolism and Intestinal Flora. *Front. Pharmacol.* **2022**, *13*, 814871. [CrossRef]
20. Domitrović, R.; Jakovac, H.; Marchesi, V.V.; Blažeković, B. Resolution of liver fibrosis by isoquinoline alkaloid berberine in CCl_4-intoxicated mice is mediated by suppression of oxidative stress and upregulation of MMP-2 expression. *J. Med. Food* **2013**, *16*, 518–528. [CrossRef]
21. Eissa, L.A.; Kenawy, H.I.; El-Karef, A.; Elsherbiny, N.M.; El-Mihi, K.A. Antioxidant and anti-inflammatory activities of berberine attenuate hepatic fibrosis induced by thioacetamide injection in rats. *Chem.-Biol. Interact.* **2018**, *294*, 91–100. [CrossRef] [PubMed]
22. Tang, S.M.; Deng, X.T.; Zhou, J.; Li, Q.P.; Ge, X.X.; Miao, L. Pharmacological basis and new insights of quercetin action in respect to its anti-cancer effects. *Biomed. Pharmacother. = Biomed. Pharmacother.* **2020**, *121*, 109604. [CrossRef] [PubMed]
23. Grewal, A.K.; Singh, T.G.; Sharma, D.; Sharma, V.; Singh, M.; Rahman, M.H.; Najda, A.; Walasek-Janusz, M.; Kamel, M.; Albadrani, G.M.; et al. Mechanistic insights and perspectives involved in neuroprotective action of quercetin. *Biomed. Pharmacother. = Biomed. Pharmacother.* **2021**, *140*, 111729. [CrossRef]
24. Chen, L.; Liu, J.; Mei, G.; Chen, H.; Peng, S.; Zhao, Y.; Yao, P.; Tang, Y. Quercetin and non-alcoholic fatty liver disease: A review based on experimental data and bioinformatic analysis. *Food Chem. Toxicol. Int. J. Publ. Br. Ind. Biol. Res. Assoc.* **2021**, *154*, 112314. [CrossRef]
25. Li, X.; Jin, Q.; Yao, Q.; Xu, B.; Li, Z.; Tu, C. Quercetin attenuates the activation of hepatic stellate cells and liver fibrosis in mice through modulation of HMGB1-TLR2/4-NF-κB signaling pathways. *Toxicol. Lett.* **2016**, *261*, 1–12. [CrossRef] [PubMed]
26. Wu, L.; Zhang, Q.; Mo, W.; Feng, J.; Li, S.; Li, J.; Liu, T.; Xu, S.; Wang, W.; Lu, X.; et al. Quercetin prevents hepatic fibrosis by inhibiting hepatic stellate cell activation and reducing autophagy via the TGF-β1/Smads and PI3K/Akt pathways. *Sci. Rep.* **2017**, *7*, 9289. [CrossRef]
27. Li, X.; Jin, Q.; Yao, Q.; Xu, B.; Li, L.; Zhang, S.; Tu, C. The Flavonoid Quercetin Ameliorates Liver Inflammation and Fibrosis by Regulating Hepatic Macrophages Activation and Polarization in Mice. *Front. Pharmacol.* **2018**, *9*, 72. [CrossRef]
28. Hao, M.; Li, Y.; Liu, L.; Yuan, X.; Gao, Y.; Guan, Z.; Li, W. The design and synthesis of a novel compound of berberine and baicalein that inhibits the efficacy of lipid accumulation in 3T3-L1 adipocytes. *Bioorg. Med. Chem.* **2017**, *25*, 5506–5512. [CrossRef]
29. Jia, D.; Dou, Y.; Li, Z.; Zhou, X.; Gao, Y.; Chen, K.; Cong, W.; Ma, M.; Wu, Z.; Li, W. Design, synthesis and evaluation of a baicalin and berberine hybrid compound as therapeutic agent for ulcerative colitis. *Bioorg. Med. Chem.* **2020**, *28*, 115697. [CrossRef]
30. Shailajan, S.; Patil, Y.; Joshi, M.; Menon, S.; Mhatre, M. Simultaneous Quantification of Pharmacological Markers Quercetin and Berberine Using High-Performance Thin-Layer Chromatography (HPTLC) and High-Performance Liquid Chromatography (HPLC) from a Polyherbal Formulation Pushyanuga Churna. *J. AOAC Int.* **2019**, *102*, 1003–1013. [CrossRef] [PubMed]
31. Urasaki, Y.; Le, T.T. Functional Complementation of Anti-Adipogenic Phytonutrients for Obesity Prevention and Management. *Nutrients* **2022**, *14*, 4325. [CrossRef]
32. Tang, Y.; Su, H.; Wang, H.; Lu, F.; Nie, K.; Wang, Z.; Huang, W.; Dong, H. The effect and mechanism of Jiao-tai-wan in the treatment of diabetes mellitus with depression based on network pharmacology and experimental analysis. *Mol. Med.* **2021**, *27*, 154. [CrossRef] [PubMed]
33. Kisseleva, T.; Brenner, D. Molecular and cellular mechanisms of liver fibrosis and its regression. *Nat. Rev. Gastroenterol. Hepatol.* **2021**, *18*, 151–166. [CrossRef]
34. Cano-Martínez, A.; Bautista-Pérez, R.; Castrejón-Téllez, V.; Carreón-Torres, E.; Pérez-Torres, I.; Díaz-Díaz, E.; Flores-Estrada, J.; Guarner-Lans, V.; Rubio-Ruíz, M.E. Resveratrol and Quercetin as Regulators of Inflammatory and Purinergic Receptors to Attenuate Liver Damage Associated to Metabolic Syndrome. *Int. J. Mol. Sci.* **2021**, *22*, 8939. [CrossRef]
35. Sun, X.; Zhang, X.; Hu, H.; Lu, Y.; Chen, J.; Yasuda, K.; Wang, H. Berberine inhibits hepatic stellate cell proliferation and prevents experimental liver fibrosis. *Biol. Pharm. Bull.* **2009**, *32*, 1533–1537. [CrossRef]
36. Bessone, F.; Razori, M.V.; Roma, M.G. Molecular pathways of nonalcoholic fatty liver disease development and progression. *Cell. Mol. Life Sci. CMLS* **2019**, *76*, 99–128. [CrossRef] [PubMed]
37. Mountford, S.; Effenberger, M.; Noll-Puchta, H.; Griessmair, L.; Ringleb, A.; Haas, S.; Denk, G.; Reiter, F.P.; Mayr, D.; Dinarello, C.A.; et al. Modulation of Liver Inflammation and Fibrosis by Interleukin-37. *Front. Immunol.* **2021**, *12*, 603649. [CrossRef] [PubMed]
38. Meng, X.M.; Nikolic-Paterson, D.J.; Lan, H.Y. TGF-β: The master regulator of fibrosis. *Nat. Rev. Nephrol.* **2016**, *12*, 325–338. [CrossRef]
39. Ahamed, J.; Laurence, J. Role of Platelet-Derived Transforming Growth Factor-β1 and Reactive Oxygen Species in Radiation-Induced Organ Fibrosis. *Antioxid. Redox Signal.* **2017**, *27*, 977–988. [CrossRef]

40. Meng, D.; Li, Z.; Wang, G.; Ling, L.; Wu, Y.; Zhang, C. Carvedilol attenuates liver fibrosis by suppressing autophagy and promoting apoptosis in hepatic stellate cells. *Biomed. Pharmacother. = Biomed. Pharmacother.* **2018**, *108*, 1617–1627. [CrossRef]
41. Koda, Y.; Teratani, T.; Chu, P.S.; Hagihara, Y.; Mikami, Y.; Harada, Y.; Tsujikawa, H.; Miyamoto, K.; Suzuki, T.; Taniki, N.; et al. CD8$^{(+)}$ tissue-resident memory T cells promote liver fibrosis resolution by inducing apoptosis of hepatic stellate cells. *Nat. Commun.* **2021**, *12*, 4474. [CrossRef]
42. Zhou, W.C.; Zhang, Q.B.; Qiao, L. Pathogenesis of liver cirrhosis. *World J. Gastroenterol.* **2014**, *20*, 7312–7324. [CrossRef] [PubMed]
43. Challa, T.D.; Wueest, S.; Lucchini, F.C.; Dedual, M.; Modica, S.; Borsigova, M.; Wolfrum, C.; Blüher, M.; Konrad, D. Liver ASK1 protects from non-alcoholic fatty liver disease and fibrosis. *EMBO Mol. Med.* **2019**, *11*, e10124. [CrossRef] [PubMed]
44. Zhao, X.; Chen, J.; Sun, H.; Zhang, Y.; Zou, D. New insights into fibrosis from the ECM degradation perspective: The macrophage-MMP-ECM interaction. *Cell Biosci.* **2022**, *12*, 117. [CrossRef]
45. Roehlen, N.; Crouchet, E.; Baumert, T.F. Liver Fibrosis: Mechanistic Concepts and Therapeutic Perspectives. *Cells* **2020**, *9*, 875. [CrossRef]
46. Nie, Q.; Li, M.; Huang, C.; Yuan, Y.; Liang, Q.; Ma, X.; Qiu, T.; Li, J. The clinical efficacy and safety of berberine in the treatment of non-alcoholic fatty liver disease: A meta-analysis and systematic review. *J. Transl. Med.* **2024**, *22*, 225. [CrossRef] [PubMed]
47. Fabregat, I.; Caballero-Diaz, D. Transforming Growth Factor-beta-Induced Cell Plasticity in Liver Fibrosis and Hepatocarcinogenesis. *Front. Oncol.* **2018**, *8*, 357. [CrossRef]
48. Wang, S.Y.; Duan, K.M.; Li, Y.; Mei, Y.; Sheng, H.; Liu, H.; Mei, X.; Ouyang, W.; Zhou, H.H.; Liu, Z.Q. Effect of quercetin on P-glycoprotein transport ability in Chinese healthy subjects. *Eur. J. Clin. Nutr.* **2013**, *67*, 390–394. [CrossRef]
49. Kim, M.K.; Choo, H.; Chong, Y. Water-soluble and cleavable quercetin-amino acid conjugates as safe modulators for P-glycoprotein-based multidrug resistance. *J. Med. Chem.* **2014**, *57*, 7216–7233. [CrossRef]
50. Zhao, Q.; Wei, J.; Zhang, H. Effects of quercetin on the pharmacokinetics of losartan and its metabolite EXP3174 in rats. *Xenobiotica* **2019**, *49*, 563–568. [CrossRef]
51. Zhang, Z.; Shang, J.; Yang, Q.; Dai, Z.; Liang, Y.; Lai, C.; Feng, T.; Zhong, D.; Zou, H.; Sun, L.; et al. Exosomes derived from human adipose mesenchymal stem cells ameliorate hepatic fibrosis by inhibiting PI3K/Akt/mTOR pathway and remodeling choline metabolism. *J. Nanobiotechnol.* **2023**, *21*, 29. [CrossRef]
52. Yu, H.; Zhang, L.; Chen, P.; Liang, X.; Cao, A.; Han, J.; Wu, X.; Zheng, Y.; Qin, Y.; Xue, M. Dietary Bile Acids Enhance Growth, and Alleviate Hepatic Fibrosis Induced by a High Starch Diet via AKT/FOXO1 and cAMP/AMPK/SREBP1 Pathway in Micropterus salmoides. *Front. Physiol.* **2019**, *10*, 1430. [CrossRef] [PubMed]
53. Lee, S.; Usman, T.O.; Yamauchi, J.; Chhetri, G.; Wang, X.; Coudriet, G.M.; Zhu, C.; Gao, J.; McConnell, R.; Krantz, K.; et al. Myeloid FoxO1 depletion attenuates hepatic inflammation and prevents nonalcoholic steatohepatitis. *J. Clin. Investig.* **2022**, *132*, e154333. [CrossRef]
54. Gao, H.; Zhou, L.; Zhong, Y.; Ding, Z.; Lin, S.; Hou, X.; Zhou, X.; Shao, J.; Yang, F.; Zou, X.; et al. Kindlin-2 haploinsufficiency protects against fatty liver by targeting Foxo1 in mice. *Nat. Commun.* **2022**, *13*, 1025. [CrossRef] [PubMed]
55. Crowley, L.C.; Marfell, B.J.; Scott, A.P.; Waterhouse, N.J. Quantitation of Apoptosis and Necrosis by Annexin V Binding, Propidium Iodide Uptake, and Flow Cytometry. *Cold Spring Harb. Protoc.* **2016**, *2016*, pdb-prot087288. [CrossRef] [PubMed]
56. Larter, C.Z.; Yeh, M.M. Animal models of NASH: Getting both pathology and metabolic context right. *J. Gastroenterol. Hepatol.* **2008**, *23*, 1635–1648. [CrossRef]
57. Ma, X.; Zhang, T.; Luo, Z.; Li, X.; Lin, M.; Li, R.; Du, P.; Yu, X.; Ma, C.; Yan, P.; et al. Functional nano-vector boost anti-atherosclerosis efficacy of berberine in Apoe$^{(-/-)}$ mice. *Acta Pharm. Sin. B* **2020**, *10*, 1769–1783. [CrossRef]

International Journal of *Molecular Sciences*

MDPI

Brief Report

Synergistic Effects of a Novel Combination of Natural Compounds Prevent H_2O_2-Induced Oxidative Stress in Red Blood Cells

Giuditta Benincasa [1,*], Paola Bontempo [2], Ugo Trama [3,†] and Claudio Napoli [1,†]

1 Department of Advanced Medical and Surgical Sciences (DAMSS), University of Campania "Luigi Vanvitelli", 80138 Naples, Italy; direzione.immunoematologia@unicampania.it
2 Department of Precision Medicine, University of Campania Luigi Vanvitelli, Via L. De Crecchio 7, 80138 Naples, Italy; paola.bontempo@unicampania.it
3 Regional Pharmaceutical Unit, Campania Region, 80143 Naples, Italy; ugo.trama@regione.campania.it
* Correspondence: giuditta.benincasa@unicampania.it; Tel.: +39-081-566-7916
† These authors contributed equally to this work.

Abstract: Novel strategies to prevent the "storage lesions" of red blood cells (RBCs) are needed to prevent the risk of adverse effects after blood transfusion. One option could be the supplementation of stored blood bags with natural compounds that may increase the basal load of antioxidant protection and the shelf life of RBCs. In this pilot study, we investigated for the first time potential synergistic effects of a triple combination of well-known anti-oxidant compounds curcumin (curc), vitamin E (vit E), and vitamin C (vit C). Briefly, we established an ex vivo model of H_2O_2-induced oxidative stress and measured the hemolysis ratio (HR) (%) and thiobarbituric acid reactive substances (TBARS) levels in RBCs with or without pre-exposure for 30 min with increasing concentrations of curc, vit E, and vit C and then exposed to 10 mM H_2O_2. for 60 min. Exposure of RBCs to a triple combination of curc, vit E, and vit C at the highest concentration (100 μM) completely prevented H_2O_2-induced hemolysis. Surprisingly, we found that pre-treatment of RBCs with curc 100 μM alone completely prevented hemolysis as compared to vit E and vit C alone or in combination at the same concentration. On the other hand, pre-treatment with the triple combination of curc, vit E, and vit C 100 μM was required to totally prevent lipid peroxidation, as compared to curc 100 μM alone, supporting their synergistic effects in preventing RBCs membrane peroxidation. Further experiments are ongoing to investigate the anti-aging effects of the triple combination of curc, vit E, and vit C on cold-stored bags.

Keywords: red blood cells; hydrogen peroxide; oxidative stress; natural compounds

1. Introduction

The transfusion of red blood cells (RBCs) is a lifesaving treatment for severe anemia, trauma, or major surgery [1,2]. RBC-related "storage lesions" represent morphological, functional, and metabolic changes that RBCs progressively undergo upon collection, processing, and refrigerated storage in blood banks for clinical use [3–5]. Often, adverse events are caused by the transfusion of aged RBCs that are closer to their expiration date (42 days), wherein the effects of RBC storage lesions can contribute to poorer outcomes as compared to outcomes using fresher RBCs [6]. In order to address this unsolved clinical need, novel options to prevent storage lesions and extend the shelf life of RBCs are required.

One approach to prevent storage lesions could be the supplementation of stored blood bags with natural compounds that may increase the basal load of antioxidant protection and shelf life of RBCs. Well-known anti-oxidant natural compounds, including vitamins [7–9], curcumin [7], quercetin [10,11], L-carnitine [12], açaì extract [13], and anthocyanin-rich extract (*Callistemon citrinus*) [14], have been widely investigated in accelerated aging models on human RBCs. However, there is a paucity of data about possible synergistic effects of specific combinations of natural compounds that may prevent oxidative stress better than one compound alone.

Here, we evaluated for the first time the possible synergistic effects of a triple combination of well-known anti-oxidant natural compounds including curcumin (curc), L-ascorbic acid (vit C), and alpha-tocopherol (vit E) entering cells via related transmembrane porters. To this aim, we used an accelerated aging model of RBCs based on H_2O_2-induced oxidative stress and measured two markers of aging, the hemolysis ratio (HR) (%) and thiobarbituric acid reactive substances (TBARS) levels, which are basic biological toxicity tests widely used to evaluate the activity of natural compounds because they are rapid, reproducible, and less expensive than other tests, such as cell culturing. Due to the lack of methodological standardization in the literature, we also offer an original framework useful for easily establishing an H_2O_2-induced oxidative stress model in human RBCs.

2. Results

2.1. Optimal Concentration of H_2O_2-Induced Oxidative Stress

In order to establish the optimal conditions of H_2O_2-induced oxidative stress, hemolysis assays were conducted upon exposure of RBCs to six increasing concentrations of peroxide. Data showed that hemolysis occurred in a concentration-dependent manner. In detail, we show a representative original plate in which we could observe the color change from red to dark brown in each well after incubation with increasing concentrations of H_2O_2 (from 4 to 50 mM) (Figure 1A). Moreover, a representative original picture shows the colour of supernatants that were collected after centrifuge that progressively changed from red towards dark brown in a direct proportional manner to increasing H_2O_2 concentration (Figure 1B). Except for 4 mM H_2O_2, RBC suspensions exposed to 6, 8, 10, 20, and 50 mM H_2O_2 had significantly increased HRs (%) as compared to that of the negative control (PBS) ($p < 0.05$) (Figure 1C). For the next experiments, we chose 10 mM H_2O_2 as a hemolysis-inducing low dose of peroxide to test the synergistic effects of natural compounds.

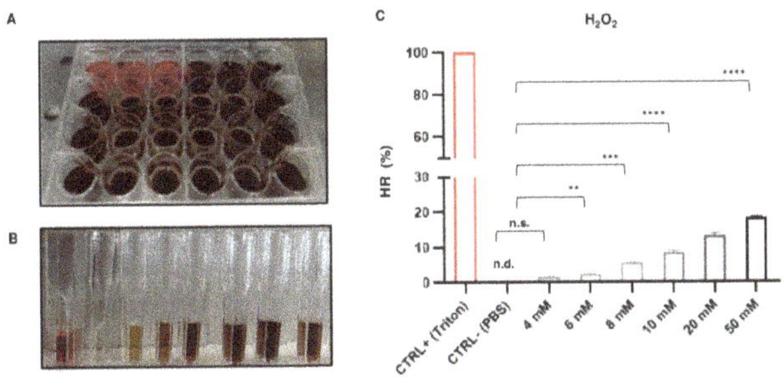

Figure 1. Selection of the H_2O_2 hemolysing dose. (**A**) A representative original image of a plate containing suspensions of RBCs without (red wells) or with (dark brown) H_2O_2. (**B**) A representative original image of cuvettes containing supernatants that were used to measure absorbance (**A**) at 540 nm after exposure to increasing concentrations of H_2O_2. From right to left in the image, contents of these cuvettes correspond to the bar graph in (**C**). (**C**) A bar graph showing the hemolysis ratios (HRs) (%) of RBCs obtained at increasing concentrations of H_2O_2. Values are the mean of independent experiments using RBCs from $n = 3$ blood donors assayed in triplicate (repeated-measures one-way ANOVA). ** $p < 0.01$; *** $p < 0.001$; **** $p < 0.0001$; n.d., not detectable; n.s., not significant.

2.2. Evaluation of Synergistic Anti-Oxidant Effects of Curc, Vit E, and Vit C

In order to exclude any possibility of hemolysis upon exposure to the triple combination of natural compounds, we preliminarily evaluated the response of RBCs towards incubation with curc, vit E, and vit C ranging from 1 to 100 μM. As shown in Figure 2, we did not observe any sign of hemolysis as proven by the fact that the colour of supernatants in the cuvettes remained light yellow (no oxidative stress) after 1 h and 30 min of incubation at 37 °C in both lower (1–20 μM) and higher (50–70 μM) concentrations.

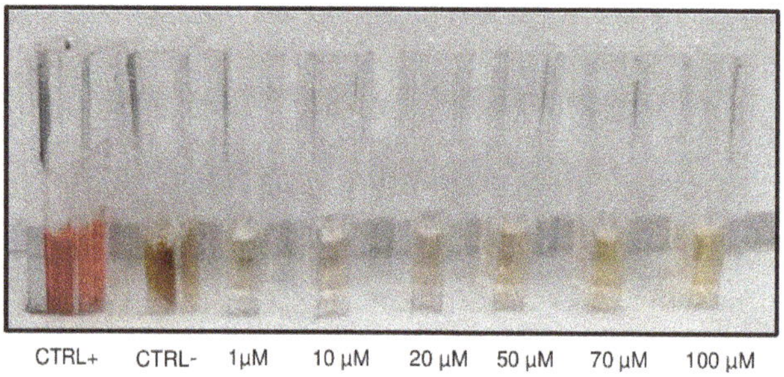

Figure 2. Preliminary evaluation of potential hemolytic effects. A representative original image of cuvettes containing supernatants from RBCs exposed to increasing concentrations of a triple combination of curc, vit E, and vit C.

After it was proven that the triple combination of curc, vit E, and vit C did not provoke hemolysis, we evaluated its ability to counteract the oxidative stress induced by peroxide. Therefore, we incubated RBCs with a triple combination of curc, vit E, and vit C at increasing concentrations (1–100 μM) for 30 min, and successively co-incubated 10 mM of H_2O_2 for 60 min. As the final step, we measured the HRs (%) of RBCs. The experimental workflow is shown in Figure 3.

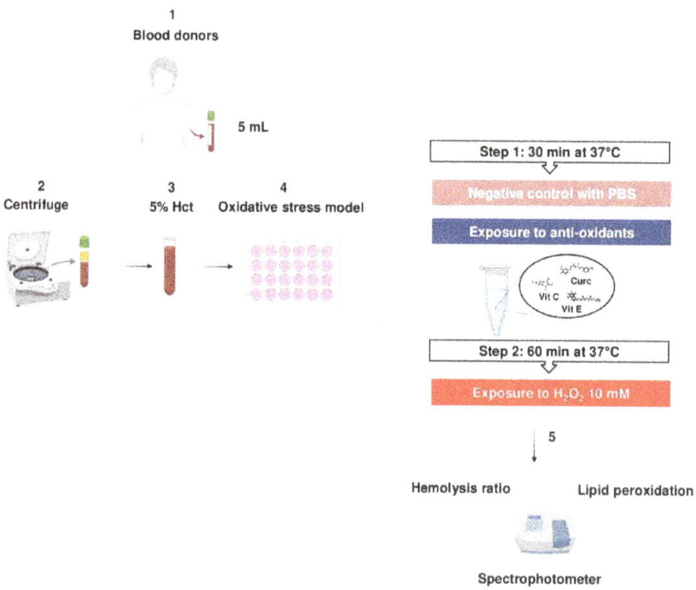

Figure 3. Experimental flowchart. Major details are provided in the Section 4.

In our experimental conditions, we observed that the HR (%) significantly decreased in a concentration-dependent manner. The ability to prevent hemolysis is observable in Figure 4A, which shows the colour of supernatants in cuvettes that progressively changed from dark brown (proven oxidative stress) towards light yellow (no oxidative stress). In detail, the triple combination of curc, vit E, and vit C at 20 µM (1-way ANOVA, $p < 0.001$), 50 µM (1-way ANOVA, $p < 0.001$), 70 µM (1-way ANOVA, $p < 0.001$), and 100 µM (1-way ANOVA, $p < 0.001$) significantly decreased the HR (%) of RBCs as compared to the H_2O_2-negative control (Figure 4B), whereas the triple combination of curc, vit E, and vit C at 1 µM and 10 µM did not prevent hemolysis as compared to the H_2O_2-negative control (1-way ANOVA, $p > 0.05$).

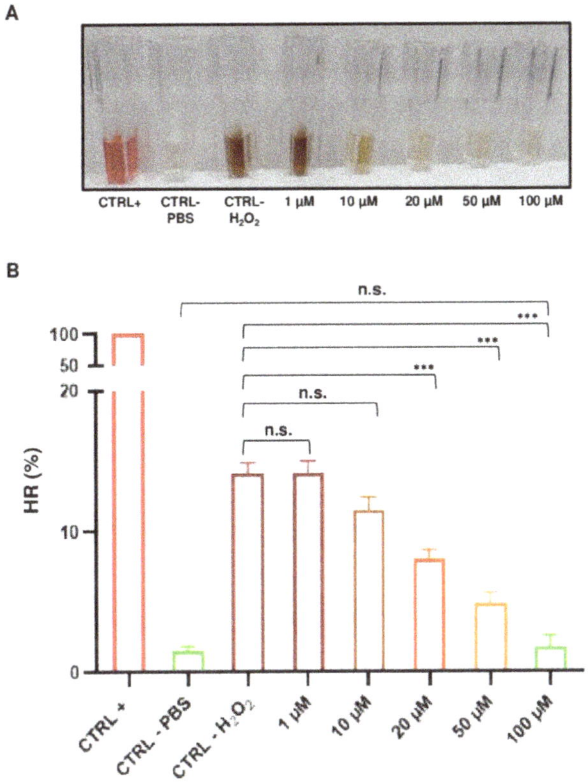

Figure 4. Synergistic anti-hemolytic effects of the triple combination of natural compounds at increasing concentrations. (**A**) A representative original image of cuvettes containing supernatants that were used to measure absorbance (**A**) at 540 nm after exposure to increasing concentrations of H_2O_2. (**B**) A bar graph showing relative hemolysis ratios (HRs) (%). Values are the mean of independent experiments using blood from $n = 3$ donors assayed in triplicate (repeated-measures 1-way ANOVA). *** $p < 0.001$; n.s., not significant.

Interestingly, we noted that the highest concentration of triple combination solution (100 µM) completely prevented oxidative stress as demonstrated by the fact that there was no significant difference between the HR (%) of RBCs pre-treated with the triple solution and that of the PBS negative control (1-way ANOVA, $p > 0.05$). Taken together, these data showed that starting from a concentration of 20 µM, a solution of curc, vit E, and vit C can prevent oxidative stress on RBCs in a dose-dependent manner, with total prevention reached at the highest concentration tested (100 μM).

Starting from previous evidence on the use of higher concentrations of vitamins to evaluate their anti-aging effects in a model of cold-stored blood bags [8,9], we did not further consider 20 μM and 50 μM concentrations, but performed the following experiments using only pre-treatments at the highest concentration (100 μM). This choice is explained by our interest in evaluating the anti-aging effects of triple combinations of curc, vit E, and vit C using a model of cold-stored blood bags in future experimental workflows.

Therefore, we measured the anti-hemolytic effects of RBCs pre-treatments with a double combination of vit C and vit E (100 μM) or single natural compounds of vit C or vit E (100 μM). We observed that pre-treatment with a combination of vit C and vit E significantly reduced the HR (%) compared to the negative control (1-way ANOVA, $p < 0.0001$), but did not totally prevent hemolysis (Figure 5A). Interestingly, pre-treatment of RBCs with curc alone was able to completely prevent hemolysis as compared to vit C and vit E alone (Figure 5B). This was proven by the fact that there was no significant difference between the HR (%) of RBCs pre-treated with curc 100 μM and that of the PBS negative control (1-way ANOVA, $p > 0.05$). These data support previous evidence that curc pre-treatment may be useful in maintaining the integrity of the RBC membrane [7].

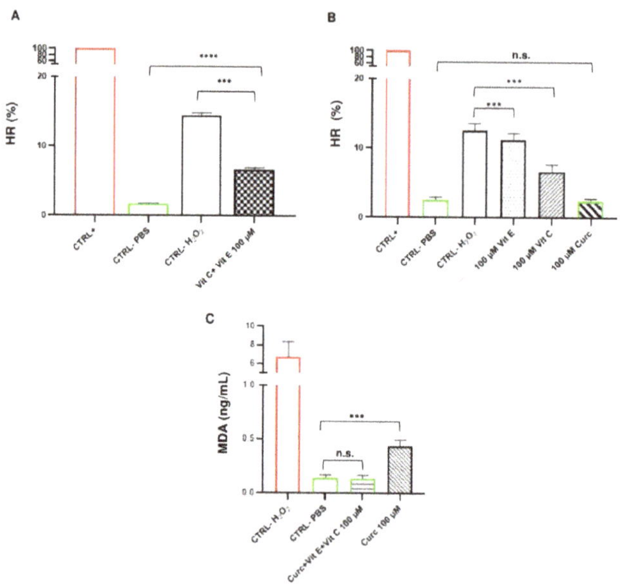

Figure 5. Evaluation of hemolysis and lipid peroxidation at the highest concentration of natural compounds. Bar graphs (**A**,**B**) show the hemolysis ratios (HRs) (%) of RBCs. Values are the mean of independent experiments using blood from $n = 3$ donors assayed in triplicate (repeated-measures 1-way ANOVA). (**C**) A bar graph showing MDA levels (ng/mL) obtained by pre-treatment with curc, vit E, and vit C in combination as compared to that of curc alone. Values are the mean of independent experiments using blood from $n = 3$ donors assayed in duplicate (repeated-measures 1-way ANOVA). *** $p < 0.001$; **** $p < 0.0001$. n.s., not significant.

Next, we chose to evaluate the triple combination of natural compounds versus curc alone in preventing lipid peroxidation of RBCs. In contrast to hemolysis, pre-treatment with curc 100 μM alone did not completely prevent lipid peroxidation. Indeed, MDA levels were statistically higher than those of the PBS negative control (1-way ANOVA, $p < 0.001$) (Figure 5C). Interestingly, the triple combination completely prevented lipid peroxidation, as shown by the fact that MDA levels were statistically similar to that of the negative control (PBS) (1-way ANOVA, $p > 0.05$) (Figure 5C). These data support that curc, vit E, and vit C may cooperate to prevent lipid peroxidation of RBC membranes with beneficial effects that are superior to those of curcumin alone.

3. Discussion

Storage lesions are heterogeneous metabolic changes that occur in cold-stored blood bags leading to unavoidable aging-induced RBC membrane loss. In turn, loss of RBC integrity results in hemolysis and the formation of microparticles, which may contribute to complications associated with transfusion. As blood transfusion is one of the most common life-saving medical therapies, it is of paramount importance to better understand storage lesion-associated mechanisms and identify novel anti-aging treatments that may advance the current paradigm of transfusion medicine. In this context, we designed a preliminary study in order to evaluate for the first time the anti-oxidant effects of a triple combination of well-known anti-oxidant natural compounds, curc, vit E, and vit C, on RBCs by measuring hemolysis and lipid peroxidation as the main biomarkers of RBC-related membrane damages. Indeed, this pilot study was based on an accelerated model of aging and is part of our ongoing experiments directly on cold-stored blood bags.

The major findings of the present pilot study are as follows. (1) The exposure of RBCs to the triple combination of curc, vit E, and vit C at the highest concentration tested (100 μM) completely prevented hemolysis induced by H_2O_2 treatment. (2) The exposure of RBCs to curc alone accounted for the highest anti-hemolytic effects as compared to pre-treatment with a combination of vit C and vit E or pre-treatment with a single natural compound. (3) The exposure of RBCs to the triple combination of curc, vit E, and vit C at the highest concentration tested (100 μM) completely prevented lipid peroxidation as compared to the exposure of RBCs to curc alone. Taken together, these data support the possible synergistic effects of curc, vit E, and vit C in preventing storage lesions in RBCs. Our focus on the highest concentration of the triple combination of natural compounds arose from our ongoing experiments evaluating its possible anti-aging effects in a model of cold-stored blood bags (about 450 mL), and therefore required higher concentrations of natural compounds. The use of higher concentrations of natural compounds in a long-term experimental model is also supported by previous studies highlighting that vit C and an analogue of vitamin E (Trolox) had lasting anti-oxidant effects starting from a concentration of 125 μM (until 3125 μM) [8,9]. In detail, previous long-term studies showed that exposure to Trolox, a water-soluble analogue of vit E, significantly prevented hemolysis and lipid peroxidation in RBCs alone [8] or in combination with vit C [9]. However, a complete prevention of oxidative stress was not observed at the tested concentrations. Moreover, pre-treatment of RBCs with curc mitigated oxidative injury, membrane deformability, and elasticity better than vit C alone did [7].

Although many studies have evaluated the antioxidant effects of vitamins [7–9], curcumin [7], quercetin [10,11], L-carnitine [12], açaì extract [13], and *Callistemon citrinus* [14] for preventing storage lesions in RBCs or additional detrimental insults to human health [15–17], there is a paucity of data about the effects of such natural compounds in specific combinations that may yield synergistic anti-oxidant effects. This is the first ex vivo study evaluating the synergistic effects of a triple combination of curc, vit E, and vit C in preventing oxidative stress in RBCs using a range of both lower and higher concentrations. In this pilot study, we were surprised that the exposure of RBCs to curcumin alone had the ability to completely prevent hemolysis in the same manner the triple combination did, and that it had a higher anti-hemolytic effect as compared to vit C and vit E, alone or in combination. Curcumin, a yellow pigment commonly used as a spice and food coloring, is widely used as a nutraceutical compound owing to its anti-inflammatory properties [18]. One explanation could be that curcumin is effective in protecting RBCs from oxidative stress events at the level of cell membrane transport. Specifically, band 3 protein, which consists of a membrane domain-mediating anion exchange and a cytoplasmic domain, mainly contributes to protein–protein interactions by coupling the lipid bilayer to the underlying cytoskeleton through cysteine -SH groups [19]. On the other hand, the exposure of RBCs to curcumin alone was not able to completely prevent lipid peroxidation compared to exposure to the triple combination of natural compounds supporting synergistic effects among them. Vitamins C and E are both naturally occurring free radical scavengers. It is well known that vit C can protect membranes against peroxidative damage and enhances the effects of vit E by reducing tocopheroxyl radicals in human RBCs [8].

Although this study was not aimed at identifying the underlying mechanisms of this synergy, we hypothesize that the simultaneous presence of curc, vit E, and vit C amplifies the antioxidant effects of each individual compound forming a synergistic barrier for the lipid bilayer via neutralizing free radicals through the donation of electrons or hydrogen atoms and iron (II) chelating ability. Natural compounds act via complex molecular networks regulating cell senescence, inflammation, and the structural integrity of the membrane, which are key contributors to aging [7,18,20]. Among the potential molecular players regulating synergistic circuits in blood bags, we may hypothesize a role for the nuclear erythroid-2-related factor (NRF2), which is a key transcription factor during erythroid development affecting the expression of several antioxidant proteins, delaying cell senescence and preventing age-related diseases [21]. Additionally, the nuclear factor kappa-light-chain-enhancer of activated B cells (NF-kB) transcription factor is vital for promoting the survival of RBCs by delaying their apoptosis [22]. It is evident that more efforts should be made to evaluate which specific mechanisms of action can have anti-aging effects on RBCs in long-term models of storage lesions.

Although this was an explorative study of small size, our results warrant further evaluation of the antioxidant effects of the triple combination of natural compounds directly in bags stored in blood biobanks [8,9]. Moreover, we provide an easy standardized methodological workflow aimed at helping researchers explore these novel beneficial anti-oxidants' effects on RBCs or other hemoderivatives using a wide range of natural compounds.

4. Materials and Methods

4.1. Study Population

As previously described [1,23–26], we enrolled periodic blood donors including non-smoker young males (30–35 years) at the Clinical Immunology, Immunohematology and Transfusion Medicine department of the University of Campania, "Luigi Vanvitelli" (Naples, Italy). RBCs were drawn from donor volunteers in accordance with guidelines from the Italian National Blood Centre (Blood Transfusion Service for donated blood).

4.2. Isolation of RBCs

Peripheral blood (5 mL) was collected in heparinised tubes and used immediately. It was centrifuged at 1500 g for 10 min at 4 °C in a clinical centrifuge, and then both the supernatant and buffy coat were gently removed. The erythrocyte pellet was washed two times with phosphate-buffered saline (PBS) (0.9% NaCl in 10 mM sodium phosphate buffer, pH 7.4). After the second wash, packed erythrocytes were gently resuspended with PBS to give a 5% hematocrit. Importantly, suspensions of RBCs were pre-incubated at 37 °C for 10 min in the presence of 1 mM of sodium azide (NaN_3) to inhibit catalase activity. This was a crucial step in establishing the oxidative stress model, as proven by the change in colour of RBCs from red to dark brown (Figure 1A), as clearly reported in previous protocols [27].

4.3. *Establishment of the H_2O_2-Induced Oxidative Stress Model*

4.3.1. Reagents

Curcumin (#HY-N0005), vitamin E (#HY-W020044), and vitamin C (HY-B0166) were purchased from MCE Med Chem Express (Princeton, NJ, USA) and 30% H_2O_2 was purchased from Sigma-Aldrich (St. Louis, MO, USA) (#BCCD8661).

4.3.2. Selection of the Suitable H_2O_2 Concentration

RBCs and enzymes are susceptible to oxidative damage, resulting in the peroxidation of membrane lipids and the release of hemoglobin (hemolysis). We chose H_2O_2 as the water-soluble oxidant because it can quickly permeate the membrane and partition in the cytosol leading to lipid membrane and enzyme oxidation [7,28]. Hemolysis assays were conducted as previously reported [7]. First, each 5% hematocrit was divided into ten aliquots: nine aliquots (after pre-exposure to NaN_3) were treated with increasing concentrations of H_2O_2 (4 mM, 6 mM, 8 mM, 10 mM, 20 mM, and 50 mM) and one aliquot (without pre-exposure to NaN_3) was treated with PBS (normal control). After 60 min at 37 °C, RBCs were immersed for 60 s in an ice bath and then centrifuged at 1500 g for 5 min at 4 °C. Supernatants were saved and their absorbance (A) was recorded at 540 nm to determine the release of hemoglobin. The hemolysis percentage was calculated using the formula:

$$\text{Hemolysis Ratio (HR) (\%)} = \text{A of sample} - \text{A of normal control (PBS)}/\text{A of positive control (Triton 1X)} \times 100$$

4.3.3. Pre-Exposure to Natural Compounds and H_2O_2-Induced Oxidative Stress

RBC suspensions (5% hematocrit in PBS, 300 μL) were divided into three groups: the normal group (treated with PBS), the treatment group (triple combination of curc, vit E, and vit C) and the negative control group (treated with an H_2O_2-related hemolysing dose of 10 mM). For both the normal and negative control groups, suspensions of RBCs were pre-incubated with 100 μL PBS and 6 μL DMSO. In the treatment group, suspensions of RBCs were pre-incubated with 100 μL of a solution containing the triple combination of curc, vit E, and vit C at increasing concentrations (1 μM, 10 μM, 20 μM, 50 μM, 70 μM and 100 μM as the final concentration). In a subsequent experimental set, suspensions of RBCs were pre-incubated with 100 μL of a solution containing a single natural compound or specific double combinations at selected concentrations. Plates were incubated at 37 °C for 30 min. Subsequently, 400 μL of PBS was added to the normal group and 400 μL H_2O_2 solutions (pH 7.4) were added to the treatment and negative control groups. RBCs were then incubated at 37 °C for 1 h. After incubation, cell pellets were collected, immersed for 60 s in an ice bath, and then centrifuged at 1500 g for 5 min at 4 °C. Supernatants were saved and their A was recorded at 540 nm to determine the HR (%), as previously described.

4.3.4. Lipid Peroxidation

Thiobarbituric acid reactive substances (TBARS) known as malondialdehyde (MDA) were measured using the Malondialdehyde (MDA) Colorimetric Assay Kit (Cell Samples) Elabscience (Houston, TX, USA) (#E-BC-K028-M), according to the manufacturer's instructions. MDA is a widely used marker of lipid peroxidation [29]. The absorbance of pink chromophore produced during the reaction of thiobarbituric acid with malondialdehyde was measured at 535 nm.

4.3.5. Statistical Analysis

Statistical procedures were performed with SPSS (version 20). Analysis of the dependent variable (absorbance at 540 nm) was performed with a repeated-measures 1-way ANOVA with "treatment" as a within-subject factor. The normality distribution of residuals was determined with Kolmogorov-Smirnov and Shapiro-Wilk tests. Sphericity was tested with the Mauchly test. If sphericity was violated (Mauchly test < 0.05) the Greenhouse–Geisser correction was applied. The alpha was 0.05. Post-hoc comparisons were performed with dependent *t*-tests corrected with the Bonferroni procedure.

Author Contributions: Conceptualization, G.B. and C.N.; methodology, P.B. and G.B.; validation, G.B., P.B., U.T. and C.N.; formal analysis, G.B.; investigation, G.B. and P.B.; resources, G.B.; data curation, G.B.; writing—original draft preparation, G.B.; writing—review and editing, G.B., P.B. and C.N.; visualization, G.B., P.B. and C.N.; supervision, G.B., P.B., U.T. and C.N.; project administration, G.B. and C.N.; funding acquisition, C.N. All authors have read and agreed to the published version of this manuscript.

Funding: This research was funded by PRIN2020XMLP45 and PNRR-AGE-IT-PE0000015 from the "Italian Ministry of University and Research (MUR)" (PI C.N.) and Ricerca Corrente (RC) 2019 from the "Italian Ministry of Health" (PI C.N.).

Institutional Review Board Statement: This study was conducted in accordance with the Declaration of Helsinki and approved by the Ethics Committee of the University of Campania "L. Vanvitelli" (protocol code 295/2018).

Informed Consent Statement: Informed consent was obtained from all subjects involved in this study.

Data Availability Statement: Data are contained within the article. Raw data and materials are available from the authors upon reasonable request.

Conflicts of Interest: The authors declare no conflicts of interest. The funders had no role in the design of this study; in the collection, analyses, or interpretation of data; in the writing of this manuscript; or in the decision to publish these results.

References

1. De Pascale, M.R.; Belsito, A.; Sommese, L.; Signoriello, S.; Sorriento, A.; Vasco, M.; Schiano, C.; Fiorito, C.; Durevole, G.; Casale, M.; et al. Blood transfusions and adverse acute events: A retrospective study from 214 transfusion-dependent pediatric patients comparing transfused blood components by apheresis or by whole blood. *Ann. Ist. Super. Sanita* **2019**, *55*, 351–356. [CrossRef]
2. Franchini, M.; Marano, G.; Mengoli, C.; Pupella, S.; Vaglio, S.; Muñoz, M.; Liumbruno, G.M. Red blood cell transfusion policy: A critical literature review. *Blood Transfus.* **2017**, *15*, 307–317. [CrossRef]
3. Bardyn, M.; Tissot, J.D.; Prudent, M. Oxidative stress and antioxidant defenses during blood processing and storage of erythrocyte concentrates. *Transfus. Clin. Biol.* **2018**, *25*, 96–100. [CrossRef] [PubMed]
4. Yoshida, T.; Prudent, M.; D'alessandro, A. Red blood cell storage lesion: Causes and potential clinical consequences. *Blood Transfus.* **2019**, *17*, 27–52. [CrossRef]
5. Bardyn, M.; Allard, J.; Crettaz, D.; Rappaz, B.; Turcatti, G.; Tissot, J.D.; Prudent, M. Image- and fluorescence-based test shows oxidant-dependent damages in red blood cells and enables screening of potential protective molecules. *Int. J. Mol. Sci.* **2021**, *22*, 4293. [CrossRef]
6. D'Alessandro, A.; Zimring, J.C.; Busch, M. Chronological storage age and metabolic age of stored red blood cells: Are they the same? *Transfusion* **2019**, *59*, 1620–1623. [CrossRef]
7. Yang, Q.; Noviana, M.; Zhao, Y.; Chen, D.; Wang, X. Effect of curcumin extract against oxidative stress on both structure and deformation capability of red blood cell. *J. Biomech.* **2019**, *95*, 109301. [CrossRef] [PubMed]
8. Antosik, A.; Czubak, K.; Cichon, N.; Nowak, P.; Zbikowska, H. Vitamin E analogue protects red blood cells against storage-induced oxidative damage. *Transfus. Med. Hemother.* **2018**, *45*, 347–354. [CrossRef] [PubMed]
9. Czubak, K.; Antosik, A.; Cichon, N.; Zbikowska, H.M. Vitamin C and Trolox decrease oxidative stress and hemolysis in cold-stored human red blood cells. *Redox Rep.* **2017**, *22*, 445–450. [CrossRef] [PubMed]

10. Remigante, A.; Spinelli, S.; Straface, E.; Gambardella, L.; Caruso, D.; Falliti, G.; Dossena, S.; Marino, A.; Morabito, R. Antioxidant activity of quercetin in a H_2O_2-induced oxidative stress model in red blood cells: Functional Role of Band 3 Protein. *Int. J. Mol. Sci.* **2022**, *23*, 10991. [CrossRef]
11. Remigante, A.; Spinelli, S.; Basile, N.; Caruso, D.; Falliti, G.; Dossena, S.; Marino, A.; Morabito, R. Oxidation stress as a mechanism of aging in human erythrocytes: Protective effect of quercetin. *Int. J. Mol. Sci.* **2022**, *23*, 7781. [CrossRef] [PubMed]
12. Ravikumar, S.; Prabhu, S.; Vani, R. Effects of L-carnitine on the erythrocytes of stored human blood. *Transfus. Med.* **2020**, *30*, 215–225. [CrossRef]
13. Remigante, A.; Spinelli, S.; Straface, E.; Gambardella, L.; Caruso, D.; Falliti, G.; Dossena, S.; Marino, A.; Morabito, R. Açaì (*Euterpe oleracea*) extract protects human erythrocytes from age-related oxidative stress. *Cells* **2022**, *11*, 2391. [CrossRef]
14. Remigante, A.; Spinelli, S.; Patanè, G.T.; Barreca, D.; Straface, E.; Gambardella, L.; Bozzuto, G.; Caruso, D.; Falliti, G.; Dossena, S.; et al. AAPH-induced oxidative damage reduced anion exchanger 1 (SLC4A1/AE1) activity in human red blood cells: Protective effect of an anthocyanin-rich extract. *Front. Physiol.* **2023**, *14*, 1303815. [CrossRef] [PubMed]
15. Maresca, V.; Capasso, L.; Rigano, D.; Stornaiuolo, M.; Sirignano, C.; Piacente, S.; Cerulli, A.; Marallo, N.; Basile, A.; Nebbioso, A.; et al. Health-promoting effects, phytochemical constituents and molecular genetic profile of the purple carrot 'Purple Sun' (*Daucus carota* L.). *Nutrients* **2024**, *16*, 2505. [CrossRef] [PubMed]
16. Maresca, V.; Badalamenti, N.; Ilardi, V.; Bruno, M.; Bontempo, P.; Basile, A. Chemical composition of thymus leucotrichus var. creticus essential oil and its protective effects on both damage and oxidative stress in leptodictyum riparium hedw. induced by cadmium. *Plants* **2022**, *11*, 3529. [CrossRef] [PubMed]
17. Bontempo, P.; Capasso, L.; De Masi, L.; Nebbioso, A.; Rigano, D. Therapeutic potential of natural compounds acting through epigenetic mechanisms in cardiovascular diseases: Current findings and future directions. *Nutrients* **2024**, *16*, 2399. [CrossRef]
18. Benincasa, G.; Coscioni, E.; Napoli, C. Cardiovascular risk factors and molecular routes underlying endothelial dysfunction: Novel opportunities for primary prevention. *Biochem. Pharmacol.* **2022**, *202*, 115108. [CrossRef]
19. Morabito, R.; Falliti, G.; Geraci, A.; Spada, G.L.; Marino, A. Curcumin protects -SH groups and sulphate transport after oxidative damage in human erythrocytes. *Cell. Physiol. Biochem.* **2015**, *36*, 345–357. [CrossRef]
20. Bjørklund, G.; Shanaida, M.; Lysiuk, R.; Butnariu, M.; Peana, M.; Sarac, I.; Strus, O.; Smetanina, K.; Chirumbolo, S. Natural Compounds and Products from an Anti-Aging Perspective. *Molecules* **2022**, *27*, 7084. [CrossRef]
21. Yang, L.; Chen, Y.; He, S.; Yu, D. The crucial role of NRF2 in erythropoiesis and anemia: Mechanisms and therapeutic opportunities. *Arch. Biochem. Biophys.* **2024**, *754*, 109948. [CrossRef] [PubMed]
22. Ghashghaeinia, M.; Toulany, M.; Saki, M.; Bobbala, D.; Fehrenbacher, B.; Rupec, R.; Rodemann, H.P.; Ghoreschi, K.; Röcken, M.; Schaller, M.; et al. The NFκB pathway inhibitors Bay 11-7082 and parthenolide induce programmed cell death in anucleated Erythrocytes. *Cell. Physiol. Biochem.* **2011**, *27*, 45–54. [CrossRef] [PubMed]
23. Vasco, M.; Costa, D.; Scognamiglio, M.; Signoriello, G.; Alfano, R.; Magnussen, K.; Napoli, C. Evidences on overweight of regular blood donors in a center of Southern Italy. *Clin. Epidemiol. Glob. Health* **2020**, *8*, 758–763. [CrossRef]
24. Vasco, M.; Signoriello, G.; Scognamiglio, M.; Moccia, G.; Filauri, P.; Sansone, A.; Matarazzo, I.; Fiorito, C.; Grimaldi, V.; Viglietti, M.; et al. Reduced levels of hepcidin associated with lower ferritin concentration and increased number of previous donations in periodic blood donors: A pilot study. *Transfus. Clin. Biol.* **2023**, *30*, 319–323. [CrossRef]
25. Belsito, A.; Costa, D.; Signoriello, S.; Fiorito, C.; Tartaglione, I.; Casale, M.; Perrotta, S.; Magnussen, K.; Napoli, C. Clinical outcome of transfusions with extended red blood cell matching in β-thalassemia patients: A single-center experience. *Transfus. Apher. Sci.* **2019**, *58*, 65–71. [CrossRef]
26. Albano, L.; Vasco, M.; Benincasa, G.; Sommese, L.; Alfano, R.; Napoli, C. Pathogenic mechanisms and current epidemiological status of HEV infection in asymptomatic blood donors and patients with chronic diseases. *Eur. J. Clin. Microbiol. Infect. Dis.* **2019**, *38*, 1203–1209. [CrossRef]
27. Qasim, N.; Mahmood, R. Diminution of Oxidative damage to human erythrocytes and lymphocytes by creatine: Possible role of creatine in blood. *PLoS ONE* **2015**, *10*, e0141975. [CrossRef] [PubMed]
28. Ambrosio, G.; Oriente, A.; Napoli, C.; Palumbo, G.; Chiariello, P.; Marone, G.; Condorelli, M.; Chiariello, M.; Triggiani, M. Oxygen radicals inhibit human plasma acetylhydrolase, the enzyme that catabolizes platelet-activating factor. *J. Clin. Investig.* **1994**, *93*, 2408–2416. [CrossRef] [PubMed]
29. Napoli, C.; Mancini, F.P.; Corso, G.; Malorni, A.; Crescenzi, E.; Postiglione, A.; Palumbo, G. A simple and rapid purification procedure minimizes spontaneous oxidative modifications of low density lipoprotein and lipoprotein (a). *J. Biochem.* **1997**, *121*, 1096–1101. [CrossRef] [PubMed]

International Journal of *Molecular Sciences*

MDPI

Review

Harnessing *Moringa oleifera* for Immune Modulation in Cancer: Molecular Mechanisms and Therapeutic Potential

Mounir Tilaoui [1,*], Jamal El Karroumi [2], Hassan Ait Mouse [3] and Abdelmajid Zyad [3]

1 Laboratory of Biotechnology, Environment, Agrifood, and Health, Department of Biology, Faculty of Sciences Dhar El Mahraz, Sidi Mohamed Ben Abdellah University, Fez 30000, Morocco

2 Laboratory of Biomolecular Chemistry, Natural Substances and Reactivity, Faculty of Sciences Semlalia, Cadi Ayyad University, Marrakech 40000, Morocco; jamalelkarroumi@gmail.com

3 Laboratory of Agro-Industrial and Medical Biotechnology, Team of Experimental Oncology and Natural Substances, Cellular and Molecular Immuno-Pharmacology, Faculty of Sciences and Technology, Sultan Moulay Slimane University, BP 523, Beni-Mellal 23000, Morocco; aitmouse@gmail.com (H.A.M.); a.zyad@usms.ma (A.Z.)

* Correspondence: mounir.tilaoui@usmba.ac.ma

Abstract

Moringa oleifera, widely recognized as the horseradish tree or drumstick tree, is classified within the *Moringaceae* family, which comprises 13 species predominantly distributed across tropical and subtropical regions. The plant possesses a variety of therapeutic, nutritional, and beneficial health properties, including its potential to enhance the immune system. The present work provides extensive bibliographic research addressing the chemical composition of *Moringa oleifera* and its immunomodulatory properties with a focus on the cellular and molecular mechanisms involved in the regulation of immune function, which is crucial in unchecked cell proliferation and metastasis. The chemical composition of *Moringa oleifera*, including kaempferol, chlorogenic acid, quercetin, and niazimicin, varies between different biological parts of the plant (seeds, leaves, roots, and stems). The presence of these various chemical compounds contributes to the plant's effect on the immune response via different pathways. Several studies indicate that *Moringa oleifera* mitigates inflammation by suppressing key pro-inflammatory mediators, such as TNF-α, IL-1β, inducible nitric oxide synthase (iNOS), prostaglandin E2 (PGE-2), and cyclooxygenase-2 (COX-2), while simultaneously enhancing anti-inflammatory mediators through activation of PPAR-γ. Furthermore, the immunomodulatory properties and possible application in health promotion and disease prevention, especially in cancer therapy, are discussed. Studies indicate that *Moringa oleifera* can modulate the tumor microenvironment (TME) by reducing Treg polarization, enhancing NK cell cytotoxicity, and prompting the proliferation and clonal expansion of $CD8^+$ and $CD4^+$ T lymphocytes. Together, *Moringa oleifera* could be considered for the treatment of conditions related to immune dysregulation, such as cancer.

Keywords: *Moringa oleifera*; chemical composition; immune system; cancer; immunomodulation; molecular mechanisms

1. Introduction

The treatment of immune-mediated inflammatory diseases, such as asthma, allergies and autoimmune diseases due to disorders of the immune system requires the modulation of the innate and adaptive immune responses [1]. Immunomodulation refers to the regulation or alteration of the immune system's activity in response to a given internal or external stimulus, often achieved through the administration of specific compounds or drugs [2]. There are numerous chemical immunomodulators, including immunosuppressive and immunostimulatory agents such as dexamethasone, fluticasone, prednisolone, prednisone, and hydrocortisone, which have been used to treat and modulate various inflammatory diseases.

Recombinant therapeutic proteins have gained increasing importance as promising treatments for immunodeficiency, infectious diseases, and cancer [3–5]. Cyclosporin A is an example of a fungal cyclic peptide widely used as an immunosuppressant drug to prevent rejection in organ-transplant recipients and to treat patients with autoimmune diseases. Nevertheless, most of these conventional drugs have been associated with several adverse effects [6,7]. Duodenal mucosa and gastric mucosal epithelium damage are the commonest adverse effects caused by non-steroidal anti-inflammatory drugs [8–11]. Other side effects, notably increased skin fragility and reduced bone marrow function, have been linked to the use of immunosuppressive drugs such as corticosteroids [12–15].

Cyclosporin A is known to cause a variety of adverse metabolic side effects and toxicities, including hypomagnesemia, hypertension, hypercalciuria, hyperkalemia, hepatotoxicity, nephrotoxicity, and gingival hypertrophy [16]. Therefore, new, safer, and more effective drugs are required as alternatives to current treatments.

Natural products are still a significant source of leads for new immunomodulatory drugs. Many plants were reported to exhibit immunomodulatory activities and to modulate immune functions. The present work provides extensive bibliographic research addressing the chemical composition of *Moringa oleifera* and its effect on the immune system, exploring its cellular activities and its molecular impacts on immunoglobulins, regulatory pathways, and cytokines.

2. Methods/Literature Search Strategy

This comprehensive narrative review was conducted to explore the immunomodulatory and therapeutic potential of *Moringa oleifera*, especially on cancer therapy, with an emphasis on molecular mechanisms, and the phytochemical composition of its various parts (leaves, seeds, pods, flowers, and roots). Relevant studies published primarily between 2020 and 2025 were conducted across databases of PubMed/Medline, Scopus, Web of Science, Google Scholar, and Cochrane Library to identify pertinent articles. The search terms used included "*Moringa oleifera*" "chemical composition" "immune system" "immune modulation" "ant-inflammatory" "cytokines" "immunotherapy" "cancer" and "molecular mechanism". Boolean operators (AND, OR) were used to effectively optimize the combination of these keywords for the literature research.

Studies were selected based on their relevance to preclinical, mechanistic, and clinical evidence, prioritizing peer-reviewed reports. Articles lacking methodological rigor, sufficient experimental detail, or unrelated to the topic were excluded. Although a formal risk-of-bias assessment was not conducted, priority was given to high-quality experimental and clinical studies to ensure the robustness, reliability, and validity of the synthesized findings.

3. Chemical Composition of *Moringa oleifera*

Each component of the plant, the leaf, root, bark, seed, flower, and pod, contains important nutritional resources [17]. Over 100 chemical constituents have been identified in *Moringa oleifera*, including proteins, fats, vitamins, minerals, carbohydrates, and dietary fibers [18]. Moreover, a multitude of phytochemicals have been identified in various parts of this plant, each offering diverse health-promoting properties [19,20].

https://doi.org/10.3390/ijms27010263

3.1. Seeds

The seeds are spherical and enclosed within a semi-permeable brown seed coat, characterized by three papery white wings. A single tree can produce approximately 15,000 to 25,000 seeds per year [21,22], each with an average mass of 0.3 g, of which about 25% is attributed to the seed coat [23]. The oil extracted from *Moringa oleifera* seeds, globally known as Ben oil, has a yield of approximately 40% [24]. According to Corbett et al., the seed oil of *Moringa oleifera* is of the oleic type with a high content of oleic acid [25]. This characteristic is particularly advantageous in the context of growing preference for replacing polyunsaturated vegetable oils with those rich in monounsaturated fatty acids, which exhibit high oxidative and thermal stability, even under high-temperature conditions such as frying. *Moringa oleifera* seed oil contains all the major fatty acids present in olive oil and is also distinguished by its notable concentration of behenic acid (C22:0), ranging from 5 to 6%. In addition, it contains small amounts of gadoleic acid (C20:1) and lignoceric acid (C24:0), along with trace levels of erucic acid (C22:1) and cerotic acid (C26:0). Figure 1 illustrates the main fatty acids found in the oil of *Moringa oleifera* seeds [26]. A study has shown that *Moringa* oil is rich in tocopherols, which are fat-soluble vitamins and powerful natural antioxidants [27].

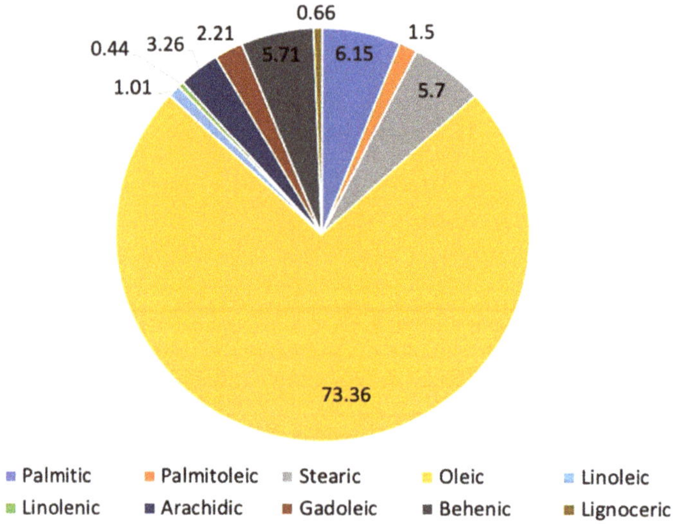

Figure 1. Major fatty acids identified in *Moringa oleifera* seeds.

3.2. Leaves

Nutritional analyses have shown that *Moringa oleifera* leaves are rich in vitamins, minerals, and proteins [28]. Through the process of hydrodistillation, the leaves of *Moringa oleifera* yielded a pale-yellow oil, accounting for 0.05% of the dry mass [29]. The first studies on *Moringa oleifera* date back to the 1970s. Recent comprehensive phytochemical analyses have identified more than 90 to over 100 compounds in *Moringa oleifera* leaves, including phenols, flavonoids, amino acids, and sugars, highlighting its diverse bioactive profile [20,30]. In 1982, several sugars such as L-arabinose, D-galactose, glucuronic acid, L-rhamnose, D-mannose, and D-xylose were identified in the following proportions: 14.5:11.3:3:2:1:1 [31]. Subsequently, in 1994, two nitrile glycosides, niazirin and niazirinin, were isolated by Shaheen Faizi et al. [32]. The same authors successfully identified glucosinolates, which, upon enzymatic hydrolysis by myrosianse, can produce isothiocyanates such as 4-[(α-L-rhamnosyloxy) benzyl]isothiocyanate, known for their various biological

https://doi.org/10.3390/ijms27010263

activities [33–36]. *Moringa oleifera* is known for its antioxidant activities, which are attributed to the presence of several flavonoids such as rutin, kaempferol, and quercetin [37] (Figure 2).

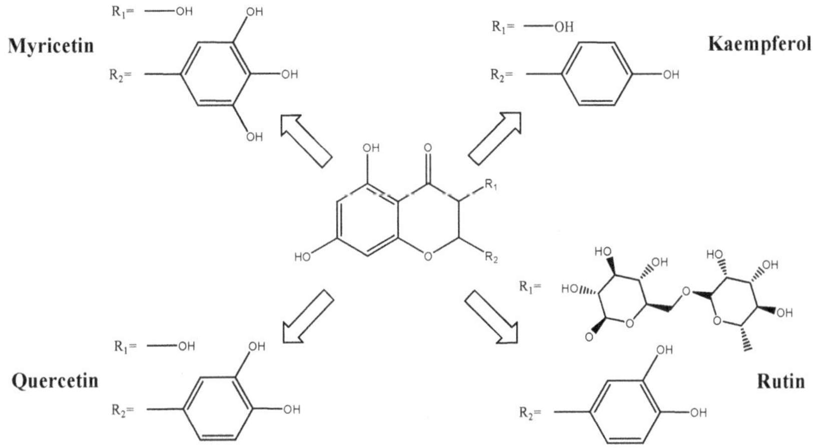

Figure 2. Chemical structures of representative flavonoids from *Moringa oleifera* leaves.

Recent investigations suggest that isothiocyanate and astragalin serve as key chemical markers for standardizing *Moringa oleifera* extracts [38], given their well-documented anti-inflammatory and antioxidant properties [39]. To quantify isothiocyanate, a specific HPLC-based cyclocondensation method was developed and applied to analyze extracts from various plant parts, such as leaves, immature seeds, mature seeds, and pods. The leaves of *Moringa oleifera* are particularly abundant in bioactive compounds, among which phenolic acids represent a prominent class [40]. Several studies have been conducted to identify and quantify these compounds with *Moringa oleifera* leaf extract. Among the most frequently reported phenolic acids are gallic acid, ferulic acid, and o-coumaric acid [41]. Figure 3 illustrates the major phenolic acids identified in *Moringa oleifera* leaves.

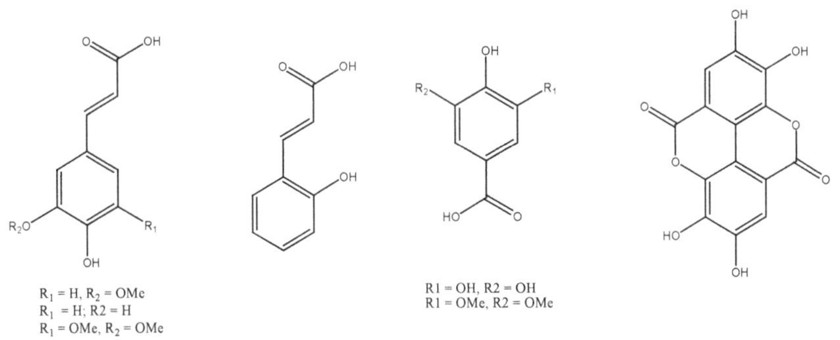

Figure 3. Structures of key phenolic acids present in *Moringa oleifera* leaves.

3.3. Roots

While the leaves of *Moringa oleifera* are the most commonly utilized, the roots also exhibit notable nutritional and therapeutic properties [19,42]. The root composition includes carbohydrates, proteins, minerals, and dietary fiber, along with important levels of antioxidant compounds such as phenolics and flavonoids [43]. Although these bioactive

https://doi.org/10.3390/ijms27010263

constituents are present in smaller concentrations compared to the leaves, the roots still provide essential vitamins, notably vitamin C, vitamin E, and multiple B-complex vitamins.

3.4. Stems

In addition to the roots, other parts of *Moringa oleifera*, including the stem, bark, and flowers, also exhibit a rich profile of bioactive phytochemicals [44]. While their chemical composition largely overlaps with that of leaves, these plant components represent significant sources of alkaloids, flavonoids, and glycosides. Moreover, compounds including steroids, terpenoids, and phenolic acids have also been identified [45]. It is worth noting that the chemical composition may be influenced by several factors, such as the plant's geographical origin, age, genetic variability, and the extraction techniques employed [46–49].

4. The Impact of *Moringa oleifera* on the Immune System

4.1. *Moringa oleifera* Effect on Innate and Adaptive Immunity

Moringa oleifera is a widely known medicinal plant indigenous to the Indian subcontinent. It has been used for centuries in Ayurvedic medicine for its healing, nutritional, rejuvenating, and preventive properties [30,50,51]. Recent investigations have underscored its potential in modulating immune system activities [52–56]. The plant harbors various bioactive compounds, including flavonoids, polyphenols, and ascorbic acid, which contribute to its immunomodulatory effects [56,57].

Evidence suggests that *Moringa oleifera* extracts can increase the production of immune cells such as lymphocytes and macrophages [58–61]. This is important, as these cells play a critical role in fighting infections within the body. Notably, *Moringa oleifera* exhibits anti-inflammatory activity that reduces inflammation, thereby promoting improved immune function [62]. Overall, the research indicates that *Moringa oleifera* exerts beneficial effects on both innate and adaptive immune systems (Tables 1 and 2). To better elucidate the mechanisms by which *Moringa oleifera* modulates immunity, we investigated the associated cellular responses and cytokine signaling in both innate and adaptive immunity.

Table 1. Effect of *Moringa oleifera* on innate immunity.

Plant Part Used	Form of Application	Model/System	Key Mechanistic Insights	Ref
Leaves	Leaf polysaccharide fraction	Lewis lung carcinoma (C57BL/6 mice, oral). In vitro bone marrow-derived macrophages	Activation of macrophage through toll-like receptor reprograms TAMs from M2 to M1 through TLR4, ↑ CXCL9/10, ↑ T-cell infiltration	[58,63]
Leaves	Oral administration	Rabbits exposed to heat stress	Enhancement of NK cell activity through up-regulation of perforin/granzyme secretion	[64–66]
Leaves	Purified protein fraction	Murine BMSCs (In vitro), followed by allogenic MLR and in vivo IgE measurement in mice	Activation and maturation of DCs via upregulation of co-stimulatory molecules and the secretion of cytokines and chemokines: ↑ CD80/CD86/MHCII; ↑ IL12 and TNF-α secretion; OX40L-TIM-4-CCL17/22 axis drives Th2 polarization; enhanced T-cell proliferation	[60,67,68]
Leaves	Oral administration	Diabetic-induced damage in male Wistar rats	Reduction in pro-inflammatory cytokines (TNF-α, IL6) and oxidative stress	[68]
Leaves	Topical ethanolic extract (10%) Moringa tea (oral drinking water)	Wistar rats with experimental burn wounds Restraint-stressed mice	Modulation of neutrophil migration/chemotaxis and enhancing phagocytosis	[59,69]

https://doi.org/10.3390/ijms27010263

Table 1. *Cont.*

Plant Part Used	Form of Application	Model/System	Key Mechanistic Insights	Ref
Leaves	Oral administration of ethanolic leaf extract at 10, 30 and 100 mg/Kg for 7 days	*Staphylococcus aureus*-challenged mice (mouse peritoneal macrophage)	↑ Macrophage phagocytic activity and capacity; ↑ Neutrophile percentage	[70]
Leaves	Purified polysaccharide fraction	RAW 264.7 macrophages (in vitro)	Enhances the gene expression of antimicrobial peptides such as Mop3 through ↑ pinocytosis; ↑ ROS/NO; ↑ IL6; TNF-α and iNOS expression	[71]
Pods	Boiled extract	RAW 264.7 murine macrophage cell line	Inhibits pro-inflammatory cytokines (IL6, TNF-α); suppresses iNOS and COX-2 expression; decreases NO production	[72]
Leaves	Ethyl acetate fraction	LPS-stimulated RAW 264.7 macrophages	Regulation of genes encoding pro-inflammatory cytokines through the regulation of MAPK and NF-κB signaling cascades	[73]
Seeds	Purified peptide (sequence KETTTIVR) administered orally	Dextran Sulfate Sodium-induced colitis in mice	Modulates gut microbiota and metabolomic profiles, inhibits JAK-STAT signaling, and protects the intestinal barrier	[74]
Leaves	Ethanolic extract; oral and in vitro exposure	Swiss albino mice peritoneal macrophages; RAW 264.7	Increase phagocytosis activity through activation of phagocytic receptors and engulfment efficiency	[70]
Roots	Hot water extract, ethanolic extract	LPS-induced RAW 264.7 macrophages murine macrophage cell line	Inhibits NO and TNF-α production; decreases iNOS mRNA production	[75]
Pods	Freeze-dried pod polyphenol extract	RAW 264.7 macrophages (LPS-stimulation)	↓ NO, ↓ TNF-α, strong anti-inflammatory effect	[76]
Leaves	Ethyl acetate extract (oral and in vitro)	RAW 264.7 macrophages	Inhibits NF-κB, suppresses pro-inflammatory cytokines (TNF-α, IL6); ↓ NO production	[77]
Pods	Digested boiled pod extract	Caco-2 cells (intestinal epithelial)	Suppressed inflammatory mediators; promoted epithelial anti-inflammatory defense	[78]
Leaves (aqueous and ethanolic extracts)	Oral administration and in vitro exposure	Various animal models (Wistar rats and BALB/c mice) and the RAW 264.7 macrophage cell line	Anti-inflammatory and antioxidant activities via free radical scavenging and inhibition of pro-inflammatory cytokines such as IL-1β, IL-6 and TNF-α	[68,77,79,80]

BMSCs, Bone marrow-derived mesenchymal stem cells; CCL, C-C motif chemokine ligand; CXCL, C-X-C motif chemokine ligand; DCs, Dendritic cells; iNOS, Inducible nitric oxide synthase; JAK, Janus kinase; LPS, Lipopolysaccharide; MAPK, Mitogen-activated protein Kinase; NF-κB, Nuclear factor kappa-light-chain-enhancer of activated B cells; OX40, OX40 ligand; STAT, Signal transducer and activator of transcription; TIM, T cell immunoglobulin and mucin domain containing protein 4. ↑ upregulate/increase expression; ↓ downregulate/decrease expression.

https://doi.org/10.3390/ijms27010263

Table 2. Effect of *Moringa oleifera* on adaptive immunity.

Plant Part Used	Form of Application	Model/System	Key Mechanistic Insights	Ref
Leaves	Purified leaf protein	In vitro, Human T-lymphoblastic leukemia (Jurkat cells)	Modulation of T cell activation and proliferation through the regulation of T cell receptor signaling cascades	[60,81,82]
Leaves	Aqueous extract	In vitro, murine splenocyte culture	B-cell activation	[73]
Leaves	Methanolic extract	SRBC immunized Wistar rats	Antibody production via B cell receptor stimulation	[83–85]
Leaves	Ethanolic extract	Breast cancer cells (MD-MA-231)	↓ of NF-κB and p65; Transcription factor modulation; ↓ cancer cell viability	[86]
Leaves	Topical application	Atopic dermatitis mouse model	Promote T-helper cell differentiation through modulation of Th1, Th2, Th17 and Treg cells responses	[61,67,87–89]
Leaves Leaf proteins	Oral administration Protein extract	Rabbits exposed to heat stress Murine BMDCs	Induction of Foxp3 expression and promotion of Treg cells	[60,64]
Pods	Pod meal (dietary inclusion)	Broiler chickens	Enhanced growth performance; improved cell-mediated immunity	[90]
Leaves	Leaf protein fraction or extract	BALB/c mice (intraperitoneal sensitization + oral exposure, allergy model); BMDCs (in vitro) and IgE induction after DCs transfer (in vivo)	Regulation of antigen presentation and processing	[60,88,91]
Leaves	Methanolic extract	Wistar albino rats	Stimulates neutrophil and lymphocyte function; Enhances humoral immunity; increases antigen-specific antibody secretion and WBC levels	[85]
Seeds	MicroRNA-enriched extract	Human PBMCs (HIV$^+$)	Modulates T-cell differentiation and memory T-cell subsets; reduces HIV replication	[92]
Leaves	Protein extract (oral and in vitro)	BALB/c mice; Murine BMDCs	Induces IgE production and Th2 polarization; activates DC; modulates humoral immunity by enhancing the production of antigen-specific antibodies	[60,93–95]

BMDCs, Bone marrow-derived dendritic cells; DCs, Dendritic cells; Foxp3, Forkhead box P3 gene; PBMCs, Peripheral blood mononuclear cells; SRBC, Sheep red blood cells; WBC, White blood cells. ↓ downregulate/decrease expression.

4.2. Cellular Responses and Cytokine Pathways in Innate and Adaptive Immunity

Recent evidence identifies *Moringa oleifera* as a source of bioactive compounds with immunomodulatory effect [54,58,60,81]. *In vivo*, *Moringa oleifera* leaf extract enhanced natural killer cell activity and stimulated cytokine production [64], highlighting its potential to modulate innate immune surveillance with potential translational relevance for immune-based therapies.

The modulatory properties of *Moringa oleifera* on immune cell activation were confirmed by Zhuping Dong et al. [96]. The result shows through an in vitro experiment that the polysaccharide MOP-1 (glucose, galactose, and arabinose) enhances the proliferation and activation of immune cells through increasing the production of reactive oxygen species, nitric oxide, along with pro-inflammatory mediators IL-6 and TNF-α through upregulation of their mRNA expression in RAW 264.7 cells (Figure 4) [96].

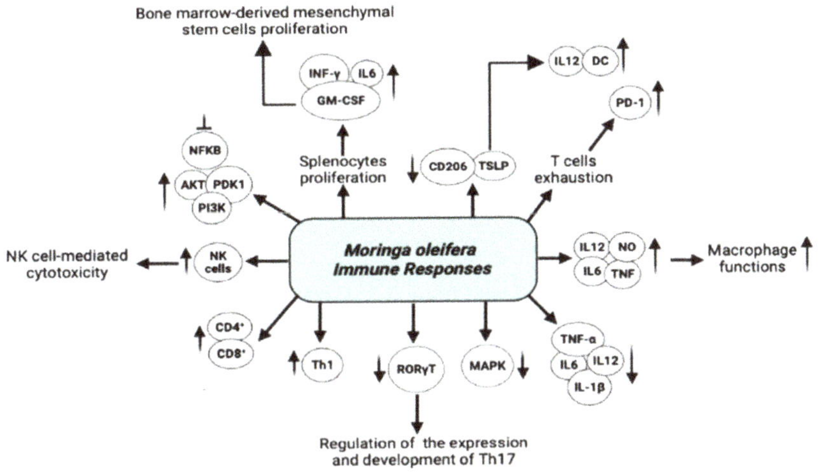

Figure 4. *Moringa oleifera* regulates diverse factors to modulate immune responses. *Moringa oleifera* modulates immunity through interconnected cellular and molecular pathways. Innate immunity is enhanced via neutrophile chemotaxis, macrophage phagocytic and secretory functions, NK cell-mediated cytotoxicity and the activation of bone marrow-derived mesenchymal stem cells that contribute to immune repair and cytokine balance. Upregulation of GM-CSF and NO further amplifies myeloid and effector cell activity. Adaptive immunity is regulated through $CD4^+$ T-cell differentiation, balancing Treg/Th17 ratios, enhancing Th2 functions via RORγT, leading to increased IL-10 and TGF-β and reduced IL-6, IL-1beta and TNF-α. Key signaling pathways, including NF-κB, MAPK, PI3K/Akt, PDK1, STAT, TSLP, and PD-1 are modulated to maintain immune homeostasis, suppression of chronic inflammation and regulation of programmed cell death. AKT, Protein Kinase B; DC, Dendritic cell; GM-CSF, Granulocyte-Macrophage Colony-Stimulating Factor; IFN, interferon; MAPK, Mitogen-activated protein Kinase; NFκB, Nuclear factor kappa-light-chain-enhancer of activated B cells; NK, Natural killer; NO, Nitric oxide; PD-1, Programmed cell death protein 1; PDK1, 3-Phosphoinositide-dependent protein kinase-1; PI3K, Phosphoinositide 3-Kinase; RORγT, Retinoic acid receptor-related orphan receptor gamma t; STAT, Signal transducer and activator of transcription; TNF, Tumor necrosis factor; TSLP, Thymic stromal lymphopoietin. ⟶ Promote/lead to; ↑ upregulate/increase expression; ↓ downregulate/decrease expression. ⊥ inhibition.

The *Moringa oleifera* roots polysaccharide MRP-1 was also reported to significantly potentiate anti-inflammatory effects observed in both in vitro and in vivo models via attenuating the release of nitric oxide and iNOS mRNA expression induced by TNF-α, suggesting its potential use as a natural treatment for inflammatory diseases (Figure 4) [97]. Building on the characterization of cellular responses and cytokine pathways, we next explored how these mechanisms orchestrate multilevel modulation of innate and adaptive immunity.

4.3. Multilevel Modulation of Innate and Adaptive Immunity

4.3.1. Gut Microbiota Modulation

Moringa oleifera promotes a balanced gut microbiota, enhancing immune homeostasis, reducing inflammation, and improving immune function [56]. It has also been demonstrated in animal experimental models that *Moringa oleifera* polysaccharides participate in the homeostasis of the gut microbiota by enhancing the proliferation of beneficial bacterial populations in the human gut, including *Lactobacillus* species, and a decrease in the abundance of pathogenic bacteria in C57BL/6 mice. *Moringa oleifera* polysaccharides

(MOPs) modulate gut microbiota through prebiotic activity, selectively enriching beneficial bacteria, including *Bacteroides* and *Lactobacillus*, while inhibiting pathogenic bacteria, thereby enhancing microbial diversity and homeostasis [98–100]. Fermentation of MOPs by gut microbiota produces short-chain fatty acids (SCFAs) such as butyrate, acetate and propionate, which act via G-protein-coupled receptors (GPR41/43) to strengthen intestinal barrier integrity, regulate mucosal immunity, and suppress pro-inflammatory NF-κB signaling pathways [101–103]. Furthermore, MOPs influence the gut-associated lymphoid tissue (GALT), enhancing a balanced Treg/Th17 cell ratio and promoting anti-inflammatory cytokines production (e.g., IL-10, TGF-β and IL-35) while decreasing pro-inflammatory cytokines (e.g., IL-6, TNF-α) [104–107].

Together, these mechanisms form an integrated bidirectional gut-immune network, linking microbial communities to systemic immune homeostasis and highlighting the therapeutic potential of MOPs in managing inflammatory and immune-related disorders [98,106–108]. The composition of gut microbiota plays a central role in shaping immune responses and preserving systemic homeostasis. *Moringa oleifera* supports this balance through promoting beneficial microbes, which in turn enhances anti-inflammatory signaling and immune homeostasis.

4.3.2. Immune Modulation and Anti-Inflammatory Balance

Moringa oleifera promotes immune homeostasis through enhancing anti-inflammatory cytokines while suppressing proinflammatory mediators. Its bioactive compounds act via key signaling pathways, including MAPK, NF-κB and PI3K/AKT/mTOR [35,109–111], thereby mitigating tissue damage and improving outcomes in inflammatory disorders. Additionally, *Moringa oleifera* polysaccharides were found to modulate the immune system, to increase the levels of anti-inflammatory cytokines, and reduce the inflammatory bowel disease [102]. Moreover, administration of *Moringa* leaf extract at a concentration of 100 mg/kg to mice enhances the production of eosinophil cells, immature neutrophils, and lymphocytes while reducing the amount of segmented neutrophils and monocytes [70]. Another study investigates the impact of *Moringa oleifera* leaf extract on proliferation capacity, physiological parameters, immune responses, and the expression of immunity-related genes in prawns exposed to both *Vibrio anguillarum* and ammonia stress [112]. The results showed that administration of 0.5% *Moringa oleifera* in prawns -after being exposed to ammonia stress- increases the levels of haemolymph SOD, glutathione peroxidase, NO, iNOS, and HSP70 upregulation in hepatopancreas while reducing the levels of haemolymph catalase, peroxiredoxin 5 in hepatopancreas, and nuclear factor-kappa-B-inhibitor alpha (IκB-α) expression compared to the control group. These findings indicate that *Moringa oleifera* leaf extract has a protective effect against ammonia-induced stress [112]. Similarly, Z Abidin et al. demonstrated that feeding whiteleg shrimp with 2.5 g of *Moringa oleifera* leaves' extract significantly enhanced the innate immunity of the shrimp by increasing the production of total hemocyte count, phenoloxidase, phagocytic rate, phagocytic index, and superoxide anion. Also, the extract improved the shrimp's resistance against *Vibrio alginolyticus*, a common pathogen in shrimp farming [113]. These findings suggest that consumption of moringa in the diet may increase the upregulation of certain genes associated with the immune system involved in catalytic activity, antimicrobial defense, antioxidant enzyme activity, and blood clotting protein production.

In agreement with the previous studies, *Moringa oleifera* leaves ethanolic extract exhibits immunomodulatory effects in broiler chicks at a dose of 200 mg/L in drinking water by increasing differential and total leukocyte counts, enhancing both the phagocytic index and phagocytic activity, elevating serum hemagglutination inhibition antibody levels against Newcastle disease virus, and reducing the mortality rate. In addition, it promotes

https://doi.org/10.3390/ijms27010263

the immune system function of the Bursa of Fabricius, spleen, and thymus indices, and stimulates lymphocytes proliferation induced by the *Moringa oleifera* leaves ethanolic extract [114]. These findings into multilevel immune regulation provide a foundation for assessing how *Moringa oleifera* engages these mechanisms to exert immunomodulatory and anti-inflammatory effects.

4.4. Immunomodulatory and Anti-Inflammatory Role of Moringa oleifera

Moringa oleifera has gained increasing attention as a bioactive plant for its promising immunomodulatory and anti-inflammatory properties [79,80,102,115,116], with accumulating evidence demonstrating its ability to regulate essential cytokine-mediated pathways [79,102,116–118], modulate inflammatory signaling cascades [74,102,116,119,120], restore immune homeostasis, and improve overall immune functions [50,56,82,102,107,116,121,122]. These effects have been observed across various experimental models and clinical settings [123–127], underscoring its promise as a natural therapeutic intervention for the prevention and management of chronic inflammatory and immune-mediated conditions.

The protective effects of *Moringa oleifera* are mediated through multiple molecular pathways. Recent studies have elucidated that its phytoconstituents exhibit multifaceted mechanisms, including the suppression of pro-inflammatory mediators such as TNF-α, IL-1β, iNOS, PGE-2, and COX-2 [76,79,111,128], alongside the activation of anti-inflammatory signaling pathways through PPAR-γ, NF-κB and MAPK [107,119,129–131]. Beyond innate immune regulation, *Moringa oleifera* influences adaptive immunity by enhancing antibody responses and modulation the PI3K/AKT/mTOR axis [34,81,119,132,133], an important regulator of cellular homeostasis and immune metabolism. Supported by these mechanistic observations, an important in vivo study examined the potential protective properties of *Moringa oleifera* ethanolic leaf extract (MOE) in *Oreochromis niloticus* to mitigate oxidative stress and immune dysfunction [72] caused by abamectin, an insecticide and acaricide known for its toxic activities on aquatic organisms, also shown to have immunosuppressive activity on *Oreochromis niloticus* by reducing the levels of lysozyme, nitric oxide, and IgM. The study demonstrated that the consumption of MOE dietary supplements ameliorated the adverse impact of *Agaricus blazei* Murrill (ABM), a mushroom known for its immunomodulatory activities, on lysozyme, nitric oxide, and IgM levels in the ABM + MOE-treated *Oreochromis niloticus* group [72].

The immunomodulatory effect of MOE could be attributed to cinnamic acid, naringenin, as well as additional phenolic and flavonoid compounds [72]. Importantly, administration of MOE in the treated group with ABM has been demonstrated to suppress the release of pro-inflammatory cytokines, including TNF-α and IL-1β, while upregulating the expression of genes involved in anti-inflammatory pathways, such as TGF-β and IL-10β. Additionally, *Moringa oleifera* pod extract inhibits the activation of mitogen-activated protein kinases (MAPKs) signaling cascades, thereby further contributing to the attenuation of inflammatory responses [72].

The protective effect of MOE on liver diseases [134–137] is probably due to the presence of quercetin [138], which plays a pivotal role in the prevention of liver inflammation through inhibiting NF-kB/TLR/NLRP3, inactivation of autophagy-mediated cell apoptosis (under ER stress) through the mTOR pathway [139–143] (Figure 5). Alterations in the mTOR signaling pathway have been associated with various human pathologies, notably liver diseases, and have also been correlated with the inhibition of cancer cell growth through suppression of the AKT/PI3K/mTOR axis [110,144] (Figure 5). Another recent study conducted by Adewale et al. demonstrates that the ethanolic extract of *Moringa oleifera* leaves exhibits an immunomodulatory effect by triggering programmed cell death pathway (apoptosis) in Jurkat cancer cells through up-regulation of CCR7, especially when those cancer cells were stimulated with CD3 and CD54 or CD28 [145]. Oral administration of aqueous extract of red *Moringa oleifera* leaves at 42 mg/kg body weight (BW) increased the number of $CD4^+CD62L^+$ and $CD8^+CD62L^+$ T cells in Salmonella typhimurium-infected BABL/c mice [145]. In a related study, the anti-inflammatory and immunostimulant effects of *Moringa oleifera* ethanolic extract were determined. The extract leads to a decrease in TNF-α on benzene-induced leukemia in Wistar rats, demonstrating its promising immunotherapy for leukemia. It has been reported that the reduction in serum levels of tumor necrosis factor alpha (TNF-α) following systemic chemotherapy in lymphoproliferative malignancies (leukemia and lymphoma) in patients is linked with response rates [146,147].

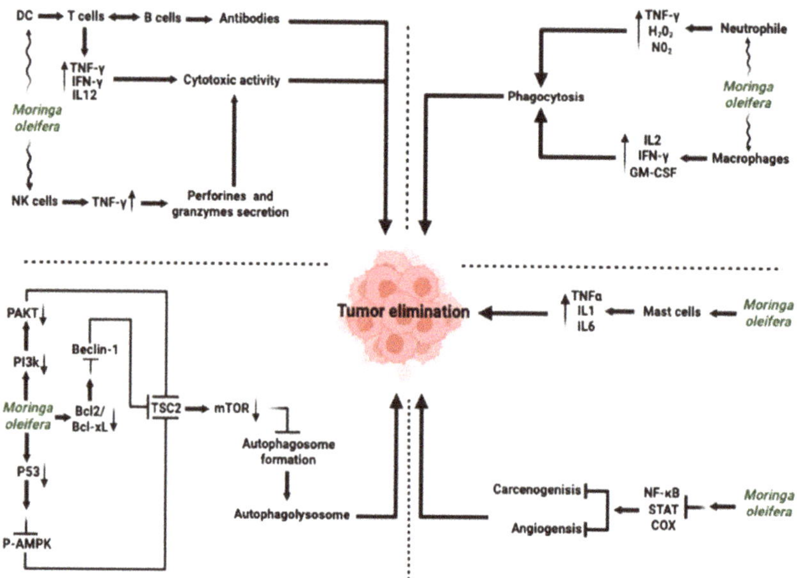

Figure 5. *Moringa oleifera* and its role in the modulation of cancer immunity. *Moringa oleifera* enhances innate immunity through stimulating neutrophil activation and macrophage functions, driving TNF-γ, H_2O_2 and NO_2 release, and enhancing phagocytosis with IL-2, IFN-γ and GM-CSF release, thereby leading to tumor elimination. In parallel, *Moringa oleifera* modulates mast cell activity, upregulating TNF-α, IL-1 and IL-2, which contribute to antitumor responses. Furthermore, *Moringa oleifera* promotes DC maturation, T and B cell activation, contributing to antibody production, IL-12 release, and NK cell-mediated cytotoxicity via perforin and granzyme secretion. In addition, H_2O_2 suppresses pro-tumorigenic cyclooxygenase activity and modulate PI3K/AKT/mTOR signaling cascades through p-AKT, p-AMPK, and TSC2, and engages p53 and Beclin-1 to trigger, apoptosis, autophagolysosome through inhibition of autophagosome formation, and inhibition of carcinogenesis, with concomitant downregulation of Bcl-2 and Bcl-xL. Antioxidant and redox regulation through H_2O_2 and NO_2 balance reduces oxidative stress, while modulation of NF-κB, STAT, and MAPK pathways limits chronic inflammation and supports cytotoxic T-cell and antibody-mediated responses against cancer cells. Overall, H_2O_2 orchestrates innate and adaptive immunity, modulates signaling pathways and balances oxidative stress to promote effective tumor elimination. COX, Cyclooxygenase; DC, Dendritic cell; GM-CSF, Granulocyte-Macrophage Colony-Stimulating Factor; IFN, interferon; mTOR, Mechanistic Target of Rapamycin; NF-κB, Nuclear factor kappa-light-chain-enhancer of activated B cells; NK, Natural killer; P-AMPK, Phosphorylated AMP-Activated protein Kinase; PAKT, Phosphorylated protein kinase B (Akt); PI3K, Phosphoinositide 3-Kinase; STAT, Signal transducer and activator of transcription; TNF, Tumor necrosis factor; TSC2, Tuberous sclerosis complex 2. ⟶ Promote/lead to; ↑ upregulate/increase expression; ↓ downregulate/decrease expression. ⊥ inhibition.

https://doi.org/10.3390/ijms27010263

Similarly, MOEs have been shown to exert immunostimulatory effects in cyclophosphamide-induced immunosuppressed Swiss albino mice [148], supporting their potential to modulate immune responses under immunosuppressive conditions. This study found that gavage with seed and root extracts at a dose of 2 g/kg BW for seven consecutive days increased the levels of white blood cells, relative spleen and thymus weights, and enhanced immunity by supporting the proliferation and activation of lymphocytes [148]. The study indicated that the extracts decrease the delayed type of hypersensitivity response, enhance the concentration of serum IgM, and increase relative spleen weight, which was associated with a higher increase in white blood cells. Also, it was demonstrated that TNF-α and IL-2 concentrations were up-regulated (Figure 5) in the treated group with *Moringa oleifera* extracts. The study concluded that *Moringa oleifera* extracts could improve patients with common variable prognosis with immunodeficiency disorders through the stimulation of the activity of hematopoietic cells, bone marrow, lymphocytes, cellular and humoral immunity [148]. These immunomodulatory and anti-inflammatory properties form the basis for investigating the molecular mechanisms by which *Moringa oleifera* impacts cancer progression.

5. *Moringa oleifera* and Cancer Progression: Molecular Mechanisms and Therapeutic Potential

Moringa oleifera exhibits broad cytotoxic and antitumor activities through multiple mechanisms, including induction of apoptosis, cell-cycle arrest, modulation of oxidative stress via Nrf2 activation, and suppression of pro-oncogenic signaling pathways NF-κB/MAPK and inhibition of oncogenic kinases like CDK2 and Src [34,81,149,150]. Furthermore, its bioactive compounds, especially moringa isothiocyanate (MIC-1; also called moringin) and niazimicin, modulate the tumor microenvironment (TME) by increasing cytotoxic T lymphocyte and NK cell activity while decreasing Treg-mediated immunosuppression [34,110,120,151,152].

5.1. *Apoptosis, Proliferation, and Cell-Cycle Control in Moringa oleifera-Mediated Cancer Progression*

Moringa oleifera extract from leaves, seeds, and oils along with the purified MIC-1 have been shown to reduce tumor cell viability and clonogenicity through triggering caspase-dependent apoptosis and arresting cells at the G1/S checkpoint. In renal cancer cells, MIC-1 suppresses growth and migration by inhibiting PTP1B-dependent Src/Ras/Raf/ERK signaling, enhances the pro-apoptotic Bax/Bcl-2 ratio in xenograft models while inhibiting ERK phosphorylation (Figure 5), establishing a mechanistic bridge between upstream phosphatase targeting and downstream caspase-driven apoptosis [34]. Furthermore, phenolic fractions from *Moringa oleifera* leaves induce activation of caspases 3/7/9, alongside mitochondrial ROS production and AIF translocation in melanoma cancer cells, suggesting simultaneous caspase-dependent and independent apoptosis signaling pathways [153].

In liver cancer, more particularly hepatocellular carcinoma (HepG2), phenolic compounds from *Moringa oleifera* fruit and leaves initiate ROS-mediated cellular signaling cascade, trigger caspase-3 activity and drive mitochondrial pathway-dependent apoptosis [152]. Concurrently, another study combining in-silico CDK2 docking with in vitro validation in ER-positive breast cancer reported that *Moringa oleifera* phytochemicals, including chlorogenic acid, niazirin, kaempferol, quercetin and ellagic acid act as putative CDK2 inhibitors, supporting the inhibition of cell growth through promoting cell cycle arrest at G1/S in hormone-receptor-positive disease [154]. Beyond its role in epithelial cancers, *Moringa oleifera* leaf protein decreases proliferation, induces apoptosis and promotes cell-cycle arrest in Jurkat T-ALL cells via MAPK/AKT pathway modulation (Figure 5), underscoring its relevance in hematologic malignancies [81].

https://doi.org/10.3390/ijms27010263

The antiproliferative and pro-apoptotic signature is consistently observed across additional cancer cell lines, with A549 (NSCLC) cells displaying caspase 9/3 activation and Bax/Bcl-2 shifts in response to *Moringa oleifera*-based extracts [155]. Similarly, colon cancer cells respond to glucosinolate-rich lead fractions with decreased proliferation [35,156], and Dalton's lymphoma cells demonstrate both in vitro and in vivo growth inhibition with upregulated Bax, cytochrome-c release and cleaved caspase 3 along with decreased Bcl-2 expression, accompanied by cell cycle arrest at G2/M, apoptosis induction, and importantly, significant improvement in hematological parameters in mice bearing Dalton's Lymphoma tumors [157]. Collectively, these studies converge on (1) caspase cascade engagement, (2) suppression of pro-survival signals such as ERK, and (3) CDK-axis inhibition at G1/S, providing a mechanistic basis for *Moringa oleifera* constituents as adjuncts to cytotoxic antitumor or targeted therapies, while emphasizing the need for standardized preparations to ensure reproducibility across experimental models.

5.2. *Moringa oleifera in Cancer: Dual Regulation of Oncogenic Signaling and Oxidative Stress*

Recent investigations have highlighted the dual modulatory effects of *Moringa oleifera* constituents on oncologic signaling and redox homeostasis, elucidating their potential as adjuncts in cancer therapy. Notably, the MIC-1 has been shown to inhibit migration and proliferation in renal cancer cells through modulation of PTP1B-dependent Src/Ras/Raf/ERK signaling [34]. This inhibition leads to reduced phosphorylation of Src (Tyr416), K-Rad, Raf and ERK1/2 [34], thus impairing tumor cell growth and migration. Concurrently, MIC-1 triggers the Nrf2/ARE pathway, upregulating the expression of cytoprotective genes such as NQO1 and enhancing cellular defense against oxidative stress [34]. This dual activity may potentially modulate tumor cell adaptation under stress conditions. In addition, in vivo studies further support these findings, showing that MIC-1 treatment results in increased Bax/Bcl-2 ratios in tumor tissues, a pivotal indicator of the induction of apoptosis [34]. Also, it has been reported that *Moringa oleifera* extracts modulate key oncogenic and stress-related signaling pathways implicated in multiple cancer types, such as breast, hepatocellular, melanoma, colon, NSCLC (Non-Small Cell Lung Cancer) and hematologic malignancies. These effects are mediated through the coordinated regulation of Nrf2-Keap1, TLR4/NF-κB and Wnt/β-catenin, thereby orchestrating oxidative stress and inflammatory responses, processes that are crucial to tumor growth, survival and therapy resistance [152].

5.3. *Inhibition of Angiogenesis and Metastasis Progression*

Recent studies have elucidated the potential role of *Moringa oleifera* extracts to modulate angiogenesis and metastasis processes (Figure 5), which are essential for tumor progression [158–160]. Although direct evidence of anti-angiogenic effect remains limited, several in vitro and in vivo investigations have demonstrated inhibition of cell migration and downregulation of metastasis-associated pathways. In a recent study, Raju and Kei (2022) reported that *Moringa oleifera* leaf extract exhibits significant anti-angiogenic activity, with the 100% aqueous extract showing an 81.33% reduction in vascular formation in the chick chorioallantoic membrane assay [161], suggesting the presence of bioactive compounds capable of inhibiting angiogenesis, a key process in tumor growth and metastasis. Furthermore, studies on *Moringa oleifera*-silver nanoparticles (MO-AgNPs) have been shown to inhibit metastasis through modulating pathways involving Snail and TGF-β. These findings suggest that MO-AgNPs may serve as a promising and effective targeted strategy to inhibit tumor metastasis [162].

Emerging evidence suggests that *Moringa oleifera* possesses significant capacity to modulate immune-regulatory networks within the TME. Notably, a recent study demonstrated

https://doi.org/10.3390/ijms27010263

that *Moringa oleifera* leaf protein inhibited the proliferation of T-lymphoblastic leukemia cells by modulating immune-related signaling pathways, specifically the MAPK (p38 and ERK) and AKT axis (Figure 5), while simultaneously inducing apoptosis, cell-cycle arrest, and autophagy [81]. Additionally, recent investigations have further revealed the ability of *Moringa oleifera*-derived polysaccharides to reprogram the TME through targeting innate immune cells. In a Lewis lung carcinoma model, *Moringa oleifera* leaf polysaccharides were shown to remodel tumor-associated macrophages from a pro-tumoral M2 phenotype toward an antitumoral M1 phenotype. This repolarization was mediated via TLR4 signaling, leading to upregulated chemokine expression CXCL9 and CXCL10 and increased intra-tumoral T-cell infiltration, thereby exerting powerful immune-activating effects [58]. In parallel, *Moringa oleifera* seed oil extract has demonstrated antioxidant, anti-inflammatory, and antitumor activities in Ehrlich ascites carcinoma-bearing Swiss albino mice treated at a dose of 500 mg/Kg [80], where treatment improved systemic oxidative stress markers through reducing malondialdehyde levels while elevating the activities of essential antioxidant enzymes, superoxide dismutase and catalase, indicating significant oxidative homeostasis amelioration. Furthermore, *Moringa oleifera* seed oil extract attenuated proinflammatory responses in tumor-bearing mice; collectively, these findings underscore its therapeutic potential to restore redox balance and mitigate the inflammatory reactions within the TME [80]. Nevertheless, caution must be taken regarding the use of *Moringa oleifera* preparations in combination with standard chemotherapy regimens, as it has been shown that in an obesity-associated triple-negative breast cancer model (MDA-MB-231 xenografts in C57BL/6J), the concomitant oral administration of *Moringa oleifera* seed extract concentrate (high-fat diet supplemented with 0.6% *w/w Moringa* concentrate) with conventional chemotherapy drug—comprising doxorubicin (2 mg/kg) and cyclophosphamide (100 mg/kg)—not only negated its angiogenesis-reducing benefit but also aggravated tumor progression, underscoring possible context-specific pro-tumor effects or adverse reactions [163]. These outcomes highlight the need for mechanistic and pharmacokinetic/pharmacodynamic clarity when exploring *Moringa oleifera* in combination with chemotherapy.

5.4. From Bench to Bedside: Addressing Translation Challenges in Moringa oleifera-Based Cancer Therapy

To advance *Moringa oleifera* as a reliable adjunct in cancer therapy, critical translational gaps must first be addressed. (i) Standardization and characterization of *Moringa oleifera* preparations. Defining Good Manufacturing Practice (GMP)-grade materials with rigorously quantified active constituents such as MIC-1, niazimicin, and principal flavonoids, alongside comprehensive impurity profiling and batch-level certificates to ensure validity and reproducibility in both preclinical and clinical investigations. Recent studies have highlighted variability in bioactive compounds across different *Moringa oleifera* sources, emphasizing the urgent need for standardized preparations [152]. (ii) A thorough understanding of pharmacodynamic, pharmacokinetic and target engagement is required. Determining human-relevant exposure levels of MIC-1 (both free and conjugated forms), evaluating tissue distribution, and confirming interaction with key molecular targets, including PTP1B and Nrf2, are crucial for predicting therapeutic outcomes and clinical efficacy [164–166]. Computational and experimental studies indicate that *Moringa oleifera* phytochemicals may modulate oncogenic and pro-survival proteins, such as Bcl-2 [81,144,167–170], providing mechanistic avenues that support translational potential. (iii) Context specificity should be rigorously addressed. The effect of *Moringa oleifera* can vary depending on tumor subtype, metabolic status (e.g., obesity), and concomitant treatments. Preclinical evidence demonstrates that *Moringa oleifera* modulates important signaling pathways, including YAP/TAZ, Wnt/β-catenin, Nrf2-Keap1 and TLR4/NF-κB, which play divergent roles depending on the tumor cellular context, microenvironment, and cancer-related immune-metabolic axis [152]. Consequently, tailoring interventions to the specific disease context is crucial to minimize adverse and prevent counterproductive outcomes, thereby enhancing therapeutic efficacy and advancing the development of targeted cancer therapies. (iv) Assessment of immuno-oncology endpoints should extend beyond conventional chemotherapeutic regimens. Evaluating *Moringa oleifera*'s impact on the TME, including Treg/CD8$^+$ balance, NK activity and M1/M2 macrophage polarization in immunocompetent models, with or without checkpoint blockade, is essential to define its immunomodulatory potential. Emerging evidence shows that *Moringa oleifera* extracts can enhance immune activation, suggesting promising opportunities for combination with immunotherapeutic strategies [81]. (v) Early-phase clinical studies are essential to translate *Moringa oleifera* findings to humans. Phase 0 or early pharmacokinetic trials, using biomarkers such as NQO1 activation, ERK signaling and cytokine profiles can guide dosing, pharmacodynamic response, and safety before disease-focused randomized controlled trials. Despite limited clinical data on cancer research, pharmacokinetics, antioxidant and anti-inflammatory activities provide a strong rationale for systemic investigation in oncology [20].

Together, addressing these translational priorities, including standardized *Moringa oleifera* preparations, rigorous pharmacological characterization, context-specific evaluation, and early clinical investigation, will enable the rational incorporation of *Moringa oleifera* into cancer therapy while ensuring mechanistic understanding, reproducibility, and safety.

6. Clinical Trials on the Immunomodulatory Properties of *Moringa oleifera* in Disease Prevention and Cancer Therapy

Moringa oleifera has emerged as a promising candidate due to its rich phytochemical composition and demonstrated ability to modulate both innate and adaptive immune responses through interconnected mechanisms. Recent clinical studies have begun to provide evidence supporting its immunostimulatory effects in humans, reflected by improved immune biomarkers, hematological parameters, and disease-specific outcomes [123–127]. These trials provide a translational perspective, bridging preclinical research findings with potential clinical applications for immune enhancement, disease prevention and management. Extending on its systemic immunomodulatory properties, clinical studies have demonstrated that *Moringa oleifera* improved immune responses in patients with gum disease [171] (CTRI/2022/02/040594), increases CD4 levels and nutritional status in HIV-positive patients [127,172], and improves immunological and hematological parameters in patients undergoing highly active antiretroviral therapy [173]. Moreover, clinical investigations have highlighted the potential of *Moringa oleifera* as an adjunct in cancer immunotherapy. In particular, an aqueous leaf extract exhibited significant anticancer activities by inducing apoptosis and inhibiting tumor progression without affecting normal physiological functions [174].

Another important study showed that *Moringa oleifera* leaf polysaccharides possess the ability to modulate the TME through inducing a phenotypic switch from immunosuppressive M2 macrophages to pro-inflammatory M1 phenotypes via TLR4 targeting [58], suggesting their promising role in cancer immunomodulation. In addition, the combination of *Moringa oleifera* with other phytochemicals, including saffron or green tea, has been demonstrated to enhance both anticancer efficacy and immunomodulatory activity, highlighting the therapeutic potential of synergistic strategies in cancer prevention and therapy [54].

Overall, these clinical findings underscore the translational potential of *Moringa oleifera* for disease prevention, immune modulation, and as an adjunct in cancer immunotherapy. While clinical trials demonstrate the therapeutic benefits of *Moringa oleifera*,

https://doi.org/10.3390/ijms27010263

investigating its long-term safety and toxicity remains critical to support its sustained therapeutic application.

7. Long-Term Safety and Toxicity of *Moringa oleifera*

The long-term safety and toxicity profile of *Moringa oleifera* has been extensively investigated in both preclinical and clinical studies, with most evidence supporting its well-established tolerability. A consolidated overview of these findings is presented in Table 3, which summarizes the available evidence across plant parts, experimental models, dosing regimens, and key safety outcomes.

Table 3. Long-term safety and toxicity profile of *Moringa oleifera*: insights from preclinical and clinical studies.

Plant Part	Study Model	Dose/Duration	Key Findings/Safety Outcomes	Ref
Seeds (hydroalcoholic extract)	Rats (14-day toxicity study)	100–2000 mg/kg/day	No significant toxicological effects; normal hematological and biochemical parameters	[175]
Leaf capsules in T2DM patients	Humans: adults with T2DM (therapy naïve)	Nutritional dose, 4 weeks	Well tolerated; no hypoglycemia; renal (BUN, creatinine) and hepatic (AST, ALT) markers remained normal	[176]
Leaves (hydroethanolic extract)	Female ICR mice	Acute: 2000 mg/kg, single dose. Sub-acute: 125–1000 mg/kg daily for 28 days.	Acute: LD_{50} > 2000 mg/kg; signs of liver and kidney damage (increased AST, CK and creatinine; hepatic degeneration; renal necrosis). Sub-acute: moderate hepato-nephrotoxicity (hepatic and renal necrosis, sinusoidal dilatation; glomerulonephritis). Lower doses (125–500 mg/kg): relatively safe with minimal adverse effects.	[177]
Leaf powder	Humans: adults	400 mg capsules, 6x daily (2.4 g/day); 12 weeks	Well tolerated; no adverse effects; kidney parameters (creatinine, urea), liver enzymes (AST, ALT), and hematological markers remained normal; decreased inflammatory parameters (CRP) and improved lipid profile (LDL-C, total cholesterol)	[124]
Leaf powder (added to RUSF)	Humans: children < 5 years with moderate malnutrition	Daily supplementation: 5 weeks for the *Moringa* group and 4 weeks for the placebo group.	No renal/hepatic toxicity reported; well tolerated	[178]
Leaf powder (10% fortified porridge)	Humans (observational study): children with cerebral palsy	Daily intake (3 months)	No serious adverse reactions or increased morbidity; improved nutritional (vitamin A and protein status) and immune markers	[179]

ALT, Alanine aminotransferase; AST, Aspartate aminotransferase; BUN, Blood urea nitrogen; CK, Creatine Kinase; CRP, C-reactive protein; ICR mice, Institute of Cancer Research (mouse strain); LDL-C, Low-density lipoprotein cholesterol; RUSF, Ready-to-use supplementary food; T2DM, Type 2 Diabetes Mellitus.

Preclinical studies on animal models have provided insights into the safety profile of *Moringa oleifera*. For instance, a 13-week repeated-dose study of an optimized aqueous leaf extract revealed no serious toxicity but recommended caution with long-term doses

https://doi.org/10.3390/ijms27010263

above 500 mg/kg in rodents, establishing a no-observed-adverse-effect level (NOAEL) close to 500 mg/kg [180]. In parallel, a 28-day repeated-dose toxicity study in rats administered orally with aqueous leaf extract exhibited no mortality or organ toxicity at 100–500 mg/kg/day, although a minor reduction in body weight was observed at the highest doses, findings that align with an acute safety margin up to 2000 mg/kg [181,182]. Moreover, disease-model studies, notably in gout, also produced favorable outcomes, further strengthening the evidence for systemic safety [183]. Reproductive and developmental toxicity assessments likewise have shown an absence of adverse effects at nutritionally relevant doses, thereby supporting its long-term safety [184–186]. Evidence on *Moringa oleifera* seeds and roots suggests a more restricted safety compared to leaves. A 14-day oral toxicity evaluation of a hydro-alcoholic seed extract did not report any significant toxicological effects, supporting its short-term tolerability [62,175]. In contrast, seed oil administered at 200–800 mg/kg in murine malaria models not only reduced parasitemia but also induced renal and hepatic lesions at higher doses, indicating a narrower therapeutic window for seed oils compared to *Moringa oleifera* leaf preparations [187]. Likewise, a 21-day oral administration of ethanol extract (150–600 mg/kg/day) in rats showed no nephrotoxic or hepatotoxicity on standard biochemical parameters, although the authors recommended longer-term evaluations due to the generally higher caution warranted for roots compared with leaves [188]. These findings emphasize the critical role of plant-part specificity, with leaves demonstrating broadly safer, whereas seed and root derivatives may require more cautious dosing and further long-term evaluation. Additional preclinical studies on related species, including *Moringa stenopetala* herbal tea blends, also support low toxicity in rats [189], though direct extrapolation to *Moringa oleifera* should be approached cautiously.

Clinical trials further corroborate these findings, indicating that oral supplementation with whole leaf powders or extracts is generally safe, well-tolerated and without consistent adverse effects in humans, while highlighting the need for standardized products and longer follow-up [62,190]. Both randomized controlled and observational studies have consistently reported no significant adverse effects, even at relatively high doses over extended periods. These observations are further supported by systematic reviews indicating that *Moringa oleifera* is well-tolerated, does not produce significant adverse effects, and serious toxicity [190–193]. Importantly, *Moringa oleifera* supplementation did not compromise hematological, renal, or hepatic functions, highlighting its favorable systemic safety profile [194–196]. Nevertheless, caution is warranted with *Moringa oleifera* parts consumption, especially the bark and roots. Overuse of these parts may carry toxicity risk due to the presence of potentially toxic bioactive compounds such as alkaloids [197–200]. Moreover, pharmacokinetic evidence indicates the possible interaction of *Moringa oleifera* with drugs metabolized by cytochrome P450 enzymes, including antihyperglycemic drugs, necessitating the need for careful use in patients receiving such therapies [192]. Rare clinical case reports have identified cutaneous reactions and potential renal/hepatic perturbations, as well as possible drug interactions, emphasizing the necessity for cautious dosing and pharmacovigilance, particularly in vulnerable populations [201]. Taken together, although *Moringa oleifera* demonstrates a generally favorable safety profile in both preclinical and clinical studies, its long-term use should be approached with consideration of potential drug–nutrient interactions, doses, and the specific parts of the plant employed.

8. Conclusions and Future Directions

Moringa oleifera has long been known to exhibit medicinal properties, and recent scientific investigations have shown its efficacy in modulating the immune system. These research findings suggest that *Moringa oleifera* plays an essential role in maintaining homeostasis between pro-inflammatory and anti-inflammatory cytokine responses and regu-

https://doi.org/10.3390/ijms27010263

lating the functions of $CD4^+$ T-cells subsets by targeting transcription factors and their signal transducers. These results highlight its promise in managing T cell-mediated autoimmune disorders and coordinating innate and adaptive immune responses. Its extracts also enhance hematopoietic activity and strengthen both cellular and humoral immunity, suggesting benefits for immunodeficiency disorders, while supporting gut microbial balance to reinforce immune homeostasis and anti-inflammatory signaling. Beyond immune regulation, *Moringa oleifera* exhibits potent anticancer activity through inducing apoptosis, cell-cycle arrest, activation of antioxidant pathways such as Nrf2, and suppressing oncogenic signaling. Evidence also indicates its capacity to inhibit angiogenesis, metastasis, and modulate immune networks within the tumor microenvironment.

To meet accepted standards, it is important to conduct high-throughput screening, analyses of the bioactive compounds, and identify the chemical targets when studying the immunomodulatory and anti-cancer activities of *Moringa oleifera* extracts. Those extracts should be subject to qualitative and quantitative analysis through validated analytical procedures to define phytochemical levels in *Moringa oleifera* extracts.

Several bioactive compounds from *Moringa oleifera*, such as isothiocyanate, flavonoids and polysaccharides (e.g., MOP-2, MRP-1) have been shown to modulate immune signaling pathways and exhibit immunomodulatory and anti-cancer effects. Nevertheless, it is necessary to conduct further mechanistic investigations with a comprehensive update on the current state of research, through multi-scale approaches spanning cellular and molecular mechanisms, host–microbiome interactions, and systemic physiology, to better understand how these compounds regulate innate, adaptive immune system responses and tumor biology. A deeper understanding of how *Moringa oleifera* modulates λδ T cells, natural killer T cells, mucosal-associated invariant T cells, and their roles in immune-regulatory circuits within the TME will be critical to advance *Moringa oleifera* from traditional medicine toward evidence-based immunotherapy and cancer interventions.

Despite the promising therapeutic effects of *Moringa oleifera*, critical gaps remain regarding its long-term safety, potential drug–nutrient interactions, optimal dosing, and the specific plant parts employed. Accordingly, rigorous standardized preclinical testing, encompassing toxicological, pharmacokinetic, and bioavailability studies, will be essential to ensure both safety and effective translation into clinical use.

Author Contributions: Conceptualization, M.T.; formal analysis, M.T. and J.E.K.; writing—original draft preparation, M.T. and J.E.K.; writing—review and editing, M.T.; visualization and critically reviewed the study design, H.A.M. and A.Z. All authors have read and agreed to the published version of the manuscript.

Funding: This research received no external funding.

Institutional Review Board Statement: Not applicable.

Informed Consent Statement: Not applicable.

Data Availability Statement: No new data were created or analyzed in this study. Data sharing is not applicable to this article.

Conflicts of Interest: The authors declare no conflicts of interest.

Abbreviations

The following abbreviations are used in this manuscript:

CCR7	C-C Chemokine Receptor Type 7
ER	Endoplasmic Reticulum
HSP70	Heat Shock Protein 70
iNOS	Inducible Nitric Oxide Synthase

https://doi.org/10.3390/ijms27010263

MOP	*Moringa oleifera* Polysaccharide
Mop3	Motif 3 Protein
MRP-1	Moringa Root Polysaccharide 1
NLRP3	NOD (Nucleotide-binding Oligomerization) LRR (Leucine-Rich Repeat) Pyrin Domain-containing Protein 3
NQO1	NAD(P)H: Quinone Oxidoreductase 1
PGE-2	Prostaglandin E2
PTP1B	Protein Tyrosine Phosphatase 1B
SOD	Superoxide Dismutase
TLR	Toll-Like Receptor
TME	Tumor Microenvironment

References

1. Pant, T.; Uche, N.; Juric, M.; Zielonka, J.; Bai, X. Regulation of Immunomodulatory Networks by Nrf2-Activation in Immune Cells: Redox Control and Therapeutic Potential in Inflammatory Diseases. *Redox Biol.* **2024**, *70*, 103077. [CrossRef]
2. Strzelec, M.; Detka, J.; Mieszczak, P.; Sobocińska, M.K.; Majka, M. Immunomodulation—A General Review of the Current State-of-the-Art and New Therapeutic Strategies for Targeting the Immune System. *Front. Immunol.* **2023**, *14*, 1127704. [CrossRef]
3. Ma, Y.; Kroemer, G. The Cancer-Immune Dialogue in the Context of Stress. *Nat. Rev. Immunol.* **2024**, *24*, 264–281. [CrossRef]
4. Carnet Le Provost, K.; Kepp, O.; Kroemer, G.; Bezu, L. Trial Watch: Dexmedetomidine in Cancer Therapy. *OncoImmunology* **2024**, *13*, 2327143. [CrossRef]
5. Walczak, J.; Iwaszkiewicz-Grześ, D.; Cholewiński, G. Approaches Towards Better Immunosuppressive Agents. *Curr. Top. Med. Chem.* **2024**, *24*, 1230–1263. [CrossRef] [PubMed]
6. La-Beck, N.M.; Owoso, J. Updates and Emerging Trends in the Management of Immune-Related Adverse Events Associated with Immune Checkpoint Inhibitor Therapy. *Asia-Pac. J. Oncol. Nurs.* **2024**, *11*, 100549. [CrossRef] [PubMed]
7. Ganguli, S.; Debnath, T. Drugs Used for Immunomodulation. In *Essentials of Pharmacodynamics and Drug Action*; Chakraverty, R., Mathur, R., Chakraborty, P., Eds.; Springer Nature: Singapore, 2024; pp. 229–240. ISBN 978-981-97-2776-6.
8. Sohail, R.; Mathew, M.; Patel, K.K.; Reddy, S.A.; Haider, Z.; Naria, M.; Habib, A.; Abdin, Z.U.; Razzaq Chaudhry, W.; Akbar, A. Effects of Non-Steroidal Anti-Inflammatory Drugs (NSAIDs) and Gastroprotective NSAIDs on the Gastrointestinal Tract: A Narrative Review. *Cureus* **2023**, *15*, e37080. [CrossRef] [PubMed]
9. Kawai, T.; Suzuki, C.; Honda, Y.; Fernandez, J.L. Long-Term Safety and Effectiveness of Vonoprazan for Prevention of Gastric and Duodenal Ulcer Recurrence in Patients on Nonsteroidal Anti-Inflammatory Drugs in Japan: A 12-Month Post-Marketing Surveillance Study. *Expert Opin. Drug Saf.* **2023**, *22*, 425–431. [CrossRef]
10. Chakik, R.M.; Alqahtani, N.I.; Al-Hagawi, Y.; Nasser Alsharif, S.; Alqahtani, A.S.; Hadi Asiri, D.; Al-Mani, S.Y. Diagnosis of Small Intestinal Diaphragms and Strictures Induced by Non-Steroidal Anti-Inflammatory Drugs Through Intraoperative Enteroscopy: A Case Study from Saudi Arabia. *Cureus* **2024**, *16*, e59752. [CrossRef]
11. Choe, Y.; Park, J.M.; Kim, J.S.; Cho, Y.K.; Kim, B.-W.; Choi, M.-G.; Kim, N.J. Drugs Effective for Nonsteroidal Anti-Inflammatory Drugs or Aspirin-Induced Small Bowel Injuries: A Systematic Review and Meta-Analysis of Randomized Controlled Trials. *J. Clin. Gastroenterol.* **2024**, *58*, 1003–1010. [CrossRef]
12. Berge, L.A.M.; Andreassen, B.K.; Stenehjem, J.S.; Heir, T.; Karlstad, Ø.; Juzeniene, A.; Ghiasvand, R.; Larsen, I.K.; Green, A.C.; Veierød, M.B.; et al. Use of Immunomodulating Drugs and Risk of Cutaneous Melanoma: A Nationwide Nested Case-Control Study. *Clin. Epidemiol.* **2020**, *12*, 1389–1401. [CrossRef] [PubMed]
13. Rahman, A.; Haider, M.F. A Comprehensive Review on Glucocorticoids Induced Osteoporosis: A Medication Caused Disease. *Steroids* **2024**, *207*, 109440. [CrossRef]
14. Tsubosaka, M.; Maruyama, M.; Lui, E.; Kushioka, J.; Toya, M.; Gao, Q.; Shen, H.; Li, X.; Chow, S.K.-H.; Zhang, N.; et al. Preclinical Models for Studying Corticosteroid-Induced Osteonecrosis of the Femoral Head. *J. Biomed. Mater. Res. Part B Appl. Biomater.* **2024**, *112*, e35360. [CrossRef]
15. Hsu, C.H.; Hsu, C.L.; Langley, A.; Wojcik, C.; Iraganje, E.; Grygiel-Górniak, B. Glucocorticoid-Induced Osteoporosis—From Molecular Mechanism to Clinical Practice. *Drugs Ther. Perspect.* **2024**, *40*, 315–329. [CrossRef]
16. Elgert, K.D. *Immunology: Understanding the Immune System*, 2nd ed.; Wiley-Blackwell: Hoboken, NJ, USA, 2009; ISBN 978-0-470-08157-0.
17. Kamran, M.; Hussain, S.; Abid, M.A.; Syed, S.K.; Suleman, M.; Riaz, M.; Iqbal, M.; Mahmood, S.; Saba, I.; Qadir, R. Phytochemical Composition of *Moringa oleifera* Its Nutritional and Pharmacological Importance. *Postep. Biol. Komorki* **2020**, *47*, 321–334.
18. Singh, A.K.; Rana, H.K.; Tshabalala, T.; Kumar, R.; Gupta, A.; Ndhlala, A.R.; Pandey, A.K. Phytochemical, Nutraceutical and Pharmacological Attributes of a Functional Crop *Moringa oleifera* Lam: An Overview. *S. Afr. J. Bot.* **2020**, *129*, 209–220. [CrossRef]

https://doi.org/10.3390/ijms27010263

19. Camilleri, E.; Blundell, R. A Comprehensive Review of the Phytochemicals, Health Benefits, Pharmacological Safety and Medicinal Prospects of *Moringa oleifera*. *Heliyon* **2024**, *10*, e27807. [CrossRef]
20. Pareek, A.; Pant, M.; Gupta, M.M.; Kashania, P.; Ratan, Y.; Jain, V.; Pareek, A.; Chuturgoon, A.A. *Moringa oleifera*: An Updated Comprehensive Review of Its Pharmacological Activities, Ethnomedicinal, Phytopharmaceutical Formulation, Clinical, Phytochemical, and Toxicological Aspects. *Int. J. Mol. Sci.* **2023**, *24*, 2098. [CrossRef]
21. Owon, M.; Osman, M.; Ibrahim, A.; Salama, M.A.; Matthäus, B. Characterisation of Different Parts from *Moringa oleifera* Regarding Protein, Lipid Composition and Extractable Phenolic Compounds. *OCL* **2021**, *28*, 45. [CrossRef]
22. Leone, A.; Spada, A.; Battezzati, A.; Schiraldi, A.; Aristil, J.; Bertoli, S. *Moringa oleifera* Seeds and Oil: Characteristics and Uses for Human Health. *Int. J. Mol. Sci.* **2016**, *17*, 2141. [CrossRef] [PubMed]
23. Takase, M.; Essandoh, P.K.; Asare, R.K.; Nazir, K.-H. The Value Chain of *Moringa oleifera* Plant and the Process of Producing Its Biodiesel in Ghana. *Sci. World J.* **2022**, *2022*, 1827514. [CrossRef]
24. Anwar, F.; Ashraf, M.; Bhanger, M.I. Interprovenance Variation in the Composition of *Moringa oleifera* Oilseeds from Pakistan. *J. Am. Oil Chem. Soc.* **2005**, *82*, 45–51. [CrossRef]
25. Corbett, P. It's Time for an Oil Change! Opportunities for High-Oleic Vegetable Oils. *Inf.-Int. News Fats Oils Relat. Mater.* **2003**, *14*, 480–481.
26. Gharsallah, K.; Rezig, L.; Msaada, K.; Chalh, A.; Soltani, T. Chemical Composition and Profile Characterization of *Moringa oleifera* Seed Oil. *S. Afr. J. Bot.* **2021**, *137*, 475–482. [CrossRef]
27. Kaur, G.; Invally, M.; Sanzagiri, R.; Buttar, H.S. Evaluation of the Antidepressant Activity of *Moringa oleifera* Alone and in Combination with Fluoxetine. *J. Ayurveda Integr. Med.* **2015**, *6*, 273. [CrossRef] [PubMed]
28. Masitlha, E.P.; Seifu, E.; Teketay, D. Nutritional Composition and Mineral Profile of Leaves of *Moringa oleifera* Provenances Grown in Gaborone, Botswana. *Food Prod. Process Nutr.* **2024**, *6*, 3. [CrossRef]
29. Marrufo, T.; Nazzaro, F.; Mancini, E.; Fratianni, F.; Coppola, R.; De Martino, L.; Agostinho, A.B.; De Feo, V. Chemical Composition and Biological Activity of the Essential Oil from Leaves of *Moringa oleifera* Lam. Cultivated in Mozambique. *Molecules* **2013**, *18*, 10989–11000. [CrossRef]
30. Divya, S.; Pandey, V.K.; Dixit, R.; Rustagi, S.; Suthar, T.; Atuahene, D.; Nagy, V.; Ungai, D.; Ahmed, A.E.M.; Kovács, B.; et al. Exploring the Phytochemical, Pharmacological and Nutritional Properties of *Moringa oleifera*: A Comprehensive Review. *Nutrients* **2024**, *16*, 3423. [CrossRef]
31. Bhattacharya, S.B.; Das, A.K.; Banerji, N. Chemical Investigations on the Gum Exudate from Sajna (*Moringa oleifera*). *Carbohydr. Res.* **1982**, *102*, 253–262. [CrossRef]
32. Faizi, S.; Siddiqui, B.S.; Saleem, R.; Siddiqui, S.; Aftab, K.; Gilani, A.H. Isolation and Structure Elucidation of New Nitrile and Mustard Oil Glycosides from *Moringa oleifera* and Their Effect on Blood Pressure. *J. Nat. Prod.* **1994**, *57*, 1256–1261. [CrossRef]
33. Waterman, C.; Rojas-Silva, P.; Tumer, T.B.; Kuhn, P.; Richard, A.J.; Wicks, S.; Stephens, J.M.; Wang, Z.; Mynatt, R.; Cefalu, W.; et al. Isothiocyanate-rich *Moringa oleifera* Extract Reduces Weight Gain, Insulin Resistance, and Hepatic Gluconeogenesis in Mice. *Mol. Nutr. Food Res.* **2015**, *59*, 1013–1024. [CrossRef]
34. Xie, J.; Qian, Y.-Y.; Yang, Y.; Peng, L.-J.; Mao, J.-Y.; Yang, M.-R.; Tian, Y.; Sheng, J. Isothiocyanate from *Moringa oleifera* Seeds Inhibits the Growth and Migration of Renal Cancer Cells by Regulating the PTP1B-Dependent Src/Ras/Raf/ERK Signaling Pathway. *Front. Cell Dev. Biol.* **2022**, *9*, 790618. [CrossRef]
35. Cuellar-Núñez, M.L.; Loarca-Piña, G.; Berhow, M.; Gonzalez de Mejia, E. Glucosinolate-Rich Hydrolyzed Extract from *Moringa oleifera* Leaves Decreased the Production of TNF-α and IL-1β Cytokines and Induced ROS and Apoptosis in Human Colon Cancer Cells. *J. Funct. Foods* **2020**, *75*, 104270. [CrossRef]
36. Chodur, G.M.; Olson, M.E.; Wade, K.L.; Stephenson, K.K.; Nouman, W.; Garima; Fahey, J.W. Wild and Domesticated *Moringa oleifera* Differ in Taste, Glucosinolate Composition, and Antioxidant Potential, but Not Myrosinase Activity or Protein Content. *Sci. Rep.* **2018**, *8*, 7995. [CrossRef] [PubMed]
37. Zarina; Wani, A.W.; Rawat, M.; Kaur, H.; Das, S.; Kaur, T.; Akram, N.; Faisal, Z.; Jan, S.S.; Oyshe, N.N.; et al. Medicinal Utilization and Nutritional Properties of Drumstick (*Moringa oleifera*)—A Comprehensive Review. *Food Sci. Nutr.* **2024**, *12*, 4546–4568. [CrossRef]
38. Engsuwan, J.; Waranuch, N.; Limpeanchob, N.; Ingkaninan, K. HPLC Methods for Quality Control of *Moringa oleifera* Extract Using Isothiocyanates and Astragalin as Bioactive Markers. *ScienceAsia* **2017**, *43*, 169–174. [CrossRef]
39. Riaz, A.; Rasul, A.; Hussain, G.; Zahoor, M.K.; Jabeen, F.; Subhani, Z.; Younis, T.; Ali, M.; Sarfraz, I.; Selamoglu, Z. Astragalin: A Bioactive Phytochemical with Potential Therapeutic Activities. *Adv. Pharmacol. Sci.* **2018**, *2018*, 9794625. [CrossRef]
40. Leone, A.; Fiorillo, G.; Criscuoli, F.; Ravasenghi, S.; Santagostini, L.; Fico, G.; Spadafranca, A.; Battezzati, A.; Schiraldi, A.; Pozzi, F. Nutritional Characterization and Phenolic Profiling of *Moringa oleifera* Leaves Grown in Chad, Sahrawi Refugee Camps, and Haiti. *Int. J. Mol. Sci.* **2015**, *16*, 18923–18937. [CrossRef] [PubMed]
41. Zduńska, K.; Dana, A.; Kolodziejczak, A.; Rotsztejn, H. Antioxidant Properties of Ferulic Acid and Its Possible Application. *Ski. Pharmacol. Physiol.* **2018**, *31*, 332–336. [CrossRef]

https://doi.org/10.3390/ijms27010263

42. Klimek-Szczykutowicz, M.; Gaweł-Bęben, K.; Rutka, A.; Blicharska, E.; Tatarczak-Michalewska, M.; Kulik-Siarek, K.; Kukula-Koch, W.; Malinowska, M.A.; Szopa, A. *Moringa oleifera* (Drumstick Tree)—Nutraceutical, Cosmetological and Medicinal Importance: A Review. *Front. Pharmacol.* **2024**, *15*, 1288382. [CrossRef]
43. Nikkon, F.; Hasan, S.; Salam, K.A.; Mosaddik, M.A.; Khondkar, P.; Haque, M.E.; Rahman, M. Benzylcarbamothioethionate from Root Bark of *Moringa oleifera* Lam. and Its Toxicological Evaluation. *Boletín Latinoam. Caribe Plantas Med. Aromát.* **2009**, *8*, 130–138.
44. Alhakmani, F.; Kumar, S.; Khan, S.A. Estimation of Total Phenolic Content, in–Vitro Antioxidant and Anti–Inflammatory Activity of Flowers of *Moringa oleifera*. *Asian Pac. J. Trop. Biomed.* **2013**, *3*, 623–627. [CrossRef] [PubMed]
45. Atawodi, S.E.; Atawodi, J.C.; Idakwo, G.A.; Pfundstein, B.; Haubner, R.; Wurtele, G.; Bartsch, H.; Owen, R.W. Evaluation of the Polyphenol Content and Antioxidant Properties of Methanol Extracts of the Leaves, Stem, and Root Barks of *Moringa oleifera* Lam. *J. Med. Food* **2010**, *13*, 710–716. [CrossRef] [PubMed]
46. Calva, J.; Cuenca, M.B.; León, A.; Benítez, Á. Chemical Composition, Acetylcholinesterase-Inhibitory Potential and Antioxidant Activity of Essential Oils from Three Populations of Parthenium *Hysterophorus* L. in Ecuador. *Molecules* **2025**, *30*, 2712. [CrossRef]
47. Malaspina, P.; Polito, F.; Mainetti, A.; Khedhri, S.; De Feo, V.; Cornara, L. Exploring Chemical Variability in the Essential Oil of *Artemisia Absinthium* L. in Relation to Different Phenological Stages and Geographical Location. *Chem. Biodivers.* **2025**, *20*, e00743. [CrossRef]
48. Hridoy, A.A.M.; Munny, F.J.; Shahriar, F.; Rahman, M.; Islam, F.; Kazmi, A.; Kawsar, A.; Bailey, C. Exploring the Potentials of Sajana (*Moringa oleifera* Lam.) as a Plant-Based Feed Ingredient to Sustainable and Good Aquaculture Practices: An Analysis of Growth Performance and Health Benefits. *Aquac. Res.* **2025**, *2025*, 3580123. [CrossRef]
49. Tilaoui, M.; Mouse, H.A.; Jaafari, A.; Zyad, A. Comparative Phytochemical Analysis of Essential Oils from Different Biological Parts of Artemisia Herba Alba and Their Cytotoxic Effect on Cancer Cells. *PLoS ONE* **2015**, *10*, e0131799. [CrossRef]
50. Balasubramaniam, M.; Sapuan, S.; Hashim, I.F.; Ismail, N.I.; Yaakop, A.S.; Kamaruzaman, N.A.; Mokhtar, A.M.A. The Properties and Mechanism of Action of Plant Immunomodulators in Regulation of Immune Response—A Narrative Review Focusing on *Curcuma longa* L., Panax Ginseng C. A. Meyer and *Moringa oleifera* Lam. *Heliyon* **2024**, *10*, e28261. [CrossRef] [PubMed]
51. Goel, F. Exploring the Therapeutic Role of *Moringa oleifera* in Neurodegeneration: Antioxidant, Anti-Inflammatory, and Neuroprotective Mechanisms. *Inflammopharmacology* **2025**, *33*, 3653–3669. [CrossRef]
52. Gaber, H.G.; Younis, N.A.; Saleh, S.Y. Comparative Physiological and Immunological Impacts of *Moringa oleifera* Leaf and Seed Water Supplements on African Catfish (*Clarias gariepinus*): Effects on Disease Resistance and Health Parameters. *BMC Vet. Res.* **2025**, *21*, 320. [CrossRef]
53. Kamble, M.T.; Gallardo, W.; Salin, K.R.; Pumpuang, S.; Chavan, B.R.; Bhujel, R.C.; Medhe, S.V.; Kettawan, A.; Thompson, K.D.; Pirarat, N. Effect of *Moringa oleifera* Leaf Extract on the Growth Performance, Hematology, Innate Immunity, and Disease Resistance of Nile Tilapia (*Oreochromis niloticus*) against Streptococcus Agalactiae Biotype 2. *Animals* **2024**, *14*, 953. [CrossRef]
54. Talib, W.H.; Al Junaidi, H.S.; Alshaeri, H.K.; Alasmari, M.M.; Hadi, R.W.; Alsayed, A.R.; Law, D. Immunomodulatory and Anticancer Effects of Moringa Polyherbal Infusions: Potentials for Preventive and Therapeutic Use. *Front. Immunol.* **2025**, *16*, 1597602. [CrossRef] [PubMed]
55. Rao, Q.; Hua, H.; Zhao, J. Advancements in Plant-Derived sRNAs Therapeutics: Classification, Delivery Strategies, and Therapeutic Applications. *Int. J. Mol. Sci.* **2025**, *26*, 4277. [CrossRef]
56. Mohai Ud Din, R.; Eman, S.; Zafar, M.H.; Chong, Z.; Saleh, A.A.; Husien, H.M.; Wang, M. *Moringa oleifera* as a Multifunctional Feed Additive: Synergistic Nutritional and Immunomodulatory Mechanisms in Livestock Production. *Front. Nutr.* **2025**, *12*, 1615349. [CrossRef]
57. Abdelazim, A.M.; Afifi, M.; Abu-Alghayth, M.H.; Alkadri, D.H. *Moringa oleifera*: Recent Insights for Its Biochemical and Medicinal Applications. *J. Food Biochem.* **2024**, *2024*, 1270903. [CrossRef]
58. Wang, S.; Hu, Q.; Chang, Z.; Liu, Y.; Gao, Y.; Luo, X.; Zhou, L.; Chen, Y.; Cui, Y.; Wang, Z. *Moringa oleifera* Leaf Polysaccharides Exert Anti-Lung Cancer Effects upon Targeting TLR4 to Reverse the Tumor-Associated Macrophage Phenotype and Promote T-Cell Infiltration. *Food Funct.* **2023**, *14*, 4607–4620. [CrossRef]
59. Valentina, A.S.; Nugroho, T.; Susilaningsih, N. The Effect of Ethanolic Extract of *Moringa oleifera* Leaves on the Macrophage Count and VEGF Expression on Wistar Rats with Burn Wound. *Biosci. Med. J. Biomed. Transl. Res.* **2022**, *6*, 1959–1964. [CrossRef]
60. Xi, C.; Li, W.; Liu, X.; Xie, J.; Li, S.; Tian, Y.; Song, S. The Potential Role of *Moringa oleifera* Lam. Leaf Proteins in Moringa Allergy by Functionally Activating Murine Bone Marrow-Derived Dendritic Cells and Inducing Their Differentiation toward a Th2-Polarizing Phenotype. *Nutrients* **2023**, *16*, 7. [CrossRef] [PubMed]
61. Pilotos, J.; Ibrahim, K.A.; Mowa, C.N.; Opata, M.M. *Moringa oleifera* Treatment Increases Tbet Expression in CD4+ T Cells and Remediates Immune Defects of Malnutrition in Plasmodium Chabaudi-Infected Mice. *Malar. J.* **2020**, *19*, 62. [CrossRef]
62. Azlan, U.K.; Mediani, A.; Rohani, E.R.; Tong, X.; Han, R.; Misnan, N.M.; Jam, F.A.; Bunawan, H.; Sarian, M.N.; Hamezah, H.S. A Comprehensive Review with Updated Future Perspectives on the Ethnomedicinal and Pharmacological Aspects of *Moringa oleifera*. *Molecules* **2022**, *27*, 5765. [CrossRef]

https://doi.org/10.3390/ijms27010263

63. Tran, T.H.M.; Lee, S.; Huh, J.-E.; Perumalsamy, H.; Balusamy, S.R.; Kim, Y.-J. Fermentation of *Moringa oleifera* Lam. Using Bifidobacterium Animalis Subsp. Lactis Enhances the Anti-Inflammatory Effect in RAW 264.7 Macrophages. *J. Funct. Foods* **2023**, *109*, 105752. [CrossRef]
64. Abdel-Latif, M.; Sakran, T.; Badawi, Y.K.; Abdel-Hady, D.S. Influence of *Moringa oleifera* Extract, Vitamin C, and Sodium Bicarbonate on Heat Stress-Induced HSP70 Expression and Cellular Immune Response in Rabbits. *Cell Stress Chaperones* **2018**, *23*, 975–984. [CrossRef]
65. Potestà, M.; Minutolo, A.; Gismondi, A.; Canuti, L.; Kenzo, M.; Roglia, V.; Macchi, F.; Grelli, S.; Canini, A.; Colizzi, V.; et al. Cytotoxic and Apoptotic Effects of Different Extracts of *Moringa oleifera* Lam on Lymphoid and Monocytoid Cells. *Exp. Ther. Med.* **2019**, *18*, 5–17. [CrossRef]
66. Budhy, T.I.; Adam, D.; Azis, Z.M.R.; Syahputri, V.; Yuliani, M.G.A.; Suwarto, M.F.S.; Setiawan, F. The Potential of Moringa Leaf Nanoparticles (*Moringa oleifera*) on the Expression of TNFα, IL10, and HSP 27 in Oral Cavity Cancer. *J. Multidiscip. Appl. Nat. Sci.* **2024**, *4*, 120–129. [CrossRef]
67. Choi, E.-J.; Debnath, T.; Tang, Y.; Ryu, Y.-B.; Moon, S.-H.; Kim, E.-K. Topical Application of *Moringa oleifera* Leaf Extract Ameliorates Experimentally Induced Atopic Dermatitis by the Regulation of Th1/Th2/Th17 Balance. *Biomed. Pharmacother.* **2016**, *84*, 870–877. [CrossRef] [PubMed]
68. Omodanisi, E.I.; Aboua, Y.G.; Chegou, N.N.; Oguntibeju, O.O. Hepatoprotective, Antihyperlipidemic, and Anti-Inflammatory Activity of *Moringa oleifera* in Diabetic-Induced Damage in Male Wistar Rats. *Pharmacogn. Res.* **2017**, *9*, 182.
69. Drue, G.; Minor, R.C. *Moringa oleifera* Tea Alters Neutrophil but Not Lymphocyte Levels in Blood of Acutely Stressed Mice. *Madridge J. Immunol.* **2018**, *2*, 43–48. [CrossRef]
70. Husni, E.; Badriyya, E.; Putri, L.; Aldi, Y. The Effect of Ethanol Extract of Moringa Leaf (*Moringa oleifera* Lam) Against the Activity and Capacity of Phagocytosis of Macrofag Cells and the Percentage of Leukosit Cells of White Mice. *Pharmacogn. J.* **2021**, *13*, 706–712. [CrossRef]
71. Li, C.; Dong, Z.; Zhang, B.; Huang, Q.; Liu, G.; Fu, X. Structural Characterization and Immune Enhancement Activity of a Novel Polysaccharide from *Moringa oleifera* Leaves. *Carbohydr. Polym.* **2020**, *234*, 115897. [CrossRef]
72. Muangnoi, C.; Chingsuwanrote, P.; Praengamthanachoti, P.; Svasti, S.; Tuntipopipat, S. *Moringa oleifera* Pod Inhibits Inflammatory Mediator Production by Lipopolysaccharide-Stimulated RAW 264.7 Murine Macrophage Cell Lines. *Inflammation* **2012**, *35*, 445–455. [CrossRef]
73. Rachmawati, I.; Rifa'i, M. In Vitro Immunomodulatory Activity of Aqueous Extract of *Moringa oleifera* Lam. Leaf to the CD4+, CD8+ and B220+ Cells in Mus Musculus. *J. Exp. Life Sci.* **2014**, *4*, 15–20. [CrossRef]
74. Hong, Z.-S.; Xie, J.; Wang, X.-F.; Dai, J.-J.; Mao, J.-Y.; Bai, Y.-Y.; Sheng, J.; Tian, Y. *Moringa oleifera* Lam. Peptide Remodels Intestinal Mucosal Barrier by Inhibiting JAK-STAT Activation and Modulating Gut Microbiota in Colitis. *Front. Immunol.* **2022**, *13*, 924178. [CrossRef]
75. Cui, C.; Chen, S.; Wang, X.; Yuan, G.; Jiang, F.; Chen, X.; Wang, L. Characterization of *Moringa oleifera* Roots Polysaccharide MRP-1 with Anti-Inflammatory Effect. *Int. J. Biol. Macromol.* **2019**, *132*, 844–851. [CrossRef]
76. Hunthayung, K.; Bhawamai, S. Polyphenol Compounds of Freeze-Dried *Moringa oleifera* Lam Pods and Their Anti-Inflammatory Effects on RAW 264.7 Macrophages Stimulated with Lipopolysaccharide. *Bioact. Compd. Health Dis.* **2024**, *7*, 185–198.
77. Arulselvan, P.; Tan, W.S.; Gothai, S.; Muniandy, K.; Fakurazi, S.; Esa, N.M.; Alarfaj, A.A.; Kumar, S.S. Anti-Inflammatory Potential of Ethyl Acetate Fraction of *Moringa oleifera* in Downregulating the NF-κB Signaling Pathway in Lipopolysaccharide-Stimulated Macrophages. *Molecules* **2016**, *21*, 1452. [CrossRef]
78. Praengam, K.; Muangnoi, C.; Dawilai, S.; Awatchanawong, M.; Tuntipopipat, S. Digested *Moringa oleifera* Boiled Pod Exhibits Anti-Inflammatory Activity in Caco-2 Cells. *J. Herbs Spices Med. Plants* **2015**, *21*, 148–160. [CrossRef]
79. Luetragoon, T.; Pankla Sranujit, R.; Noysang, C.; Thongsri, Y.; Potup, P.; Suphrom, N.; Nuengchamnong, N.; Usuwanthim, K. Bioactive Compounds in *Moringa oleifera* Lam. Leaves Inhibit the Pro-Inflammatory Mediators in Lipopolysaccharide-Induced Human Monocyte-Derived Macrophages. *Molecules* **2020**, *25*, 191. [CrossRef]
80. Aldayel, T.S.; Gad El Hak, H.N.; Nafie, M.S.; Saad, R.; Abdelrazek, H.M.A.; Kilany, O.E. Evaluation of Antioxidant, Anti-Inflammatory, Anticancer Activities and Molecular Docking of *Moringa oleifera* Seed Oil Extract against Experimental Model of Ehrlich Ascites Carcinoma in Swiss Female Albino Mice. *BMC Complement. Med. Ther.* **2023**, *23*, 457. [CrossRef] [PubMed]
81. Liu, X.; Yang, H.; Xie, S.; Wang, X.; Tian, Y.; Song, S. *Moringa oleifera* Leaves Protein Suppresses T-Lymphoblastic Leukemogenesis via MAPK/AKT Signaling Modulation of Apoptotic Activation and Autophagic Flux Regulation. *Front. Immunol.* **2025**, *16*, 1546189. [CrossRef]
82. Ababneh, S.K.; Khwaldeh, A.; Alsarhan, A.A.; Yousef, I.; Al-Shdefat, R.; Shoiab, A.; Ababneh, S. The Modulatory Effects of Statins, Vitamin E, and *Moringa oleifera* Extract on CD3 Expression in Ribociclib-Induced Hepatotoxicity in Rats. *Acta Inf. Med.* **2025**, *33*, 90–95. [CrossRef]
83. Poluan, J.C.; Zubair, M.S.; Ramadani, A.P.; Hayati, F. Narrative Review: Potential of Flavonoids from Moringa (*Moringa oleifera* Lam.) Leaves as Immunomodulators. *J. Farm. Galen. Galen. J. Pharm.* **2023**, *9*, 270–283. [CrossRef]

84. Swaroop, A.K.; Lalitha, C.M.V.N.; Shanmugam, M.; Subramanian, G.; Natarajan, J.; Selvaraj, J. Plant Derived Immunomodulators; A Critical Review. *Adv. Pharm. Bull.* **2022**, *12*, 712. [CrossRef]
85. Nfambi, J.; Bbosa, G.S.; Sembajwe, L.F.; Gakunga, J.; Kasolo, J.N. Immunomodulatory Activity of Methanolic Leaf Extract of *Moringa oleifera* in Wistar Albino Rats. *J. Basic Clin. Physiol. Pharmacol.* **2015**, *26*, 603–611. [CrossRef]
86. Al-Asmari, A.K.; Albalawi, S.M.; Athar, M.T.; Khan, A.Q.; Al-Shahrani, H.; Islam, M. *Moringa oleifera* as an Anti-Cancer Agent against Breast and Colorectal Cancer Cell Lines. *PLoS ONE* **2015**, *10*, e0135814. [CrossRef] [PubMed]
87. Ma, N.; Tang, Q.; Wu, W.-T.; Huang, X.-A.; Xu, Q.; Rong, G.-L.; Chen, S.; Song, J.-P. Three Constituents of *Moringa oleifera* Seeds Regulate Expression of Th17-Relevant Cytokines and Ameliorate TPA-Induced Psoriasis-Like Skin Lesions in Mice. *Molecules* **2018**, *23*, 3256. [CrossRef]
88. Zhang, J.; Liu, X.; Wang, Z.; Zhang, H.; Gao, J.; Wu, Y.; Meng, X.; Zhong, Y.; Chen, H. Potential Allergenicity Response to *Moringa oleifera* Leaf Proteins in BALB/c Mice. *Nutrients* **2022**, *14*, 4700. [CrossRef] [PubMed]
89. Wuryandari, M.R.E.; Atho'illah, M.F.; Laili, R.D.; Fatmawati, S.; Widodo, N.; Widjajanto, E.; Rifa'i, M. Lactobacillus Plantarum FNCC 0137 Fermented Red *Moringa oleifera* Exhibits Protective Effects in Mice Challenged with Salmonella Typhi via TLR3/TLR4 Inhibition and down-Regulation of Proinflammatory Cytokines. *J. Ayurveda Integr. Med.* **2022**, *13*, 100531. [CrossRef]
90. Eladia, R.E.; Ampode, K.M.B. Moringa (*Moringa oleifera* Lam.) Pod Meal: Nutrient Analysis and Its Effect on the Growth Performance and Cell-Mediated Immunity of Broiler Chickens. *J. Anim. Health Prod.* **2021**, *9*, 170–177.
91. Yasoob, T.B.; Khalid, A.R.; Zhang, Z.; Zhu, X.; Hang, S. Liver Transcriptome of Rabbits Supplemented with Oral *Moringa oleifera* Leaf Powder under Heat Stress Is Associated with Modulation of Lipid Metabolism and Up-Regulation of Genes for Thermo-Tolerance, Antioxidation, and Immunity. *Nutr. Res.* **2022**, *99*, 25–39. [CrossRef] [PubMed]
92. Minutolo, A.; Potestà, M.; Roglia, V.; Cirilli, M.; Iacovelli, F.; Cerva, C.; Fokam, J.; Desideri, A.; Andreoni, M.; Grelli, S. Plant microRNAs from *Moringa oleifera* Regulate Immune Response and HIV Infection. *Front. Pharmacol.* **2021**, *11*, 620038. [CrossRef]
93. Aldi, Y.; Ramadhan, Z.L.; Dillasamola, D.; Alianta, A.A. The Activities of *Moringa oleifera* Lam Leaf Extract on Specific Cellular Immune of Albino Male Mice Induced by the COVID-19 Antigen. *Trop. J. Nat. Prod. Res.* **2025**, *9*, 2421–2425. [CrossRef]
94. Hazarika, S.; Lal, M.; Rahaman, S.B.; Debnath, U.; Rawal, R.K. Computational Screening of *Moringa oleifera* Compounds for Immunomodulatory Targets: Identification and Validation of Core Genes. *S. Afr. J. Bot.* **2025**, *185*, 331–347. [CrossRef]
95. Lu, J.; Liu, X.; Li, W.; Xi, C.; Feng, D.; Song, S. Analysis of the Sensitization Activity of *Moringa oleifera* Leaves Protein. *Front. Nutr.* **2025**, *11*, 1509343. [CrossRef]
96. Dong, Z.; Li, C.; Huang, Q.; Zhang, B.; Fu, X.; Liu, R.H. Characterization of a Novel Polysaccharide from the Leaves of *Moringa oleifera* and Its Immunostimulatory Activity. *J. Funct. Foods* **2018**, *49*, 391–400. [CrossRef]
97. Zheng, L.; Lu, X.; Yang, S.; Zou, Y.; Zeng, F.; Xiong, S.; Cao, Y.; Zhou, W. The Anti-Inflammatory Activity of GABA-Enriched *Moringa oleifera* Leaves Produced by Fermentation with Lactobacillus Plantarum LK-1. *Front. Nutr.* **2023**, *10*, 1093036. [CrossRef] [PubMed]
98. Li, L.; Ma, L.; Wen, Y.; Xie, J.; Yan, L.; Ji, A.; Zeng, Y.; Tian, Y.; Sheng, J. Crude Polysaccharide Extracted from *Moringa oleifera* Leaves Prevents Obesity in Association with Modulating Gut Microbiota in High-Fat Diet-Fed Mice. *Front. Nutr.* **2022**, *9*, 861588. [CrossRef]
99. Husien, H.M.; Rehman, S.U.; Duan, Z.; Wang, M. Effect of *Moringa oleifera* Leaf Polysaccharide on the Composition of Intestinal Microbiota in Mice with Dextran Sulfate Sodium-Induced Ulcerative Colitis. *Front. Nutr.* **2024**, *11*, 1409026. [CrossRef] [PubMed]
100. Jia, L.; Peng, X.; Deng, Z.; Zhang, B.; Li, H. The Structural Characterization of Polysaccharides from Three Cultivars of *Moringa oleifera* Lam. Root and Their Effects on Human Intestinal Microflora. *Food Biosci.* **2023**, *52*, 102482. [CrossRef]
101. Wen, Z.; Tian, H.; Liang, Y.; Guo, Y.; Deng, M.; Liu, G.; Li, Y.; Liu, D.; Sun, B. *Moringa oleifera* Polysaccharide Regulates Colonic Microbiota and Immune Repertoire in C57BL/6 Mice. *Int. J. Biol. Macromol.* **2022**, *198*, 135–146. [CrossRef]
102. Mohamed Husien, H.; Peng, W.; Su, H.; Zhou, R.; Tao, Y.; Huang, J.; Liu, M.; Bo, R.; Li, J. *Moringa oleifera* Leaf Polysaccharide Alleviates Experimental Colitis by Inhibiting Inflammation and Maintaining Intestinal Barrier. *Front. Nutr.* **2022**, *9*, 1055791. [CrossRef]
103. Khalid, A.R.; Yasoob, T.B.; Zhang, Z.; Zhu, X.; Hang, S. Dietary *Moringa oleifera* Leaf Powder Improves Jejunal Permeability and Digestive Function by Modulating the Microbiota Composition and Mucosal Immunity in Heat Stressed Rabbits. *Env. Sci. Pollut. Res.* **2022**, *29*, 80952–80967. [CrossRef] [PubMed]
104. Tian, H.; Liang, Y.; Liu, G.; Li, Y.; Deng, M.; Liu, D.; Guo, Y.; Sun, B. *Moringa oleifera* Polysaccharides Regulates Caecal Microbiota and Small Intestinal Metabolic Profile in C57BL/6 Mice. *Int. J. Biol. Macromol.* **2021**, *182*, 595–611. [CrossRef]
105. Li, C.; Zhou, S.; Fu, X.; Huang, Q.; Chen, Q. In Vitro Digestibility and Prebiotic Activities of a Bioactive Polysaccharide from *Moringa oleifera* Leaves. *J. Food Biochem.* **2021**, *45*, e13944. [CrossRef]
106. Zhao, C.; Li, H.; Gao, C.; Tian, H.; Guo, Y.; Liu, G.; Li, Y.; Liu, D.; Sun, B. *Moringa oleifera* Leaf Polysaccharide Regulates Fecal Microbiota and Colonic Transcriptome in Calves. *Int. J. Biol. Macromol.* **2023**, *253*, 127108. [CrossRef]

https://doi.org/10.3390/ijms27010263

107. Husien, H.M.; Peng, W.; Essa, M.O.A.; Adam, S.Y.; Ur Rehman, S.; Ali, R.; Saleh, A.A.; Wang, M.; Li, J. The Anti-Inflammatory Properties of Polysaccharides Extracted from *Moringa oleifera* Leaves on IEC6 Cells Stimulated with Lipopolysaccharide In Vitro. *Animals* **2024**, *14*, 3508. [CrossRef]
108. Tian, H.; Wen, Z.; Liu, Z.; Guo, Y.; Liu, G.; Sun, B. Comprehensive Analysis of Microbiome, Metabolome and Transcriptome Revealed the Mechanisms of *Moringa oleifera* Polysaccharide on Preventing Ulcerative Colitis. *Int. J. Biol. Macromol.* **2022**, *222*, 573–586. [CrossRef]
109. Chiricosta, L.; Silvestro, S.; Pizzicannella, J.; Diomede, F.; Bramanti, P.; Trubiani, O.; Mazzon, E. Transcriptomic Analysis of Stem Cells Treated with Moringin or Cannabidiol: Analogies and Differences in Inflammation Pathways. *Int. J. Mol. Sci.* **2019**, *20*, 6039. [CrossRef]
110. Wu, Y.-Y.; Xu, Y.-M.; Lau, A.T.Y. Anti-Cancer and Medicinal Potentials of Moringa Isothiocyanate. *Molecules* **2021**, *26*, 7512. [CrossRef] [PubMed]
111. Cuellar-Núñez, M.L.; De Mejia, E.G.; Loarca-Piña, G. *Moringa oleifera* Leaves Alleviated Inflammation through Downregulation of IL-2, IL-6, and TNF-α in a Colitis-Associated Colorectal Cancer Model. *Food Res. Int.* **2021**, *144*, 110318. [CrossRef] [PubMed]
112. Kaleo, I.V.; Gao, Q.; Liu, B.; Sun, C.; Zhou, Q.; Zhang, H.; Shan, F.; Xiong, Z.; Bo, L.; Song, C. Effects of *Moringa oleifera* Leaf Extract on Growth Performance, Physiological and Immune Response, and Related Immune Gene Expression of Macrobrachium Rosenbergii with Vibrio Anguillarum and Ammonia Stress. *Fish Shellfish Immunol.* **2019**, *89*, 603–613. [CrossRef]
113. Abidin, Z.; Huang, H.-T.; Liao, Z.-H.; Chen, B.-Y.; Wu, Y.-S.; Lin, Y.-J.; Nan, F.-H. *Moringa oleifera* Leaves' Extract Enhances Nonspecific Immune Responses, Resistance against Vibrio Alginolyticus, and Growth in Whiteleg Shrimp (*Penaeus vannamei*). *Animals* **2021**, *12*, 42. [CrossRef] [PubMed]
114. Mohamed, M.A.; El-Mleeh, A.A.; Hamad, R.T.; Abu-Alya, I.S.; El-Hewaity, M.H.; Elbestawy, A.R.; Elbagory, A.M.; Sayed-Ahmed, A.S.; Abd Eldaim, M.A.; Elshabrawy, O.I. Immunostimulant Potential of *Moringa oleifera* Leaves Alcoholic Extract versus Oregano Essential Oil (OEO) against Cyclophosphamide-Induced Immunosuppression in Broilers Chicks. *Trop. Anim. Health Prod.* **2023**, *55*, 209. [CrossRef]
115. Zhang, S.; Cao, Y.; Huang, Y.; Zhang, S.; Wang, G.; Fang, X.; Bao, W.; Aqueous, M. Oleifera Leaf Extract Alleviates DSS-Induced Colitis in Mice through Suppression of Inflammation. *J. Ethnopharmacol.* **2024**, *318*, 116929. [CrossRef]
116. Sailaja, B.S.; Aita, R.; Maledatu, S.; Ribnicky, D.; Verzi, M.P.; Raskin, I. Moringa Isothiocyanate-1 Regulates Nrf2 and NF-κB Pathway in Response to LPS-Driven Sepsis and Inflammation. *PLoS ONE* **2021**, *16*, e0248691. [CrossRef]
117. Herman-Lara, E.; Rodríguez-Miranda, J.; Ávila-Manrique, S.; Dorado-López, C.; Villalva, M.; Jaime, L.; Santoyo, S.; Martínez-Sánchez, C.E. In Vitro Antioxidant, Anti-Inflammatory Activity and Bioaccessibility of Ethanolic Extracts from Mexican *Moringa oleifera* Leaf. *Foods* **2024**, *13*, 2709. [CrossRef]
118. Luetragoon, T.; Daowtak, K.; Thongsri, Y.; Potup, P.; Calder, P.C.; Usuwanthim, K. Anti-Inflammatory Potential of 3-Hydroxy-β-Ionone from *Moringa oleifera*: Decreased Transendothelial Migration of Monocytes Through an Inflamed Human Endothelial Cell Monolayer by Inhibiting the IκB-α/NF-κB Signaling Pathway. *Molecules* **2024**, *29*, 5873. [CrossRef]
119. Shahbaz, M.; Naeem, H.; Batool, M.; Imran, M.; Hussain, M.; Mujtaba, A.; Alsagaby, S.A.; Al Abdulmonem, W.; El-Ghorab, A.H.; Ghoneim, M.M.; et al. Antioxidant, Anticancer, and Anti-inflammatory Potential of Moringa Seed and Moringa Seed Oil: A Comprehensive Approach. *Food Sci. Nutr.* **2024**, *12*, 6157–6173. [CrossRef]
120. Farias-Pereira, R.; Camayoc, P.; Raskin, I. Isothiocyanate-Rich Moringa Seed Extract Activates SKN-1/Nrf2 Pathway in Caenorhabditis Elegans. *Int. J. Mol. Sci.* **2024**, *25*, 10917. [CrossRef] [PubMed]
121. Nassar, A.A.M.A.; Gharib, A.A.E.-A.; Abdelgalil, S.Y.; AbdAllah, H.M.; Elmowalid, G.A. Immunomodulatory, Antioxidant, and Growth-Promoting Activities of Dietary Fermented *Moringa oleifera* in Nile Tilapia (*Oreochromus niloticus*) with in-Vivo Protection against Aeromonas Hydrophila. *BMC Vet. Res.* **2024**, *20*, 231. [CrossRef] [PubMed]
122. Feng, Y.; Huang, J.; Gong, J.; Peng, W.; Yu, C.; Huai, Y.; Chen, C.; Bo, R.; Liu, M.; Li, J. Composition and Anti-Colitis Efficacy Analysis of Ethanol and Aqueous Extraction of *Moringa oleifera* Leaves. *Fitoterapia* **2025**, *185*, 106720. [CrossRef]
123. Luetragoon, T.; Sranujit, R.P.; Noysang, C.; Thongsri, Y.; Potup, P.; Somboonjun, J.; Maichandi, N.; Suphrom, N.; Sangouam, S.; Usuwanthim, K. Evaluation of Anti-Inflammatory Effect of *Moringa oleifera* Lam. and Cyanthillium Cinereum (Less) H. Rob. Lozenges in Volunteer Smokers. *Plants* **2021**, *10*, 1336. [CrossRef]
124. Díaz-Prieto, L.E.; Gómez-Martínez, S.; Vicente-Castro, I.; Heredia, C.; González-Romero, E.A.; Martín-Ridaura, M.D.C.; Ceinos, M.; Picón, M.J.; Marcos, A.; Nova, E. Effects of *Moringa oleifera* Lam. Supplementation on Inflammatory and Cardiometabolic Markers in Subjects with Prediabetes. *Nutrients* **2022**, *14*, 1937. [CrossRef]
125. Prabowo, N.A. *Effect of Moringa Leaf Extract on Disease Activity in Rheumatoid Arthritis Patients*; Clinicaltrials.gov: Bethesda, MD, USA, 2022.
126. Asare, B.; Selorm Segbefia, P.; Awuku-Larbi, R.; Asema Asandem, D.; Brenko, T.; Bentum-Ennin, L.; Osei, F.; Teye-Adjei, D.; Agyekum, G.; Akuffo, L. Immunomodulatory Effect of *Moringa oleifera* and *Phyllanthus niruri extracts* on Anti-HBV Cytokine Production by Human Peripheral Blood Mononuclear Cells. *Open Res. Eur.* **2025**, *5*, 149. [CrossRef]

https://doi.org/10.3390/ijms27010263

127. Gambo, A.; Moodley, I.; Babashani, M.; Babalola, T.K.; Gqaleni, N. A Double-Blind, Randomized Controlled Trial to Examine the Effect of *Moringa oleifera* Leaf Powder Supplementation on the Immune Status and Anthropometric Parameters of Adult HIV Patients on Antiretroviral Therapy in a Resource-Limited Setting. *PLoS ONE* **2021**, *16*, e0261935. [CrossRef]
128. Gandhi, G.R.; Mohana, T.; Athesh, K.; Hillary, V.E.; Vasconcelos, A.B.S.; Farias de Franca, M.N.; Montalvão, M.M.; Ceasar, S.A.; Jothi, G.; Sridharan, G.; et al. Anti-Inflammatory Natural Products Modulate Interleukins and Their Related Signaling Markers in Inflammatory Bowel Disease: A Systematic Review. *J. Pharm. Anal.* **2023**, *13*, 1408–1428. [CrossRef] [PubMed]
129. Zhang, Y.; Zhang, X.-Y.; Shi, S.-R.; Ma, C.-N.; Lin, Y.-P.; Song, W.-G.; Guo, S.-D. Natural Products in Atherosclerosis Therapy by Targeting PPARs: A Review Focusing on Lipid Metabolism and Inflammation. *Front. Cardiovasc. Med.* **2024**, *11*, 1372055. [CrossRef]
130. Mohd Sahardi, N.F.N.; Makpol, S. Suppression of Inflamm-Aging by *Moringa oleifera* and Zingiber Officinale Roscoe in the Prevention of Degenerative Diseases: A Review of Current Evidence. *Molecules* **2023**, *28*, 5867. [CrossRef] [PubMed]
131. Chiș, A.; Noubissi, P.A.; Pop, O.-L.; Mureșan, C.I.; Fokam Tagne, M.A.; Kamgang, R.; Fodor, A.; Sitar-Tăut, A.-V.; Cozma, A.; Orășan, O.H.; et al. Bioactive Compounds in *Moringa oleifera*: Mechanisms of Action, Focus on Their Anti-Inflammatory Properties. *Plants* **2023**, *13*, 20. [CrossRef]
132. Ajzashokouhi, A.H.; Rezaee, R.; Omidkhoda, N.; Karimi, G. Natural Compounds Regulate the PI3K/Akt/GSK3β Pathway in Myocardial Ischemia-Reperfusion Injury. *Cell Cycle* **2023**, *22*, 741–757. [CrossRef] [PubMed]
133. Zhang, T.; Zhao, L.; Xu, M.; Jiang, P.; Zhang, K. Moringin Alleviates DSS-Induced Ulcerative Colitis in Mice by Regulating Nrf2/NF-κB Pathway and PI3K/AKT/mTOR Pathway. *Int. Immunopharmacol.* **2024**, *134*, 112241. [CrossRef]
134. Kim, C.G.; Chang, S.N.; Park, S.M.; Hwang, B.S.; Kang, S.-A.; Kim, K.S.; Park, J.G. *Moringa oleifera* Mitigates Ethanol-Induced Oxidative Stress, Fatty Degeneration and Hepatic Steatosis by Promoting Nrf2 in Mice. *Phytomedicine* **2022**, *100*, 154037. [CrossRef]
135. Afolabi, O.A.; Akhigbe, T.M.; Akhigbe, R.E.; Alabi, B.A.; Gbolagun, O.T.; Taiwo, M.E.; Fakeye, O.O.; Yusuf, E.O. Methanolic *Moringa oleifera* Leaf Extract Protects against Epithelial Barrier Damage and Enteric Bacterial Translocation in Intestinal I/R: Possible Role of Caspase 3. *Front. Pharmacol.* **2022**, *13*, 989023. [CrossRef]
136. Laftah, A.H.; Alhelfi, N.; Al Salait, S.K.; Altemimi, A.B.; Tabandeh, M.R.; Tsakali, E.; Van Impe, J.F.; Abd El-Maksoud, A.A.; Abedelmaksoud, T.G. Mitigation of Doxorubicin-Induced Liver Toxicity in Mice Breast Cancer Model by Green Tea and *Moringa oleifera* Combination: Targeting Apoptosis, Inflammation, and Oxidative Stress. *J. Funct. Foods* **2025**, *124*, 106626. [CrossRef]
137. Reda, R.M.; Helmy, R.M.A.; Osman, A.; Ahmed, F.A.G.; Kotb, G.A.M.; El-Fattah, A.H.A. The Potential Effect of *Moringa oleifera* Ethanolic Leaf Extract against Oxidative Stress, Immune Response Disruption Induced by Abamectin Exposure in *Oreochromis niloticus*. *Env. Sci. Pollut. Res.* **2023**, *30*, 58569–58587. [CrossRef]
138. Liga, S.; Magyari-Pavel, I.Z.; Avram, Ș.; Minda, D.I.; Vlase, A.-M.; Muntean, D.; Vlase, L.; Moacă, E.-A.; Danciu, C. Comparative Analysis of *Moringa oleifera* Lam. Leaves Ethanolic Extracts: Effects of Extraction Methods on Phytochemicals, Antioxidant, Antimicrobial, and In Ovo Profile. *Plants* **2025**, *14*, 1653. [CrossRef] [PubMed]
139. Lin, J.; Li, F.; Jiao, J.; Qian, Y.; Xu, M.; Wang, F.; Sun, X.; Zhou, T.; Wu, H.; Kong, X. Quercetin, a Natural Flavonoid, Protects against Hepatic Ischemia–Reperfusion Injury via Inhibiting Caspase-8/ASC Dependent Macrophage Pyroptosis. *J. Adv. Res.* **2024**, *70*, 555–569. [CrossRef]
140. Yu, J.; Fu, R.; Buhe, A.; Xu, B. Quercetin Attenuates Lipopolysaccharide-Induced Hepatic Inflammation by Modulating Autophagy and Necroptosis. *Poult. Sci.* **2024**, *103*, 103719. [CrossRef] [PubMed]
141. Katsaros, I.; Sotiropoulou, M.; Vailas, M.; Kapetanakis, E.I.; Valsami, G.; Tsaroucha, A.; Schizas, D. Quercetin's Potential in MASLD: Investigating the Role of Autophagy and Key Molecular Pathways in Liver Steatosis and Inflammation. *Nutrients* **2024**, *16*, 3789. [CrossRef]
142. Xiong, F.; Zhang, Y.; Li, T.; Tang, Y.; Song, S.-Y.; Zhou, Q.; Wang, Y. A Detailed Overview of Quercetin: Implications for Cell Death and Liver Fibrosis Mechanisms. *Front. Pharmacol.* **2024**, *15*, 1389179. [CrossRef] [PubMed]
143. Zhang, Y.; Qu, X.; Gao, H.; Zhai, J.; Tao, L.; Sun, J.; Song, Y.; Zhang, J. Quercetin Attenuates NLRP3 Inflammasome Activation and Apoptosis to Protect INH-Induced Liver Injury via Regulating SIRT1 Pathway. *Int. Immunopharmacol.* **2020**, *85*, 106634. [CrossRef]
144. Guon, T.E.; Chung, H.S. *Moringa oleifera* Fruit Induce Apoptosis via Reactive Oxygen Species-Dependent Activation of Mitogen-Activated Protein Kinases in Human Melanoma A2058 Cells. *Oncol. Lett.* **2017**, *14*, 1703–1710. [CrossRef]
145. Adewale, G.B.; Olowoparija, S.F.; Olayemi, J.O.; Olayinka, E. Effect of Ethanol Extract of *Moringa oleifera* Leaves on T-Cell Proliferation. *West. J. Med. Biomed. Sci.* **2020**, *1*, 176–184.
146. Akanni, E.O.; Adedeji, A.L.; Akanni, R.A.; Oloke, J.K. Up-Regulation of TNF-α by Ethanol Extract of *Moringa oleifera* in Benzene-Induced Leukemic Wister Rat: A Possible Mechanism of Anticancer Property. *Res. Rev. J. Oncol. Haematol.* **2014**, *3*, 16–21.
147. Abutu, P.; Amuda, O.; Osinaya, O.; Babatunde, B. Regulatory Activity of Ethanol Leaves Extract of *Moringa oleifera* on Benzene Induced Leukemia in Wister Rat Using TNF-α Analysis. *Am. J. Med. Biol. Res.* **2019**, *7*, 6–11.
148. Van Pham, A.T.; Luong, M.H.; Dinh, H.T.T.; Mai, T.P.; Trinh, Q.V.; Luong, L.H. Immunostimulatory Effect of *Moringa oleifera* Extracts on Cyclophosphamide-Induced Immunosuppressed Mice. *J. Herbs Spices Med. Plants* **2021**, *27*, 377–385. [CrossRef]

https://doi.org/10.3390/ijms27010263

149. Asaduzzaman, A.K.M.; Hasan, I.; Chakrabortty, A.; Zaman, S.; Islam, S.S.; Ahmed, F.R.S.; Kabir, K.A.; Nurujjaman, M.; Uddin, M.B.; Alam, M.T. *Moringa oleifera* Seed Lectin Inhibits Ehrlich Ascites Carcinoma Cell Growth by Inducing Apoptosis through the Regulation of Bak and NF-κB Gene Expression. *Int. J. Biol. Macromol.* **2018**, *107*, 1936–1944. [CrossRef]
150. Ndlovu, S.S.; Chuturgoon, A.A.; Ghazi, T. *Moringa oleifera* Lam Leaf Extract Stimulates NRF2 and Attenuates ARV-Induced Toxicity in Human Liver Cells (HepG2). *Plants* **2023**, *12*, 1541. [CrossRef]
151. Sailaja, B.S.; Hassan, S.; Cohen, E.; Tmenova, I.; Farias-Pereira, R.; Verzi, M.P.; Raskin, I. Moringa Isothiocyanate-1 Inhibits LPS-Induced Inflammation in Mouse Myoblasts and Skeletal Muscle. *PLoS ONE* **2022**, *17*, e0279370. [CrossRef]
152. Villegas-Vazquez, E.Y.; Gómez-Cansino, R.; Marcelino-Pérez, G.; Jiménez-López, D.; Quintas-Granados, L.I. Unveiling the Miracle Tree: Therapeutic Potential of *Moringa oleifera* in Chronic Disease Management and Beyond. *Biomedicines* **2025**, *13*, 634. [CrossRef] [PubMed]
153. Do, B.H.; Hoang, N.S.; Nguyen, T.P.T.; Ho, N.Q.C.; Le, T.L.; Doan, C.C. Phenolic Extraction of *Moringa oleifera* Leaves Induces Caspase-Dependent and Caspase-Independent Apoptosis through the Generation of Reactive Oxygen Species and the Activation of Intrinsic Mitochondrial Pathway in Human Melanoma Cells. *Nutr. Cancer* **2021**, *73*, 869–888. [CrossRef]
154. Sultan, R.; Ahmed, A.; Wei, L.; Saeed, H.; Islam, M.; Ishaq, M. The Anticancer Potential of Chemical Constituents of *Moringa oleifera* Targeting CDK-2 Inhibition in Estrogen Receptor Positive Breast Cancer Using in-Silico and in Vitro Approches. *BMC Complement. Med. Ther.* **2023**, *23*, 396. [CrossRef]
155. Qian, D.; Zha, D.; Sang, Y.; Tao, J.; Cheng, Y. *Moringa oleifera* Mediated Green Synthesis of Gold Nanoparticles and Their Anti-Cancer Activity against A549 Cell Line of Lung Cancer through ROS/Mitochondrial Damage. *Front. Chem.* **2025**, *13*, 1521089. [CrossRef]
156. Kaavyaunni; Antony, S.; Chanthini, K.M.-P. Biogenically Synthesized *Moringa oleifera* Leaf Extract-Derived Silver Nanoparticles Exhibit Potent Antimicrobial and Anticancer Effects via Oxidative Stress and Apoptotic Pathways in AGS Gastric Cancer Cells. *Biocatal. Agric. Biotechnol.* **2025**, *67*, 103674. [CrossRef]
157. Kumar, S.; Verma, P.K.; Shukla, A.; Singh, R.K.; Patel, A.K.; Yadav, L.; Kumar, S.; Kumar, N.; Kaushalendra; Acharya, A. *Moringa oleifera* L. Leaf Extract Induces Cell Cycle Arrest and Mitochondrial Apoptosis in Dalton's Lymphoma: An in Vitro and in Vivo Study. *J. Ethnopharmacol.* **2023**, *302*, 115849. [CrossRef]
158. Al-Shalabi, R.; abdul Samad, N.; Al-Deeb, I.; Joseph, J.; Abualsoud, B.M. Assessment of Antiangiogenic and Cytotoxic Effects of *Moringa oleifera* Silver Nanoparticles Using Cell Lines. *Curr. Res. Nutr. Food Sci. J.* **2024**, *12*, 196–211. [CrossRef]
159. Choirunnisa, N.L.; Kezia, D.C.; Permana, A.Z.; Malek, N.A.N.N.; Nurkolis, F.; Permatasari, H.K. Modulation of pAKT and NF-κB Pathways by *Moringa oleifera* Silver Nanoparticles: Inducing Apoptosis in Cervical Cancer Cells. *CyTA J. Food* **2025**, *23*, 2543878. [CrossRef]
160. Abd Allah, R.S.; Ahmmed, N.H.; Walaa, A.; Heibashy, D.M. The Curative Role of Moringa Leaf Extract Nanomaterial and Low Doses of Gamma Irradiation on Ehrlich Carcinoma Bearing Mice. *Lett. Appl. NanoBioSci.* **2025**, *14*, 74.
161. Raju, N.S.C.; Kei, W.W. Anti-Angiogenic Screening of *Moringa oleifera* Leaves Extract Using Chorioallantonic Membrane Assay. *Iraqi J. Pharm. Sci.* **2022**, *31*, 225–232. [CrossRef]
162. Susanto, H.; Firdaus, S.R.A.; Sholeh, M.; Endharti, A.T.; Taufiq, A.; Malek, N.A.N.N.; Permatasari, H.K. *Moringa oleifera* Leaf Powder—Silver Nanoparticles (MOLP-AgNPs) Efficiently Inhibit Metastasis and Proliferative Signaling in HT-29 Human Colorectal Cancer Cells. *J. Agric. Food Res.* **2024**, *16*, 101149. [CrossRef]
163. Zunica, E.R.; Yang, S.; Coulter, A.; White, C.; Kirwan, J.P.; Gilmore, L.A. *Moringa oleifera* Seed Extract Concomitantly Supplemented with Chemotherapy Worsens Tumor Progression in Mice with Triple Negative Breast Cancer and Obesity. *Nutrients* **2021**, *13*, 2923. [CrossRef] [PubMed]
164. Armagan, G.; Sevgili, E.; Gürkan, F.T.; Köse, F.A.; Bilgiç, T.; Dagcı, T.; Saso, L. Regulation of the Nrf2 Pathway by Glycogen Synthase Kinase-3β in MPP+-Induced Cell Damage. *Molecules* **2019**, *24*, 1377. [CrossRef]
165. Bourebaba, L.; Komakula, S.S.B.; Weiss, C.; Adrar, N.; Marycz, K. The PTP1B Selective Inhibitor MSI-1436 Mitigates Tunicamycin-Induced ER Stress in Human Hepatocarcinoma Cell Line through XBP1 Splicing Modulation. *PLoS ONE* **2023**, *18*, e0278566. [CrossRef]
166. Xu, Q.; Wu, N.; Li, X.; Guo, C.; Li, C.; Jiang, B.; Wang, H.; Shi, D. Inhibition of PTP1B Blocks Pancreatic Cancer Progression by Targeting the PKM2/AMPK/mTOC1 Pathway. *Cell Death Dis.* **2019**, *10*, 874. [CrossRef] [PubMed]
167. Hagoel, L.; Vexler, A.; Kalich-Philosoph, L.; Earon, G.; Ron, I.; Shtabsky, A.; Marmor, S.; Lev-Ari, S. Combined Effect of *Moringa oleifera* and Ionizing Radiation on Survival and Metastatic Activity of Pancreatic Cancer Cells. *Integr. Cancer Ther.* **2019**, *18*, 1534735419828829. [CrossRef]
168. Aja, P.M.; Agu, P.C.; Ezeh, E.M.; Awoke, J.N.; Ogwoni, H.A.; Deusdedit, T.; Ekpono, E.U.; Igwenyi, I.O.; Alum, E.U.; Ugwuja, E.I.; et al. Prospect into Therapeutic Potentials of *Moringa oleifera* Phytocompounds against Cancer Upsurge: De Novo Synthesis of Test Compounds, Molecular Docking, and ADMET Studies. *Bull. Natl. Res. Cent.* **2021**, *45*, 99. [CrossRef]

https://doi.org/10.3390/ijms27010263

169. Saha, A.; Desai, B.M.A.; Biswas, P. Multi-Ligand Simultaneous Docking Analysis of *Moringa oleifera* Phytochemicals Reveals Enhanced BCL-2 Inhibition via Synergistic Action. In Proceedings of the 2024 5th International Conference on Biomedical Engineering (IBIOMED), Bali, Indonesia, 23–25 October 2024; pp. 51–56.
170. Bhadresha, K.; Thakore, V.; Brahmbhatt, J.; Upadhyay, V.; Jain, N.; Rawal, R. Anticancer Effect of *Moringa oleifera* Leaves Extract against Lung Cancer Cell Line via Induction of Apoptosis. *Adv. Cancer Biol. Metastasis* **2022**, *6*, 100072. [CrossRef]
171. ICTRP Search Portal. Available online: https://trialsearch.who.int/Trial2.aspx?TrialID=CTRI/2022/02/040594 (accessed on 26 August 2025).
172. Tshingani, K.; Donnen, P.; Mukumbi, H.; Duez, P.; Dramaix-Wilmet, M. Impact of *Moringa oleifera* Lam. Leaf Powder Supplementation versus Nutritional Counseling on the Body Mass Index and Immune Response of HIV Patients on Antiretroviral Therapy: A Single-Blind Randomized Control Trial. *BMC Complement. Altern. Med.* **2017**, *17*, 420. [CrossRef] [PubMed]
173. Aprioku, J.S.; Robinson, O.; Obianime, A.W.; Tamuno, I. Moringa Supplementation Improves Immunological Indices and Hematological Abnormalities in Seropositive Patients Receiving HAARTs. *Afr. Health Sci.* **2022**, *22*, 1–11. [CrossRef]
174. Barhoi, D.; Upadhaya, P.; Barbhuiya, S.N.; Giri, A.; Giri, S. Aqueous Extract of *Moringa oleifera* Exhibit Potential Anticancer Activity and Can Be Used as a Possible Cancer Therapeutic Agent: A Study Involving In Vitro and In Vivo Approach. *J. Am. Coll. Nutr.* **2021**, *40*, 70–85. [CrossRef]
175. Kim, Y.; Jaja-Chimedza, A.; Merrill, D.; Mendes, O.; Raskin, I. A 14-Day Repeated-Dose Oral Toxicological Evaluation of an Isothiocyanate-Enriched Hydro-Alcoholic Extract from *Moringa oleifera* Lam. Seeds in Rats. *Toxicol. Rep.* **2018**, *5*, 418–426. [CrossRef]
176. Taweerutchana, R.; Lumlerdkij, N.; Vannasaeng, S.; Akarasereenont, P.; Sriwijitkamol, A. Effect of *Moringa oleifera* Leaf Capsules on Glycemic Control in Therapy-Naïve Type 2 Diabetes Patients: A Randomized Placebo Controlled Study. *Evid. Based Complement. Altern. Med.* **2017**, *2017*, 6581390. [CrossRef]
177. Aliyu, A.; Shaari, M.R.; Ahmad Sayuti, N.S.; Reduan, F.H.; Sithambaram, S.; Mohamed Mustapha, N.; Shaari, K.; Hamzah, H.B. *Moringa oleifera* Hydorethanolic Leaf Extract Induced Acute and Sub-Acute Hepato-Nephrotoxicity in Female ICR-Mice. *Sci. Prog.* **2021**, *104*, 00368504211004272. [CrossRef]
178. Manzo, M.L.; Hallarou, M.E.; Halidou, M.D.; Maiga, D.A.; Bahwere, P.; Salimata, W.; Castetbon, K.; Wilmet-Dramaix, M.; Donnen, P. Effect of Moringa Supplementation in the Management of Moderate Malnutrition in Children under 5 Receiving Ready-to-Use Supplementary Foods in Niger: A Randomized Clinical Trial. *GSC Adv. Res. Rev.* **2021**, *8*, 071–086. [CrossRef]
179. Kajuju Malla, J.; Ochola, S.; Ogada, I.; Munyaka, A. Effect of *Moringa oleifera* Fortified Porridge Consumption on Protein and Vitamin A Status of Children with Cerebral Palsy in Nairobi, Kenya: A Randomized Controlled Trial. *PLoS Glob. Public Health* **2022**, *2*, e0001206. [CrossRef] [PubMed]
180. Silva, N.R.G.; Costa, W.K.; Ferreira, M.R.A.; Coelho, L.C.B.B.; Soares, L.A.L.; Napoleão, T.H.; Paiva, P.M.G.; de Oliveira, A.M. 13-Week Repeated-Dose Toxicity Study of Optimized Aqueous Extract of *Moringa oleifera* Leaves in Mice. *J. Ethnopharmacol.* **2024**, *335*, 118637. [CrossRef] [PubMed]
181. Adedapo, A.A.; Mogbojuri, O.M.; Emikpe, B.O. Safety Evaluations of the Aqueous Extract of the Leaves of *Moringa oleifera* in Rats. *J. Med. Plants Res.* **2009**, *3*, 586–591.
182. Iliyasu, D.; Rwuaan, J.S.; Sani, D.; Nwannenna, A.I.; Njoku, C.O.; Mustapha, A.R.; Peter, I.D. Evaluation of Safety, Proximate and Efficacy of Graded Dose of *Moringa oleifera* Aqueous Seed Extract as Supplement That Improve Live-Body Weight and Scrotal Circumference in Yankasa Ram. *Int. J. Livest. Res.* **2020**, *10*, 33–46. [CrossRef]
183. Palomino-Pacheco, M.; Rojas-Armas, J.P.; Ortiz-Sánchez, J.M.; Arroyo-Acevedo, J.L.; Justil-Guerrero, H.J.; Martínez-Heredia, J.T. Assessment of Oral Toxicity of *Moringa oleifera* Lam Aqueous Extract and Its Effect on Gout Induced in a Murine Model. *Vet. World* **2024**, *17*, 1449. [CrossRef]
184. Musa, A.H.; Gebru, G.; Debella, A.; Makonnen, E.; Asefa, M.; Woldekidan, S.; Abebe, A.; Lengiso, B.; Bashea, C. Prenatal Developmental Toxicity Study of Herbal Tea of *Moringa stenopetala* and *Mentha spicata* Leaves Formulation in Wistar Rats. *Toxicol. Rep.* **2022**, *9*, 1853–1862. [CrossRef] [PubMed]
185. Hajar, A.; Abdelmounaim, B.; Hamid, K.; Jaouad, L.; Abdelfattah, A.B.; Majda, B.; Loubna, E.Y.; Mohammed, L.; Rachida, A.; Abderrahman, C. Developmental Toxicity of *Moringa oleifera* and Its Effect on Postpartum Depression, Maternal Behavior and Lactation. *S. Afr. J. Bot.* **2024**, *171*, 257–266. [CrossRef]
186. Soliman, S.S.; Suliman, A.A.; Fathy, K.; Sedik, A.A. Ovario-Protective Effect of *Moringa oleifera* Leaf Extract against Cyclophosphamide-Induced Oxidative Ovarian Damage and Reproductive Dysfunction in Female Rats. *Sci. Rep.* **2025**, *15*, 1054.
187. Abdulahi, S.K.; Dada, E.O.; Adebayo, R.O. Histopathological Effects of Seed Oil of *Moringa oleifera* Lam. on Albino Mice Infected with Plasmodium Berghei (NK65). *Adv. J. Grad. Res.* **2022**, *11*, 71–79. [CrossRef]
188. Oghama, O.E.; Timothy, O.; Gabriel, B.O.; Abaku, S.N. 21 Days Effects of Ethanol Root Extract of *Moringa oleifera* Lam. on Kidney and Liver Functions in Wistar Rats. *Trop. J. Chem.* **2025**, *1*, 14–18. [CrossRef]

189. Musa, A.H.; Hagos, A.D.; Dimsu, G.G.; Eshetu, E.M.; Tola, M.A.; Admas, A.; Gelagle, A.A.; Tullu, B.L. Subchronic Toxicity Study of Herbal Tea of *Moringa stenopetala* (Baker f.) Cudof. and *Mentha spicata* L. Leaves Formulation in Wistar Albino Rats. *Toxicol. Rep.* **2022**, *9*, 797–805. [CrossRef]
190. Stohs, S.J.; Hartman, M.J. Review of the Safety and Efficacy of *Moringa oleifera*. *Phytother. Res.* **2015**, *29*, 796–804. [CrossRef] [PubMed]
191. Thanikachalam, P.V.; Ramesh, K.; Hydar, M.I.; Dhalapathy, V.V.; Devaraji, M. Therapeutic Potential of *Moringa oleifera* Lam. in Metabolic Disorders: A Molecular Overview. *Asian Pac. J. Trop. Biomed.* **2025**, *15*, 263. [CrossRef]
192. Nakata, H.; Nakayama, S.M.M.; Kataba, A.; Toyomaki, H.; Doya, R.; Beyene Yohannes, Y.; Ikenaka, Y.; Ishizuka, M. Does *Moringa oleifera* Affect Element Accumulation Patterns and Lead Toxicity in Sprague–Dawley Rats? *J. Funct. Foods* **2022**, *97*, 105242. [CrossRef]
193. Popoola, J.O.; Aworunse, O.S.; Oyesola, O.L.; Akinnola, O.O.; Obembe, O.O. A Systematic Review of Pharmacological Activities and Safety of *Moringa oleifera*. *J. Herbmed Pharmacol.* **2020**, *9*, 174–190. [CrossRef]
194. Abdelwanis, F.M.; Abdelaty, H.S.; Saleh, S.A. Exploring the Multifaceted Uses of *Moringa oleifera*: Nutritional, Industrial and Agricultural Innovations in Egypt. *Discov. Food* **2024**, *4*, 146. [CrossRef]
195. Elabd, H.; Soror, E.; El-Asely, A.; El-Gawad, E.A.; Abbass, A. Dietary Supplementation of Moringa Leaf Meal for Nile Tilapia *Oreochromis niloticus*: Effect on Growth and Stress Indices. *Egypt. J. Aquat. Res.* **2019**, *45*, 265–271. [CrossRef]
196. Sallam, G.R.; Abdel-Rahim, M.M.; Lotfy, A.M.; Fayed, W.M.; Shehata, A.I.; El Basuini, M.F.; Elwan, R.I.; Al-absawey, M.A.; Elhetawy, A.I.G. Long Term Dietary *Moringa oleifera* Leaf Extract to Florida Red Tilapia Oreochromis Sp Improves Performance Immunity Maturation and Reproduction in Saltwater. *Sci. Rep.* **2025**, *15*, 20261. [CrossRef] [PubMed]
197. Ofulue, O.O.; Ebomoyi, M.I. Acute Toxicity Test of Alkaloid Fraction of *Moringa oleifera* Leaf and Its Effect on Reproductive Hormones of Pregnant Wistar Rats. *J. Appl. Sci. Environ. Manag.* **2023**, *27*, 2235–2241. [CrossRef]
198. Tilaoui, M.; Ait Mouse, H.; Zyad, A. Update and New Insights on Future Cancer Drug Candidates from Plant-Based Alkaloids. *Front. Pharmacol.* **2021**, *12*, 719694. [CrossRef]
199. Ma, J.; Li, M.; Li, N.; Chan, W.Y.; Lin, G. Pyrrolizidine Alkaloid-Induced Hepatotoxicity Associated with the Formation of Reactive Metabolite-Derived Pyrrole–Protein Adducts. *Toxins* **2021**, *13*, 723. [CrossRef]
200. Yang, N.; Guo, J.; Zhang, J.; Gao, S.; Xiang, Q.; Wen, J.; Huang, Y.; Rao, C.; Chen, Y. A Toxicological Review of Alkaloids. *Drug Chem. Toxicol.* **2024**, *47*, 1267–1281. [CrossRef]
201. Sagrera, A.; Montenegro, T.; Borrego, L. Toxicodermia por *Moringa oleifera*. *Actas Dermosifiliogr.* **2021**, *112*, 953–954. [CrossRef]

https://doi.org/10.3390/ijms27010263

MDPI AG
Grosspeteranlage 5
4052 Basel
Switzerland
Tel.: +41 61 683 77 34

International Journal of Molecular Sciences Editorial Office
E-mail: ijms@mdpi.com
www.mdpi.com/journal/ijms

www.ingramcontent.com/pod-product-compliance
Lightning Source LLC
Chambersburg PA
CBHW040750250926
51850CB00007BB/1458